Gene Regulation

A CETUS–UCLA Symposium

Proceedings of the Symposium
on Gene Regulation
held March 28 - April 4, 1982,
Keystone, Colorado

UCLA Symposia on Molecular and Cellular Biology
Volume XXVI, 1982

GENE REGULATION

EDITED BY

BERT W. O'MALLEY

Department of Cell Biology
Baylor College of Medicine
Houston, Texas

SERIES EDITOR

C. FRED FOX

Department of Microbiology and Molecular Biology Institute
University of California
Los Angeles, California

MANAGING EDITOR

SANDY MALONE

UCLA Symposium
University of California
Los Angeles, California

ACADEMIC PRESS   1982
A Subsidiary of Harcourt Brace Jovanovich, Publishers
New York   London
Paris   San Diego   San Francisco   São Paulo   Sydney   Tokyo   Toronto

Academic Press Rapid Manuscript Reproduction

ACADEMIC PRESS, INC.
111 Fifth Avenue, New York, New York 10003

United Kingdom Edition published by
ACADEMIC PRESS, INC. (LONDON) LTD.
24/28 Oval Road, London NW1   7DX

Library of Congress Cataloging in Publication Data
Main entry under title:

Gene regulation.

    (UCLA symposia on molecular and cellular biology ;
v. 26)
    Includes bibliographies and index.
    1. Genetic regulation--Congresses. I. O'Malley,
Bert W. II. University of California, Los Angeles.
III. Series.
QH450.G463   1982   574.87'3223   82-20709
ISBN 0-12-525960-3

PRINTED IN THE UNITED STATES OF AMERICA

82 83 84 85   9 8 7 6 5 4 3 2 1

Contents

II. CHROMATIN STRUCTURE

III. GENE EXPRESSION

IV. CELLULAR BIOLOGY

Contributors

Numbers in parentheses indicate the chapter numbers.

Sidney Altman (15), *Department of Biology, Yale University, New Haven, Connecticut 06520*

G. Atfield (2), *Department of Immunology, The London Hospital Medical School, University of London, Turner Street, London E1 2AD, England*

Richard Axel (18), *Institute of Cancer Research, 701 West 168th Street, New York, New York 10032*

Madeline Baer (15), *Department of Biology, Yale University, New Haven, Connecticut 06520*

Melvyn Baez (6), *Department of Cell Biology, Baylor College of Medicine, Houston, Texas 77030*

John D. Baxter (17, 25), *Metabolic Research Unit, 681 HSE, University of California School of Medicine, San Francisco, California 94143*

Michael Behe (8), *Laboratory of Molecular Biology, Building 2, Room 305, National Institutes of Health, Bethesda, Maryland 20205*

Graeme I. Bell (3), *Department of Biochemistry and Biophysics, University of California, San Francisco, California 94143*

Bonnie Blomberg (1), *Center of Cancer Research, Massachusetts Institute of Technology, E17-350, Cambridge, Massachusetts 02139*

B. R. Brinkley (23), *Department of Cell Biology, Baylor College of Medicine, Texas Medical Center, Houston, Texas 77030*

Roy J. Britten (19), *Division of Biology 156-29, California Institute of Technology, Pasadena, California 91125*

H. Bud (2), *Clinical Research Centre, Northwick Park Hospital, Watford Road, Harrow, Middlesex HA1 3UJ, England*

H. Bullman (2), *Clinical Research Centre, Northwick Park Hospital, Watford Road, Harrow, Middlesex HA1 3UJ, England*

Harris Busch (13), *Department of Pharmacology, Baylor College of Medicine, 1200 Moursund Avenue, Houston, Texas 77030*

M. Busslinger (2), *Clinical Research Centre, Northwick Park Hospital, Watford Road, Harrow, Middlesex HA1 3UJ, England*

Carlos V. Cabrera (19), *Centro de Biologie Molecular, Universidad Autonoma de Madrid, Madrid-34, Spain*

Iain L. Cartwright (9), *Campus Box 1137, Department of Biology, Washington University, St. Louis, Missouri 63130*

Guy Cathala (17), *U.S.T.L. Laboratoire de Biologie Moléculaire, Place Eugène Bataillon, 34060 Montpellier Cédex, France*

James G. Chafouleas (22), *Department of Cell Biology, Baylor College of Medicine, Houston, Texas 77030*

Pierre Chambon (14), *Laboratoire de Génétique Moléculaire des Eucaryotes du CNRS, Faculté de Médecine, 11 rue Humann, 67085 Strasbourg Cédex, France*

Thirumalachary Chandra (4), *Howard Hughes Medical Institute, Baylor College of Medicine, 1200 Moursund Avenue, Houston, Texas 77030*

Subramanyam Chirala (13), *Department of Pharmacology, Baylor College of Medicine, 1200 Moursund Avenue, Houston, Texas 77030*

Lynn Cooley (16), *Department of Molecular Biophysics and Biochemistry, Yale University, P.O. Box 6666, New Haven, Connecticut 06511*

Jeffrey Corden (14), *Department of Microbiology, The Johns Hopkins University School of Medicine, Baltimore, Maryland 21205*

Charles S. Craik (3), *Department of Biochemistry and Biophysics, University of California, San Francisco, California 94143*

J. E. Darnell, Jr. (12), *The Rockefeller University, 1230 York Avenue, New York, New York 10021*

Eric H. Davidson (19), *Division of Biology 156-29, California Institute of Technology, Pasadena, California 91125*

Earl W. Davie (4), *Department of Biochemistry, University of Washington, Seattle, Washington 98195*

E. deBoer (2), *Clinical Research Centre, Northwick Park Hospital, Watford Road, Harrow, Middlesex HA1 3UJ, England*

Donald B. DeFranco (16), *Department of Biochemistry and Biophysics, University of California, San Francisco, California 94143*

William J. Deery (23), *Department of Cell Biology, Baylor College of Medicine, Texas Medical Center, Houston, Texas 77030*

Theodor Dingermann (16), *Department of Molecular Biophysics and Biochemistry, Yale University, P.O. Box 6666, New Haven, Connecticut 06511*

Naochika Domae (13), *Department of Pharmacology, Baylor College of Medicine, 1200 Moursund Avenue, Houston, Texas 77030*

Norman L. Eberhardt (17), *Metabolic Research Unit, 1143 HSW, University of California, San Francisco, California 94143*

Sarah C. R. Elgin (9, 11), *Campus Box 1137, Department of Biology, Washington University, St. Louis, Missouri 63130*

Paul Epstein (13), *Department of Pharmacology, Baylor College of Medicine, 1200 Moursund Avenue, Houston, Texas 77030*

Gary Felsenfeld (8), *Laboratory of Molecular Biology, Building 2, Room 301, National Institutes of Health, Bethesda, Maryland 20205*

J. Ferluga (2), *Department of Immunology, The London Hospital Medical School, University of London, Turner Street, London E1 2AD, England*

H. Festenstein (2), *Department of Immunology, The London Hospital Medical School, University of London, Turner Street, London E1 2AD, England*

R. A. Flavell (2), *Clinical Research Centre, Northwick Park Hospital, Watford Road, Harrow, Middlesex HA1 3UJ, England*

Robert Fletterick (3), *Department of Biochemistry and Biophysics, University of California, San Francisco, California 94143*

Constantin N. Flytzanis (19), *Division of Biology 156-29, California Institute of Technology, Pasadena, California 91125*

David L. Gard (24), *Division of Biology, California Institute of Technology, Pasadena, California 91125*

David Gardner (17), *Metabolic Research Unit, 681 HSE, University of California, San Francisco, California 94143*

Marie-Pierre Gaub (14), *Laboratoire de Génétique Moléculaire des Eucaryotes du CNRS, Faculté de Médecine, 11 rue Humann, 67085 Strasbourg Cédex, France*

L. Golden (2), *Clinical Research Centre, Northwick Park Hospital, Watford Road, Harrow, Middlesex HA1 3UJ, England*

F. G. Grosveld (2), *Clinical Research Centre, Northwick Park Hospital, Watford Road, Harrow, Middlesex HA1 3UJ, England*

G. C. Grosveld (2), *Clinical Research Centre, Northwick Park Hospital, Watford Road, Harrow, Middlesex HA1 3UJ, England*

Cecilia Guerrier-Takada (15), *Department of Biology, Yale University, New Haven, Connecticut 06520*

Arthur Gutierrez-Hartmann (17), *Metabolic Research Unit, 1143 HSW, University of California, San Francisco, California 94143*

Rene Hen (14), *Laboratoire de Génétique Moléculaire des Eucaryotes du CNRS, Faculté de Médecine, 11 rue Humann, 67085 Strasbourg Cédex, France*

Dale Henning (13), *Department of Pharmacology, Baylor College of Medicine, 1200 Moursund Avenue, Houston, Texas 77030*

Andrew A. Hobbs (20), *Department of Cell Biology, Baylor College of Medicine, Houston, Texas 77030*

J. Hurst (2), *Clinical Research Centre, Northwick Park Hospital, Watford Road, Harrow, Middlesex HA1 3UJ, England*

R. James (2), *ICRF Tumour Immunology Unit, University College, Gower Street, London WC1E 6BT, England*

Deborah L. Johnson (16), *Department of Molecular Biophysics and Biochemistry, Yale University, P.O. Box 6666, New Haven, Connecticut 06511*

Mark L. Johnson (20), *Department of Cell Biology, Baylor College of Medicine, Houston, Texas 77030*

Michael Karin (17), *Department of Microbiology, University of Southern California School of Medicine, Los Angeles, California 90033*

Claude Kédinger (14), *Laboratoire de Génétique Moléculaire des Eucaryotes du CNRS, Faculté de Médecine, 11 rue Humann, 67085 Strasbourg Cédex, France*

Vincent J. Kidd (4), *Howard Hughes Medical Institute, Baylor College of Medicine, 1200 Moursund Avenue, Houston, Texas 77030*

D. Kioussis (2), *Clinical Research Centre, Northwick Park Hospital, Watford Road, Harrow, Middlesex HA1 3UJ, England*

Kotoku Kurachi (4), *Department of Biochemistry, University of Washington, Seattle, Washington 98195*

Lisette Lagace (22), *Department of Cell Biology, Baylor College of Medicine, Houston, Texas 77030*

Eugene Lai (22), *Department of Cell Biology, Baylor College of Medicine, Houston, Texas 77030*

Nancy C. Lan (17), *Metabolic Research Unit, 681 HSE, University of California, San Francisco, California 94143*

Orgad Laub (3), *Department of Biochemistry and Biophysics, University of California, San Francisco, California 94143*

George M. Lawson (6), *Clinical Chemistry Section, Hilton Building, Mayo Clinic, Rochester, Minnesota*

Elias Lazarides (24), *Division of Biology, California Institute of Technology, Pasadena, California 91125*

L. Leben (2), *Department of Immunology, The London Hospital Medical School, University of London, Turner Street, London E1 2AD, England*

Mei-Hua Liu (13), *Department of Pharmacology, Baylor College of Medicine, 1200 Moursund Avenue, Houston, Texas 77030*

George L. Long (4), *Department of Biochemistry, University of Washington, Seattle, Washington 98195*

James McGhee (8), *Laboratory of Molecular Biology, Building 2, Room 401, National Institutes of Health, Bethesda, Maryland 20205*

Anthony R. Means (22), *Department of Cell Biology, Baylor College of Medicine, Houston, Texas 77030*

Synthia H. Mellon (17), *Metabolic Research Unit, 1143 HSW, University of California, San Francisco, California 94143*

A. L. Mellor (2), *Clinical Research Centre, Northwick Park Hospital, Watford Road, Harrow, Middlesex HA1 3UJ, England*

François Meyer (5), *CIBA-GEIGY AG, K 121-210, 4000 Basel, Switzerland*

Barry Nelkin (7), *The Johns Hopkins University School of Medicine, Oncology Center, Room 1-127, 600 North Wolfe Street, Baltimore, Maryland 21205*

J. R. Nevins (12), *The Rockefeller University, 1230 York Avenue, New York, New York 10021*

Joanne Nickol (8), *Laboratory of Molecular Biology, Building 2, Room 305, National Institutes of Health, Bethesda, Maryland 20205*

Bert W. O'Malley (6), *Department of Cell Biology, Baylor College of Medicine, Houston, Texas 77030*

Inbok Paek (18), *Institute of Cancer Research, 701 West 168th Street, New York, New York 10032*

Drew Pardoll (7), *The Johns Hopkins University School of Medicine, Oncology Center, Room 1-127, 600 North Wolfe Street, Baltimore, Maryland 21205*

James W. Posakony (19), *Department of Biochemistry and Molecular Biology, Harvard University, Cambridge, Massachusetts 02138*

Donald C. Rau (8), *Laboratory of Chemical Physics, Building 2, Room B1-03, National Institutes of Health, Bethesda, Maryland 20205*

Ramachandra Reddy (13), *Department of Pharmacology, Baylor College of Medicine, 1200 Moursund Avenue, Houston, Texas 77030*

Robin Reed (15), *Department of Biology, Yale University, New Haven, Connecticut 06520*

Jakob Reiser (5), *Institut für Molekularbiologie I, Universität Zürich, Hönggerberg, 8093 Zürich, Switzerland*

Diane M. Robins (18), *Institute of Cancer Research, 701 West 168th Street, New York, New York 10032*

Sabina Robinson (7), *The Johns Hopkins University School of Medicine, Oncology Center, Room 1-127, 600 North Wolfe Street, Baltimore, Maryland 21205*

John R. Rodgers (20), *Department of Cell Biology, Baylor College of Medicine, Houston, Texas 77030*

Jeffrey M. Rosen (20), *Department of Cell Biology, Baylor College of Medicine, Houston, Texas 77030*

Lawrence Rothblum (13), *Department of Pharmacology, Baylor College of Medicine, 1200 Moursund Avenue, Houston, Texas 77030*

William J. Rutter (3), *Department of Biochemistry and Biophysics, University of California, San Francisco, California 94143*

M. Santamaria (2), *Department of Immunology, The London Hospital Medical School, University of London, Turner Street, London E1 2AD, England*

Paolo Sassone-Corsi (14), *Laboratoire de Génétique Moléculaire des Eucaryotes du CNRS, Faculté de Médecine, 11 rue Humann, 67085 Strasbourg Cédex, France*

Jerome B. Schaack (16), *Department of Molecular Biophysics and Biochemistry, Yale University, P.O. Box 6666, New Haven, Connecticut 06511*

W. Schmidt (2), *Department of Immunology, The London Hospital Medical School, University of London, Turner Street, London E1 2AD, England*

William T. Schrader (25), *Department of Cell Biology, Baylor College of Medicine, 1200 Moursund Avenue, Houston, Texas 77030*

Wayne Schrier (13), *Department of Pharmacology, Baylor College of Medicine, 1200 Moursund Avenue, Houston, Texas 77030*

Peter H. Seeburg (18), *Genentech, Inc., 460 Point San Bruno Boulevard, South San Francisco, California 94080*

Mark Selby (17), *Department of Microbiology and Immunology, University of California, San Francisco, California 94143*

Stephen Sharp (16), *Department of Molecular Biophysics and Biochemistry, Yale University, P.O. Box 6666, New Haven, Connecticut 06511*

E. Simpson (2), *Clinical Research Centre, Northwick Park Hospital, Watford Road, Harrow, Middlesex HA1 3UJ, England*

Don Small (7), *The Johns Hopkins University School of Medicine, Oncology Center, Room 1-127, 600 North Wolfe Street, Baltimore, Maryland 21205*

Dieter Söll (16), *Department of Molecular Biophysics and Biochemistry, Yale University, P.O. Box 6666, New Haven, Connecticut 06511*

David Spector (13), *Department of Pharmacology, Baylor College of Medicine, 1200 Moursund Avenue, Houston, Texas 77030*

Stephen Sprang (3), *Department of Biochemistry and Biophysics, University of California, San Francisco, California 94143*

Joseph P. Stein (22), *Department of Endocrinology, University of Texas Health Science Center, Houston, Texas 77030*

William E. Stumph (6), *Department of Cell Biology, Baylor College of Medicine, Houston, Texas 77030*

Jamshed R. Tata (21), *National Institute for Medical Research, Mill Hill, London NW7 1AA, England*

P. M. Taylor (2), *Division of Immunology, National Institute for Medical Research, The Ridgeway, Mill Hill, London NW7 1AA, England*

Terry L. Thomas (19), *Division of Biology 156-29, California Institute of Technology, Pasadena, California 91125*

Susumu Tonegawa (1), *Center of Cancer Research, Massachusetts Institute of Technology, E17-352, Cambridge, Massachusetts 02139*

A. R. M. Townsend (2), *Division of Immunology, National Institute for Medical Research, The Ridgeway, Mill Hill, London NW7 1AA, England*

Ming-Jer Tsai (6), *Department of Cell Biology, Baylor College of Medicine, Houston, Texas 77030*

Bert Vogelstein (7), *The Johns Hopkins University School of Medicine, Oncology Center, Room 1-127, 600 North Wolfe Street, Baltimore, Maryland 21205*

E. Weiss (2), *Clinical Research Centre, Northwick Park Hospital, Watford Road, Harrow, Middlesex HA1 3UJ, England*

Charles Weissmann (5), *Institut für Molekularbiologie I, Universität Zürich, Hönggerberg, 8093 Zürich, Switzerland*

Berend Wieringa (5), *Institut für Molekularbiologie I, Universität Zürich, Hönggerberg, 8093 Zürich, Switzerland*

Savio L. C. Woo (4), *Howard Hughes Medical Institute, Baylor College of Medicine, 1200 Moursund Avenue, Houston, Texas 77030*

William Wood (8), *Laboratory of Molecular Biology, Building 2, Room 307, National Institutes of Health, Bethesda, Maryland 20205*

S. Wright (2), *Clinical Research Centre, Northwick Park Hospital, Watford Road, Harrow, Middlesex HA1 3UJ, England*

Carl Wu (10), *Laboratory of Biochemistry, National Cancer Institute, National Institutes of Health, Bethesda, Maryland 20205*

Li Yuan Yu-Lee (20), *Department of Cell Biology, Baylor College of Medicine, Houston, Texas 77030*

M. Zeevi (12), *Biotechnology General, Kiryat Weizmann, Rehovot 76326, Israel*

Preface

This volume documents the proceedings of the CETUS–UCLA Symposium "Gene Regulation," held in Keystone, Colorado in March/April 1982. It was one of the conferences of the 1982 series of the UCLA Symposia on Molecular and Cellular Biology.

The symposium related gene structure and regulatory sequences to overall genomic organization and genetic evolution. It was the first meeting to focus on regulation of eukaryotic gene expression since the maturation in recombinant DNA technology.

The success of the meeting was due to a great extent to the professionalism of Sandy Malone and her staff at the UCLA Symposia: Robert (Hank) Harwood, Maureen Kronish, and Sylvia Sledger. We wish to thank CETUS Corporation for its generous sponsorship of this meeting and gratefully acknowledge the financial support of Beckman Instruments, Inc., and Lilly Research Laboratories.

ORGANIZATION AND EXPRESSION OF MOUSE λ LIGHT
CHAIN IMMUNOGLOBULIN GENES [1]

Bonnie Blomberg and Susumu Tonegawa

Center for Cancer Research
Massachusetts Institute of Technology
Cambridge, MA. 02139

ABSTRACT   The four λ light chain constant region (C)
genes have been cloned from BALB/c mouse embryo DNA. The
$C\lambda_1$ gene segment was previously analyzed (1,2). Each $C\lambda$
gene carries its own J segment approximately 1.3 kilo-
bases to its 5' side which contrasts with both the kappa
(κ) and heavy (H) chain immunoglobulin gene systems with
a cluster of four functional joining (J) sequences 5' to
the constant gene segment(s). The four $C\lambda$ genes occur in
two clusters: $5'J_3C_3J_1C_1 3'$ and $5'J_2C_2J_4C_4 3'$. The J DNA
segments of λ_2, λ_3 and λ_4 were sequenced and compared with
that of λ_1. Sequence homology (particularly in the non-
coding regions) was greatest between J_1 and J_4 and between J_2 and J_3 which suggests, along with the similar
organization of JCJC and crosshybridization of C_1 and C_4
and of C_2 and C_3, that the two clusters are products of a
duplication event. A single variable region (V)λ gene,
5' of each JCJC cluster, was probably part of this duplication unit. We have confirmed that there are only two
Vλ genes in mouse (Vλ_1 and Vλ_2), and we have also shown
that the Vλ_1 gene segment is joined productively to $C\lambda_3$
in a λ_3 myeloma. Vλ_1 has been found associated only with
$C\lambda_3$ or $C\lambda_1$ and in most cases Vλ_2 joins with $C\lambda_2$ (the
exceptions allow us to deduce a probable organization of
the total λ locus). From these data and from the analysis of germ line and rearranged Vλ genes in myelomas, the
two Vλ genes must be interspersed by a JCJC cluster if
the looping-out and deletion model is generally used for
V-J joining. The organization of the λ locus is most

[1]This work was supported by a Post Residency Cancer
fellowship award from the American Cancer Society to B.B.
S.T. is supported in part by grants from NIH (CA-14051),
the Cancer Research Institute (NY), and the Whitaker
Health Sciences Fund (82-09). This investigation was
conducted, in part, at the Basel Institute for Immunology,
which was founded and is supported by Hoffmann - La
Roche.

1

likely: $5'V_2-J_2C_2J_4C_4-V_1-J_3C_3J_1C_1 3'$. The λ_4 gene is probably
not functional since the J_4 sequence does not contain a valid
splice site and has a 2-bp deletion in the signal heptamer
sequence 5' to J4. The signal nonamer sequence 5' to J_3
differs from that of J_1 in two consecutive base pairs and may
account for the lower level of λ_3 expression as compared with
λ_1 in mouse lymphocytes.

INTRODUCTION

The immunoglobulin genes for the light chains, κ and λ,
and the heavy chains occur in three families. In mouse, the
genes for the light chains have been placed on chromosome 6
(3,4), the heavy (H) chains on chromosome 12 (5,6) and the
light chains on chromosome 16 (7). It was proposed that the λ
germ line variable (V) region gene segments are separate from
the constant (C) gene segments for light chains (8) and likewise for heavy chains (9, 10). This assumption was demonstrated to be correct and it was found that formation of a
functional immunoglobulin (Ig) gene requires DNA linkage of V
to the J (joining) region in the case of light chain genes
(11, 1) and V, D, and J in the case of heavy chain genes (12,
13)(the D, diversity, DNA segments encode primarily the third
hypervariable region of heavy chains). For the κ light chain,
one of about two hundred V genes (14, 15) can combine with one
of four functional J segments (16, 17). The exact site of
joining may vary slightly (16, 17). Thus two somatic mechanisms
occur to generate diversity within a light chain: <u>combinatorial</u>
diversity when different combinations of V and J gene segments
assemble and <u>junctional diversity</u> when different amino acids
are generated by slippage in the actual site of V-J joining.
Likewise, increased antibody diversity is generated when one
of the V_H gene segments is joined with one of approximately
twenty D segments (18) and one of four J_H segments (13, 12).
Junctional diversity can also occur for heavy chains (13, 19).
Antibody diversity is also generated when different light
chains combine with different heavy chains. The antibody
repertoire is further expanded by somatic mutation of the V
gene segments for both light chains (20, 2, 21, 22) and heavy
chains (13, 23-26).

We wished to study the organization and molecular basis
for differential expression of the mouse λ light chain genes.
The mouse λ light chain system is simple when compared with
that of the κ and H chains in that there are only two V genes,
V_1 and V_2, (1) and the λ myeloma and serum proteins have been
well defined. The λ light chains, which comprise about 5% of
the mouse serum Ig (27-29) occur in three subtypes, λ_1 (20,
30, 31), λ_2 (32, 33), and λ_3 (34). A subtype is defined by

its C region amino acid sequence. The three subtypes, λ_1, λ_2 and λ_3 occur in the serum in the approximate ratio 8:1:1 (29, 34) and in spleen lymphocytes in the approximate ratio 1.0: 0.7: 0.3 (35). The λ_1 chains previously reported all contained V regions encoded by V_1 (1, 15) and all known λ_2 chains use V_2 (29, 34, 36). It was recently shown that λ_3 chains also use the V_1 gene (37,38). Each of the subtypes is encoded by its own constant region gene. We describe here the V, J, and C gene segments of the mouse λ locus and give their probable organization in the genome. We show the DNA sequences surrounding each of the λJ segments, suggest a mechanism for control of λ_1 and λ_3 subtype expression, and demonstrate that λ_4 is a pseudogene.

RESULTS AND DISCUSSION

<u>Description and Characterization of Four λC Genes</u>. High molecular weight kidney DNA from several mouse strains was digested with <u>Eco</u>RI endonuclease and analyzed by the Southern gel blotting procedure as previously described (37). Six mouse strains gave identical results as those shown for the BALB/c mouse strain in Figure 1A. The hybridization probe was either $C\lambda_1$ cDNA (600 base pairs complementary to $C\lambda_1$ from an Hha I/Hae III digest of the B1 plasmid of H2020 cDNA $(V+C)\lambda_1$) or $(V+C)\lambda_2$ cDNA (cDNA prepared from MOPC 315, containing $(V+C)\lambda_2$, from R. Schwartz and M. Gefter). With the $(V+C)\lambda_2$ probe we usually saw four bands, at 8.6, 4.8, 3.5, and 3.2 kilobases (kb). The 4.8 kb and 3.5 kb bands are $V\lambda_2$ and $V\lambda_1$ respectively (the two Vλ genes cross hybridize) (1). The 8.6 kb and 3.2 kb bands were candidates for the $C\lambda_2$ and $C\lambda_3$ genes. We expected that $C\lambda_2$ and $C\lambda_3$ would be detected by the $(V+C)\lambda_2$ probe since the amino acid sequences of $C\lambda_2$ and $C\lambda_3$ differ by only five out of 107 residues (34). With the $C\lambda_1$ probe we saw a band at 8.6 kb, known to contain the $C\lambda_1$ gene (1) and another band at 2.8 kb which we designated $C\lambda_4$ (37). The $C\lambda_4$ band did not cross hybridize with the $(V+C)\lambda_2$ probe and was not a candidate for the $C\lambda_2$ or $C\lambda_3$ gene. Occasionally, we detected a band at 6 kb (due to partial digestion, not shown in Figure 1A) which hybridized with both probes.

The bands at 2.8 kb, 6.0 kb, and 8.6 kb were enriched by preparative gel electrophoresis and cloned in λWES phage. A clone of the 8.6 fragment was selected by hybridization with the $(V+C)\lambda_2$ probe but was found to hybridize with both probes and was identical to our previously described clone of $C\lambda_1$, Ig25 (1)(Figure 1B). When the Ig25 cloned DNA is digested with <u>Eco</u>RI, the 8.6 kb band hybridizes to both the $C\lambda_1$ and $(V+C)\lambda_2$ probes (Figure 1B). The region of this <u>Eco</u>RI insert which hybrides to $(V+C)\lambda_2$ is 5' to that of the $C\lambda_1$ gene; when

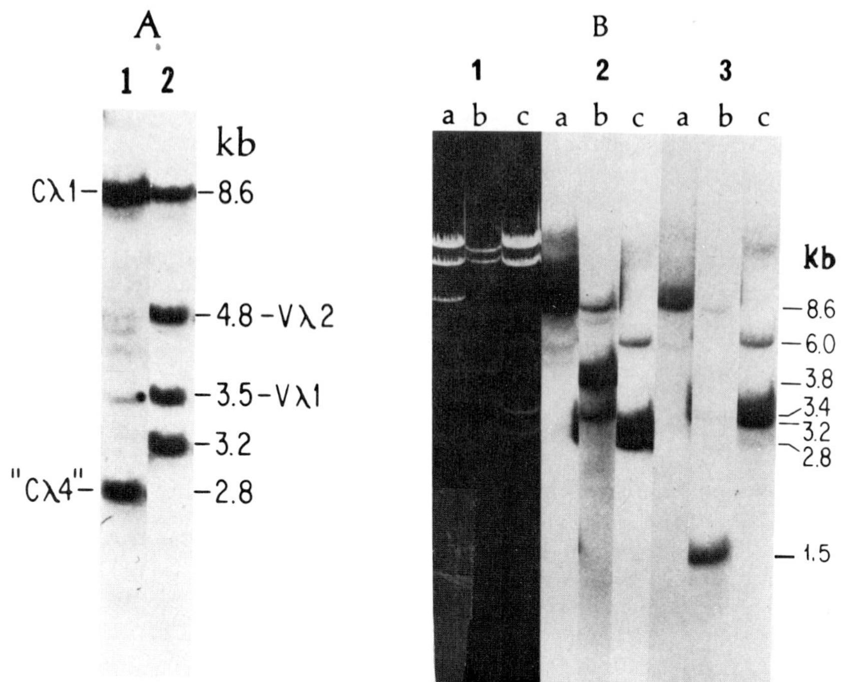

FIGURE 1A. <u>BALB/c embryo DNA digested with EcoRI and hybridized with 1) $C\lambda_1$ or 2) $(V+C)\lambda_2$.</u> Fragment sizes (in kilobases (kb)) were determined by comparison with Hind III-digested λ phage DNA as a size marker.

FIGURE 1B. <u>Cloned DNA fragments containing $C\lambda_1$- or $(V+C)\lambda_2$-hybridizing sequences.</u> Duplicate samples of cloned DNA were electrophoresed in agar, blotted according to Southern and hybridized to either $C\lambda_1$ or $(V+C)\lambda_2$. Section 1: ethidium bromide stain of gel. The top two bands in each sample are phage arms of λWES. Section 2: hybridized to $C\lambda_1$. Section 3: hybridized to $(V+C)\lambda_2$. Lane a: clone Ig25 digested with EcoRI, showing the 8.6 kb fragment hybridizing with both $C\lambda_1$, and $(V+C)\lambda_2$. Lane b: Ig25 digested with KpnI. Lane c: 10A1 digested with <u>Eco</u>RI showing a band at 2.8 kb hybridizing with $C\lambda_1$ and a band at 3.2 kb hybridizing to $(V+C)\lambda_2$. The weak partial band at 6.0 kb in clone 10A1 can be seen to hybridize with both probes.

Ig25 insert DNA is digested with KpnI, the 1.5 kb band, at the 5' side of $C\lambda_1$ (2) hybridizes with $(V+C)\lambda_2$ but not $C\lambda_1$ and the 3.8 and 3.4 kb fragments hybridize only with $C\lambda_1$, as expected. Similarly, clone 10A1, from the 6 kb fragment, when completely digested with <u>Eco</u>RI, revealed two fragments: one at 2.8 kb hybridized only to $C\lambda_1$, and the other at 3.2 kb hybridized only to $(V+C)\lambda_2$. Thus both clones contained two $C\lambda$ genes.

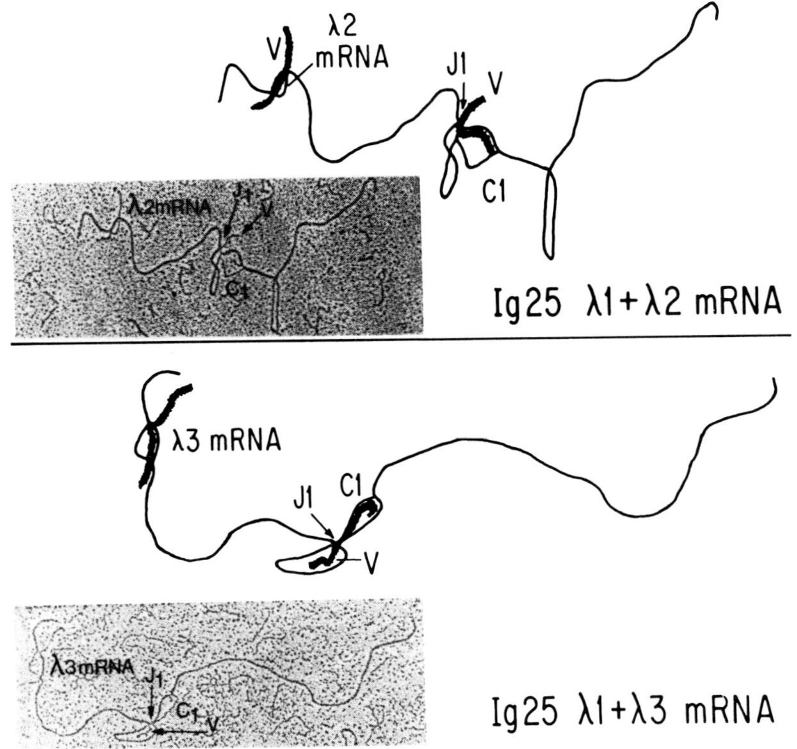

FIGURE 2. Electron micrographs and line drawings of clone Ig25 DNA. R-loops of Ig25 double-stranded insert (8.6 kb) DNA with λ_1 and λ_2 mRNA (top panel) or λ_1 and λ_3 mRNA (bottom panel).

We were able to demonstrate that the clone containing $C\lambda_1$ (Ig25) also carried $C\lambda_3$ and that $C\lambda_2$ and $C\lambda_4$ were closely linked (in clone 10A1)(37). Comparisons of the r-loop structures formed by hybridization of DNA from clone Ig25 with a mixture of λ_1 and λ_2 mRNA or λ_1 and λ_3 mRNA showed a larger, more open loop structure with λ_3 mRNA, indicating a greater degree of homology (Figure 2). Similarly, when the 3.2 kb EcoRI insert of clone 10A1 was hybridized with λ_2 or λ_3 mRNA (Figure 3), more homology was seen with λ_2 mRNA. The 2.8 kb band of clone 10A1 was identical to the 2.8 kb band cloned from embryo DNA by restriction enzyme mapping and DNA heteroduplex analysis. Each of the four $C\lambda$ gene segments was shown by r-looping to carry its own $J\lambda$ segment approximately 1.3 kb to the 5' side (also see J sequence data below).

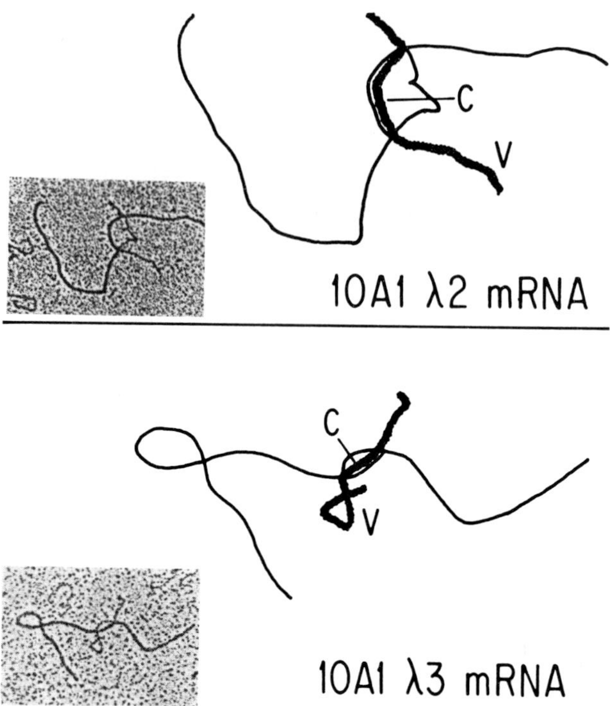

FIGURE 3.   Electron micrographs and line drawings of DNA from clone Ig10A1.   R-loops of the double-stranded 3.2 kb fragment from EcoRI-digested Ig10A1 DNA with λ_2 mRNA (top panel) or λ_3 mRNA (bottom panel).

The organization of these two Cλ gene clusters is shown in Figure 4.   The $J_3C_3J_1C_1$ organization was shown independently by Miller et al. (39).   From the similarity in organization of the two clusters and from the cross hybridization between C_2 and C_3, and C_1 and C_4, we proposed that the two clusters arose by duplication and that this duplication unit most likely also included V at the 5' side of the JCJC cluster (37).

Placement of the Two λV Genes.   We showed that $V\lambda_1$ has joined to $J\lambda_3$ in the productive rearrangement of a λ_3 producing myeloma.   (See Figure 5 A and B: CBPC-49, a λ_3 myeloma shows a rearranged band at 2.8 kb with a $V(J)\lambda_1$ probe.   This band was cloned and shown by heteroduplex and r-loop analysis to contain $V\lambda_1$, $J\lambda_3$, and the sequence 3' of $J\lambda_3$ up to the EcoRI

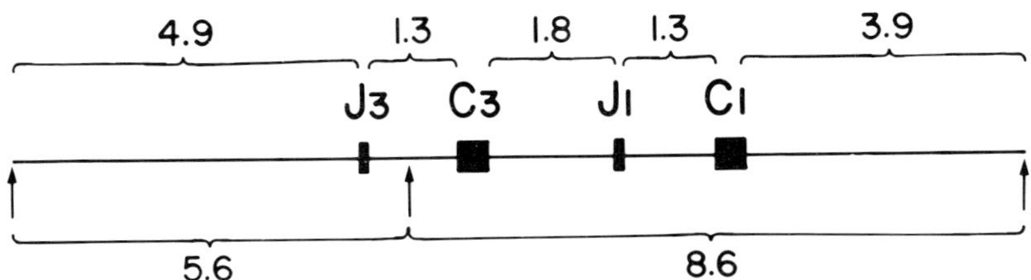

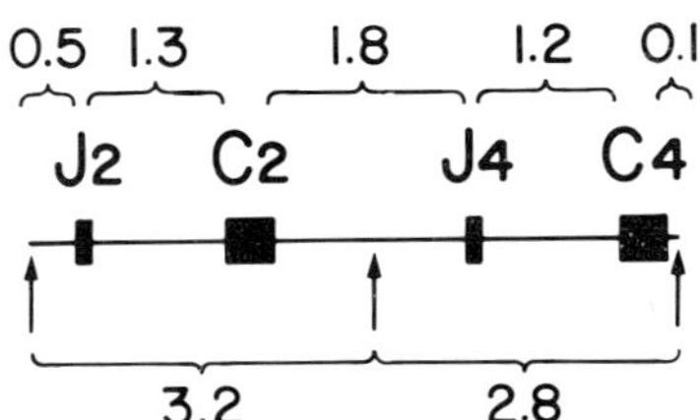

FIGURE 4. <u>Germ line configuration of mouse λ J and C gene segments.</u> Cloned DNA from <u>Eco</u>RI partial digests of BALB/c embryo DNA. The $J_3C_3J_1C_1$ gene segments are from clone Ig S8.2 (not presented) and $J_2C_2J_4C_4$ from clone Ig 10A1 (37). Distances are in kilobases and the J and C regions are designated. Arrows mark the EcoRI cleavage sites. The distance between the two clusters is unknown.

site between $J\lambda_3$ and $C\lambda_3$ (37)). Since V_1 could be shown to be used with C_3 as well as C_1, since V_2 had only been found associated with C_2, and since the V_1 and V_2 gene segments were very homologous (41, 2), we concluded that the probable evolutionary duplication unit was V-JCJC (37). DNA sequence evidence for the J_2-J_3 and J_4-J_1 homology is given below and further supports the above argument.

Further data from which we may deduce the most likely arrangement of the mouse λ locus are given in Figures 5A and 5B. Figure 5A shows a Southern blot of the myeloma Ag8.653 with the $V(J)\lambda_1$ probe where $V\lambda_2$ has undergone two rearrangements (the embryonic $V\lambda_2$ band at 4.8 kb is absent). The two rearranged bands are at 6.5 kb (V_2C_2) and 9.0 kb (9.0 kb is the size predicted for a V_2C_1 rearrangement: the distance from the 5' <u>Eco</u>RI site to the $V\lambda_2$ gene segment (in clone Ig13, ref. 42) is 3.3 kb, the complete V gene is about 0.5 kb, and the distance of $J\lambda_1$ to the 3' <u>Eco</u>RI site (in clone Ig25, ref. 1,

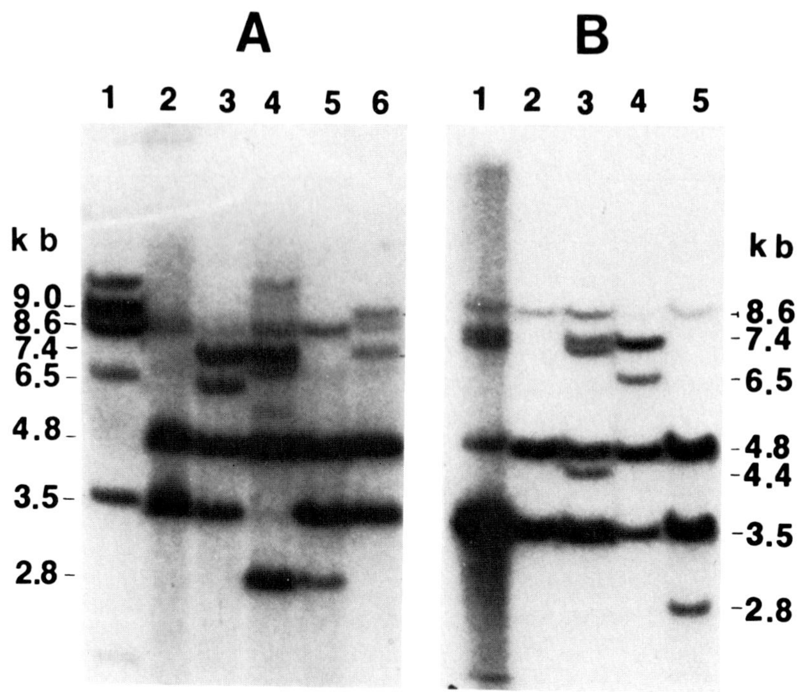

FIGURE 5 A and B. <u>BALB/c embryo, myeloma and hybridoma</u>
<u>DNA fragments from EcoRI digestion hybridized to V(J)λ_1</u>
<u>sequences.</u> The probe contains only the V and J regions of λ_1
cDNA. A. DNA from (1) Ag8.653, myeloma fusion parent for
hybridoma production, (2) BALB/c embryo, (3) MOPC-315 ($\lambda_2\alpha$), a
myeloma producing V_2C_2 light chains (rearranged band at 6.5 kb)
and which has also rearranged V_1C_1 (7.4 kb) (See ref. 40 and
37), (4) J558 ($\lambda_1\alpha$) has productively rearranged V_1C_1 and also
aberrantly rearranged V_1C_3 (2.8 kb), (5) CBPC-49, $a^1\lambda_3$ producing
myeloma (V_1C_3 at 2.8 kb), and (6) P543.6, anti-NP λ-producing
hybridoma from T. Imanishi-Kari, possibly V_2C_1 (9.0 kb) as
well as V_1C_1 (7.4 kb). B. DNA from (1) SP-2, myeloma fusion
parent for 6-2 hybridoma, (2) BALB/c embryo, (3) 6-2, hybridoma
(from H.N. Eisen) with 4.4 kb rearranged band, V_2J_3, (4) MOPC-
315, and (5) CBPC-49. Fragment sizes are in kb.

see also Figure 4 this paper) is 5.2 kb, producing a rearranged
fragment of 9.0 kb). The presence of an embryonic $V\lambda_1$ band in
Ag8.653 (in the absence of any $V\lambda_2$ band) rules out the possibility that $V\lambda_1$ lies 3' to $V\lambda_2$ without an intervening JCJC
cluster, i.e. the V genes are not clustered (V_2V_1). By a
similar argument, the J558 DNA pattern in Figure 5A eliminates

another possibility that $V\lambda_2$ lies 3' to $V\lambda_1$ without an intervening JCJC cluster (the organization is not $V_1 V_2$). In J558 two $V\lambda_1$ gene segments have rearranged, one productively to $J\lambda_1$ (at 7.4 kb) and the other to $J\lambda_3$ (at 2.8 kb); no embryonic $V\lambda_1$ band (at 3.5 kb) can be seen, and yet an embryonic $V\lambda_2$ band (at 4.8 kb) is present. The two cell types, Ag8.653 and J558, are both diploid (as demonstrated by the absence of the embryonic V_2 or V_1 bands,respectively). Therefore, provided that looping out and deletion is the predominant mechanism for V-J joining (16), the two V genes cannot exist without interspersion by a JCJC cluster, i.e. the organization is V-JCJC-V-JCJC.

We suggested above that unusual λ rearrangements may occur (as $V_2 C_1$ in Ag8.653, Figure 5A). This $V_2 C_1$ rearrangement may also have occured in a λ hybridoma (P543.6, from T. Imanishi-Kari, Figure 5A) which appears to produce a protein which shares characteristics of V_2 and C_1 both serologically and biochemically (Reilly, E., B. Blomberg, T. Imanishi-Kari, and H.N. Eisen, in preparation). Another rare λ rearrangement, in the hybridoma 6-2, is shown in Figure 5B. SP-2 is the myeloma which was fused with BALB/c spleen cells for the 6-2 hybridoma production. The rearranged band (the only non-embryo or new band) in 6-2, at 4.4 kb, is consistent only with a rearrangement of $V\lambda_2$ to $J\lambda_3$ (3.3 + 0.5 + 0.6 (the distance of $J\lambda_3$ to the 3' $\underline{EcoRI}$ site)). Recently, the amino acid sequence of the 6-2 protein was shown to contain the sequence of $V\lambda_2$, $J\lambda_3$, and $C\lambda_3$ (Elliot, B., H.N. Eisen, and L. Steiner, submitted). These rare recombinations ($V_2 C_1$ and $V_2 C_3$ but not $V_1 C_2$) in the approximately 40 λ myelomas and hybridomas studied to date suggest the organization seen in Figure 6.

<u>DNA Sequences of the Four J Regions</u>. Although V_1 is capable of joining with both J_3 and J_1, the ratio of $\lambda_1:\lambda_2: \lambda_3$ is 8:1:1 in the serum and 1.0: 0.7: 0.3 in spleen lymphocytes (35). In order to determine a possible explanation for the lower expression of λ_3 as compared with that of λ_1, to obtain further evidence for an evolutionary duplication unit of JCJC, and to establish whether the $J_4 C_4$ gene segments might be functional, we determined the DNA sequences of the λ J segments of J_2, J_3, and J_4 (Figure 7).

<u>The λ_4 gene is probably a pseudogene</u>. There has been a 2 base pair (bp) deletion within the heptamer recognition sequence (underlined) 5' of J_4 and the donor RNA splice site is missing at the 3' side of J_4. The dinucleotide GT, present at amino acid position 110 in all functional J regions and an obligatory part of all RNA splicing signals (44) is absent in all reading frames of J_4 in this position (43,45). In

Organization of Mouse λ Light Chain Genes

Chromosome 16

FIGURE 6. <u>Mouse λ light chain locus</u>. The most probable arrangement of the gene segments known to date in the λ light chain locus. Distances between V and its C gene cluster and between each C cluster (shown as -//- and ...) have not been determined.

addition, no protein representative of λ_4 has been described and all serum, hybridoma, and myeloma λ chains can be accounted for by λ_1, λ_2 and λ_3 (20, 29, 34).

<u>The $J_3C_3J_1C_1$ and $J_2C_2J_4C_4$ Sequences Arose by Duplication.</u> The J_1 and J_4 sequences and the J_2 and J_3 sequences are more alike, especially in the noncoding regions (43). The DNA sequences of J_3 and J_2 are 92% homologous in the coding region, 83% (84/101 bp) in the noncoding regions; J_1 and J_4 are 82% homologous in the coding region, 74% (115/155 bp) in the noncoding region. The sequences of J_3 and J_1 are only 77% homologous in the coding region and 30% (49/163 bp) in the noncoding regions; J_2 and J_4 are 75% homologous in the coding region and 35% (50/144 bp) in the noncoding regions. Figure 7 also confirms that each Cλ gene carries its own J sequence to its 5' side: in the case of J_1, J_2, and J_3, the DNA sequences correspond to the previously determined amino acid sequences (20, 38, 34). For J_4, no amino acid sequence is available, but a J-like sequence exists approximately 1.3 kb 5' to the C_4 segment.

<u>Differences Within the Nonamer Sequence 5' of J_3 and J_1 May Account for the Differential Expression of λ_3 and λ_1 Igs.</u> The λ_1 and λ_3 genes share the same $V\lambda_1$ gene and yet the level of expression of λ_3 is much lower than that of λ_1. The ratio of $\lambda_1 : \lambda_2 : \lambda_3$ is about 8:1:1 in the serum and 1.0: 0.7: 0.3 in the spleen lymphocytes (35). Consensus nonamer and heptamer sequences 3' of all V gene segments, 5' of all J gene segments and flanking the D gene segments have been proposed as possible recognition sites for a "recombinase" involved in V-J or V-D-J gene assembly (16, 17, 12, 13). These consensus sequences are

                                              98 99   102   106   110
                                          TrpValPheGlyGlyGlyThrLysLeuThrValLeuGly

J1   Ig25λ   AAATGCATGC-AAGGTTTTTGCATGAGTCTATATCACAGTGCTGGGTGTTCGGTGGAGGAACCAAACTCACTGTCCTAGGTGAGTGACTCCTTCCTCCT
                                                                       * *   * * *   *
         ** *   ** *   * *   **   *TrpValPheGlyGlyGlyThrArgLeuThrValLeuAsp*   *
J4   Ig10A1 AGGTACATGCAGAGTTTTTTGCATTAGACTATAT--CAGTGTTGGGTGTTCGGAGGTGGAACCAGATTGACTGTCCTAGATGAGTGACTCCTCCCTCCT

                                          PheIlePheGlySerGlyThrLysValThrValLeuGly
J3   IgS8.2 TGCTTGCCCCACAGGTTTAGGGTTGGGTTTCAGTCACTGTGGTTTATTTTCGGCAGTGGAACCAAGGTCACTGTCCTAGGTAAGTGGCTTTAATGCTTC
                                                       * *   *
          *   *   *   *   *   *TyrValPheGlyGlyGlyThrLysValThrValLeuGly*   * * * *
J2   Ig10A1 TGCTGGCCCCATAGGTTTTGGGTTGGGTTTTAGTCATTGTGTTATGTTTTCGGCGGTGGAACCAAGGTCACTGTCCTAGGTAAGTAGTTTCAAAGC

FIGURE 7.   Comparison of nucleotide sequences of germ line λ
segments and surrounding regions.   Extensions of these sequences
and sequence strategy are given in ref. 43.   *, nonidentical base
pairs in comparisons of the J_1-J_4 and J_3-J_2 sequences.   Signal
nonamer and heptamer sequences 5' to the J regions are underlined.
Amino acids encoded by nucleotide sequences are shown in italics.
The J_1 sequence was reported previously by Bernard et al. (2).

GGTTTTTGT and CACTGTG. There are two consecutive base pairs changed within the nonamer signal sequence of J_3 as compared with J_1 (Figure 7) (43,45). This difference may cause less efficient joining of V_1 to J_3 and account for a level of expression of λ_3 which is lower than that of λ_1 as seen in serum and in lymphocytes.

CONCLUDING REMARKS

We have described the organization of all of the gene segments known to date for the mouse λ light chain immunoglobulin locus. This locus is unusual and simple in that there are only two V gene segments and four C gene segments, each with its own J segment approximately 1.3 kb to its 5' side. The most likely organization of this locus is shown in Figure 6. These gene segments would account for all the expressed λ light chains, in fact it appears that the λ_4 gene is a pseudogene and not expressed at the protein level. We cloned a crosshybridizing gene segment but it was not further analyzed (37); it is possible that other C gene segments exist in mouse and are not expressed. Recently, wild mice gave 5-7 DNA fragments hybridizing to a $C\lambda_1$ probe, suggesting the presence of additional $C\lambda_1$-like genes (46). Perhaps the weakly hybridizing bands seen in Figure 1A with the $C\lambda_1$ probe are analogous to those seen in the wild mice. The human λ light chain locus has also been found to contain multiple (at least six) copies of closely linked C gene segments (47) but the J gene segments have not yet been located.

The organization of the mouse λ locus (Figure 6) is derived, initially, from the data for physical linkage of $J_3C_3J_1C_1$ and $J_2C_2J_4C_4$ within each cluster (37, 39). The placement of the V gene segments in relation to the two clusters, and the placement of one V-JCJC cluster in relation to the other was deduced from analyses of germ line and rearranged V genes (Figure 5, A and B and ref. 37) and from the fact that V_1-J_1, V_1-J_3 and V_2-J_2 associations occur preferentially in the production of λ_1, λ_2 and λ_3 light chains. The latter preference may simply reflect the proximity of a given V gene segment to its JCJC cluster.

Within a given cluster, eg. V_1-$J_3C_3J_1C_1$, the level of expression of a particular subtype, eg. λ_3 or λ_1, may reflect the efficiency of the V-J joining. The nonamer sequence, 5' of J_3 is different from that of J_1 in two consecutive base pairs (Figure 7). This difference from the consensus sequence may result in less efficient joining of V_1-J_3 as compared with V_1-J_1 and hence a lower level of expression of λ_3 in lymphocytes and serum.

Recently, we cloned the V, J and C gene segments for λ_1 from SJL, a mouse strain with a genetic defect resulting in a low level of λ_1 serum Ig and λ_1-bearing lymphocytes (28). Geckeler et al. (28) proposed that this defect resided in "one of the DNA level recognition sites involved in the transloca- tion event which places the $V\lambda_1$ and $C\lambda_1$ structural genes in a transcriptional unit". This was a reasonable proposal since it appeared that the expression of the λ locus, but not the λ structural locus itself, was affected; the defect behaved as a single gene, and the defect was cis-dominant (normal by λ-low mice gave one-half the normal level of λ_1 serum Ig). We have sequenced the SJL $J\lambda_1$ and $C\lambda_1$ gene segments and surrounding regions. The nonamer and heptamer recognition sequences are identical to those of BALB/c (the SJL nonamer sequence is like that of Mill et al. (45) and different from that of Bernard et al. (2). The sequences for the donor splice site, at the 3' side of J, and the acceptor splice site, at the 5' side of C are intact and identical to those of BALB/c. The poly A addition site is identical to that reported for BALB/c $C\lambda_1$ (40). Within the $C\lambda_1$ coding region there were changes in two bp which give two amino acid differences in the C region. Therefore from the DNA sequence data SJL does not appear defective in the potential for V-J joining or λ_1 expression. The cause of the defect may occur during or after transcription, or perhaps there is cellular suppression of the lymphocytes expressing SJL λ_1 (48).

ACKNOWLEDGMENTS

We wish to thank Michéle Courtet for excellent technical assistance.

REFERENCES

1. Brack, C., Hirama, M., Lenhard-Schuller, R. and Tonegawa, S. (1978). Cell 15, 1.
2. Bernard, O., Hozumi, N., and Tonegawa, S. (1978). Cell 15, 1133.
3. Gottlieb, P. (1974). J. Exp. Med. 140, 1432.
4. Weigert, M. and Potter, M. (1977). Immunogenetics 4, 401.
5. Hengartner, H., Meo, T., and Müller, E. (1978) Proc. Natl. Acad. Sci. USA 75, 4494.
6. Meo, T., Johnson, J., Beechey, C.V., Andrews, S.J., Peters, J., Searle, A.G. (1980). Proc. Natl. Acad. Sci. USA 77, 550.
7. D'Eustachio, P., Bothwell, A.L.M., Takaro, T.K., Baltimore, D., and Ruddle, F.H. 1981. J. Exp. Med. 153, 793.

8. Dreyer, W.J. and Bennett, J.C. (1965). Proc. Natl. Acad. Sci. 54, 864.

9. Mage, R., Young, G.O., and Alexander, C. (1971). Nature New Biol. 230, 63.

10. Riblet, R., Weigert, M., and Makela, O. (1975). Eur. J. Immunol. 5, 778.

11. Hozumi, N., and Tonegawa, S. (1976). Proc. Natl. Acad. Sci. USA 73, 3632.

12. Early, P., Huang, H., Davis, M.M., Calame, K., and Hood, L. (1980). Cell 19, 981.

13. Sakano, H., Maki, R., Kurosawa, Y., Roeder, W., and Tonegawa, S. (1980). Nature 286, 676.

14. Cohn, M., Blomberg, B., Geckeler, W., Raschke, W., Riblet, R., and Weigert, M. (1974). In "The Immune System: Genes, Receptors, Signals" (E.E. Sercarz, et al., eds.), p. 89, Academic Press, New York.

15. Weigert, M., and Riblet, R. (1977). Cold Spring Harbor Symp. Quant. Biol. 41, 837.

16. Sakano, H., Huppi, K., Heinrich, G., and Tonegawa, S. (1979). Nature 280, 288.

17. Max, E.E., Seidman, J.G., and Leder, P. (1979). Proc. Natl. Acad. Sci. USA 76, 3450.

18. Kurosawa, Y., and Tonegawa, S. (1982). J. Exp. Med. 155, 201.

19. Kurosawa, Y., von Boehmer, H., Haas, W., Sakano, H., Trauneker, A., and Tonegawa, S. (1981). Nature 290, 565.

20. Weigert, M., Cesari, I.M., Yonkovich, S.J., and Cohn, M. (1970). Nature 228, 1045.

21. Selsing, E. and Storb, U. (1981) Cell 25, 47.

22. Gershenfeld, H.K., Tsukamoto, A., Weissman, I.L., and Joho, K. (1981). Proc. Natl. Acad. Sci. USA 78, 7674.

23. Bothwell, A.L.M., Paskind, M., Reth, M., Imanishi-Kari, T., Rajewsky, K., and Baltimore, D. (1981). Cell 24, 625.

24. Gearhart, P.J., Johnson, N.D., Douglas, R., and Hood, L. (1981). Nature 291, 29.

25. Crews, S., Griffin, J., Huang, H., Calame, K., and Hood, L. (1981) Cell 25, 59.

26. Kim, S., Davis, M., Sinn, E., Patten, P., and Hood, L. (1981). Cell 27, 573.

27. McIntire, K.R. and Rouse, A.M. (1970). Fed. Proc. Fed. Am. Soc. Exp. Biol. 29, 704.

28. Geckeler, W., Faversham, J., and Cohn, M. (1978). J. Exp. Med. 148, 1122.

29. Cotner, T., and Eisen, H.N. (1978). J. Exp. Med. 148, 1388.

30. Appella, E. (1971). Proc. Natl. Acad. Sci. USA 68, 590.

31. Cesari, I.M., and Weigert, M. (1973). Proc. Natl. Acad. Sci. USA 70, 2112.

32. Dugan, E.S., Bradshaw, R.A., Simms, E.S., and Eisen, H.N.
    (1973). Biochemistry 12, 5400.
33. Blaser, K. and Eisen, H.N. (1978). Proc. Natl. Acad.
    Sci. USA 75, 1495.
34. Azuma, T., Steiner, L.A., and Eisen, H.N. (1981). Proc.
    Natl. Acad. Sci. USA 78, 569.
35. Reilly, E.B., Frackelton, A.R. Jr., and Eisen, H.N.
    (1982). Eur. J. Immunol. (in press).
36. Elliot, B.W. Jr., Steiner, L.A., and Eisen, H.N. (1981)
    (Abstract) Fed. Proc. Fed. Am. Soc. Exp. Biol. 40, 1098.
37. Blomberg, B., Traunecker, A., Eisen, H.N., and Tonegawa,
    S. (1981). Proc. Natl. Acad. Sci. USA 78, 3765.
38. Breyer, R., Sauer, R., and Eisen, H.N. (1981). In
    "Immunoglobulin Idiotypes, ICN-UCLA Symposia on Molecular
    and Cellular Biology", (C.A. Janeway, E.E. Sercarz, H.
    Wigzell, and C.F. Fox, eds.), pp. 105-110, Academic
    Press, New York.
39. Miller, J., Bothwell, A., and Storb, U. (1981). Proc.
    Natl. Acad. Sci. USA 78, 3829.
40. Bothwell, A.L.M., Paskind, M., Schwartz, R., Sonenshein,
    G.E., Gefter, M.L., and Baltimore, D. (1981). Nature
    290, 65.
41. Tonegawa, S., Maxam, A., Tizard, R., Bernard, O., Gilbert,
    W. (1978). Proc. Natl. Acad. Sci. USA 75, 1485.
42. Tonegawa, S., Brack, C., Hozumi, N., and Schuller, R.
    (1977). Proc. Natl. Acad. Sci. 74, 3518.
43. Blomberg, B. and Tonegawa, S. (1982). Proc. Natl. Acad.
    Sci. USA 79, 530.
44. Breathnach, R., Benoist, C., O'Hare, K, Gannon, F., and
    Chambon, P. Proc. Natl. Acad. Sci. USA 75, 4853.
45. Miller, J., Selsing, E., and Storb, U. (1982). Nature
    295, 428.
46. Scott, C.L., Mushinski, J.F., Potter, M., Huppi, K., and
    Weigert, M. (1981) (Abstract) Fed. Proc. Fed. Am. Soc.
    Exp. Biol. 41, 725.
47. Hieter, P.A., Hollis, G.F., Korsmeyer, S.J., Waldmann, T.
    A., and Leder, P. (1981). Nature 294, 536.
48. Takemori, T. and Rajewsky, K. 1981. Eur. J. Immunol.
    11, 618.

STRUCTURE AND EXPRESSION OF GLOBIN AND H-2 GENES

R. A. Flavell, F. G. Grosveld, G. C. Grosveld, S. Wright,
M. Busslinger, E. deBoer, D. Kioussis, A. L. Mellor, L.Golden,
E. Weiss, J. Hurst, H. Bud, H. Bullman, E. Simpson,[1] R. James,[2]
A. R. M. Townsend,[3] P. M. Taylor,[3] W. Schmidt,[4] J. Ferluga[4],
L. Leben,[4] M. Santamaria,[4] G. Atfield,[4] and H. Festenstein[4]

Laboratory of Gene Structure and Expression, National Institute
for Medical Research, The Ridgeway, Mill Hill, London NW7 1AA

ABSTRACT   The use of DNA-mediated gene transfer for the
study of the rabbit β-globin promotor, the molecular
basis of β^+thalassaemia and the specific induction of
human β-globin expression in mouse erythroid cells is
described.   In addition, we describe the isolation of
the class I mouse MHC genes from the $H-2^b$ haplotype.
We further show that an $H-2K^b$ gene when introduced into
mouse L cells renders these cells a target for both
allospecific and H-2 restricted, virus-specific
cytotoxic T cell killing.

INTRODUCTION

The globin system has served us well as a model for the
ways in which genes are organized on the chromosome and
expressed in the cells of higher eukaryotes.   In this article
we consider three aspects of globin gene transcription and
RNA processing, all studied using the techniques of DNA-
mediated gene transfer.   In the second part of the paper, we
describe the application of these techniques to the analysis
of the functioning of cell surface antigens of the major
histocompatibility complex (H-2) of the mouse.

[1]Clinical Research Centre, Northwick Park Hospital,
Watford Road, Harrow, Middlesex HA1 3UJ.

[2]ICRF Tumour Immunology Unit, University College,
Gower Street, London WC1E 6BT.

[3]Division of Immunology, National Institute for Medical
Research, The Ridgeway, Mill Hill, London NW7 1AA.

[4]Dept. of Immunology, The London Hospital Medical School,
University of London, Turner Street, London E1 2AD.

THE RABBIT β-GLOBIN PROMOTOR

For a number of years we have studied the DNA sequences
required for the transcription, <u>in vivo</u> and <u>in vitro</u>, of the
rabbit β-globin gene.   DNA sequence comparisons made by
others, have pointed to two sequences, present to the 5' side
of genes transcribed by RNA polymerase II called the "ATA" box
($^{T}_{C}$ATA$^{T}_{A}$AAG; refs. 1, 2, 3) and the "CAAT" box (GG$^{T}_{C}$CAATCT (3).
Subsequent to their discovery, considerable effort has been
expended to test the requirement of these sequences for transcription of cellular and viral genes.   Thus, deletion of the
ATA box causes a loss of specificity in the choice of the site
of initiation of RNA synthesis (4, 5, 6, 7) and usually (4, 5)
but not always (5, 6) a strong down mutation.   Deletion of
the CAAT box region is usually a strong down mutation (8, 5,
6, 7, 9), but these studies did not determine the necessary
DNA sequences within this region.   In the past nine months we
have attempted to delineate these DNA sequences, within the
-80 region which are required for transcription <u>in vivo</u>.   The
approach used is shown in Fig. 1.   A variety of deletion
mutants have been generated and then analysed in the SV40 HeLa
cell transient expression assay system (10) as described (7).
Basically deletions emanating around the CCAAT sequence at
position -75 or further upstream at -96 have been analyzed.
The mutant DNAs were introduced into HeLa cells as globin-SV40
-pBR328 plasmid recombinants.   The transcription, which occurs
from the rabbit β-globin gene promotor requires the presence
of the SV40 enhancer sequence (72bp repeat); the plasmid used
by us also contains the early region of SV40 so that T antigen
is also expressed in these cells (about 30% of the cells are
T antigen   positive two days after transformation).

To quantitate our data, we cotransformed the HeLa cells
with mutant rabbit β-globin genes and the 'wild type' human
β-globin gene; transcripts from both genes are analyzed
simultaneously in S_1 mapping experiments and the human β-globin
gene serves as an internal control.   A summary of these
experiments performed with a number of deletion mutants is
shown in Fig. 2.
The following conclusions can be drawn:
1.   Most, if not all, of the entire GGCCAATCT sequence is
required for efficient transcription <u>in vivo</u>.   Deletion of
the GG, or AATCT residues are down mutations.
2.   Deletion of a region to the 5' side of the CCAAT sequence
has an even stronger effect.   Deletion -76 to -93 has an
essentially intact''CAAT'' box yet it is a 50-fold down mutation.
Although this upstream region is less precisely defined, it is
likely that the sequences from -83 to -94 are involved.   This
region is characterized by many C residues and is therefore

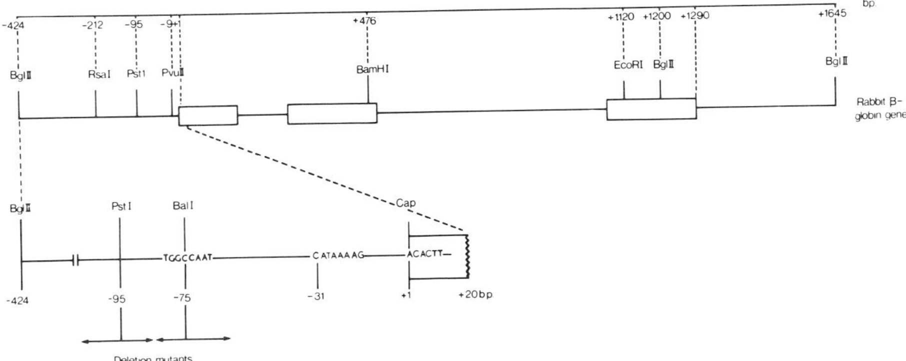

FIGURE 1.   General structure of the rabbit β-globin gene and the strategy used to generate deletion mutants.   The rabbit β-globin gene is shown together with a number of relevant restriction enzyme cleavage sites.   The bottom half of the figure shows an expanded view of the promotor region.

Transcription Level

```
         -130      -120      -110      -100       -90       -80       -70       -60       -50
     ACAGGGGTGCTGTCATCACCCAGACCTCACCCTGCAGAGCCACACCCTGGTGTTGGCCAATCTACACACGGGGTAGGGATT                              1

     ACAGGGGTGCTGTCATCACCCAGACCTCACCCTGCAGAGCCACACCCTGGTGTT       CCAATCTACACACGGGGTAGGGATT        -76 to -77      0.12

     ACAGGGGTGCTGTCATCACCCAGACCTCACCCTGCAGAGCCACACCCTGG            AATCTACACACGGGGTAGGGATT         -74 to -81      0.2

     ACAGGGGTGCTGTCATCACCCAGACCTCACCCTGCAGAGCCACACCCTGGTGTTGGCC        ACACACGGGGTAGGGATT         -69 to -73      0.2

     ACAGGGGTGCTGTCATCACCCAGACCTCACCCTGCAGAGCCACACCCTGGTG           TCTACACACGGGGTAGGGATT         -72 to -79      0.08

     ACAGGGGTGCTGTCATCACCCAGACCTCACCCTGCAGAGCCACACCCT                       GTAGGGATT             -60 to -83      0.25

     ACAGGGGTGCTGTCATCACCCAGACCTCACCCTGCAGAGCCACACCC                      GGTAGGGATT              -61 to -84      0.15

     ACAGGGGTGCTGTCATCACCCAGACCTCACCCTGCAG               CCCAATCTACACACGGGGTAGGGATT               -76 to -93      0.02

     ACAGGGGT                                                    CGGGGTAGGGATT                    -64 to -123     0.005

     ACAGGGGTGCTGTCATCAC                       CCCTGGTGTTGGCCAATCTACACACGGGGTAGGGATT               -88 to -112     0.2

     ACAGGGGTGCTGTCATCACCCAGACCT           AGCCACACCCTGGTGTTGGCC       ACACACGGGGTAGGGATT          -94 to -104     0.1

     ACAGGGGTGCTGTCATCAC                 AGAGCCACACCCTGGTGTTGGCC       ACACACGGGGTAGGGATT          -97 to -112     0.1

     ACAGGGGTGCTGTCATCACCCAGACCTCACCCTGCAGAGCCACACCCTGGTGTTGGCCAATCTACACACGGGGTAGGGATT                             0.4
                                                        GCTGCAGC

                                              CCAAGCTTGG                                                           0.7

                                            CCAAGCTTGGCCAAGCTTGG                                                   0.25

                               CCAAGCTTGGCCAAGCTTGGCCAAGCTTGGCCAAGCTTGGCCAAGCTTGG                                  0.2
```

FIGURE 2.   The transcription efficiencies of deletion and insertion mutants in the -80 region of the rabbit β-globin gene. The sequence in the -80 region of each mutant template is presented, and the gaps show the DNA residues deleted.   Transcription level was determined by scanning the autoradiographs of the S_1 mapping experiments and correcting the signal for the deletion mutants by reference to the internal human β-globin gene control.

similar to a region of the TK gene, mapping from −95 to −105,
which is required for transcription of that gene in mouse L
cells (6).

The rabbit β-globin gene promotor can therefore be
seen conceptually as a tripartite structure, the ATA box,
CAAT box and the −90 region. In our transient expression
system the enhancer sequence of SV40 is required for
expression. Why this is, and whether there is a cellular
counterpart for enhancing globin gene expression, is not
clear.

PHENOTYPIC ASSAY OF INHERITED DEFECTS IN HUMAN GENES

Structural studies of cloned genes from patients with
inherited diseases have in many cases provided a clear
explanation for the genetic defect. Thus, many such diseases
result from a single base substitution which generates a stop
codon; as a result the protein is not formed (e.g. 11, 12,13).
Yet it is likely that genetic lesions exist for which the
molecular defect is not so easy to determine, either because
the mutations detected by structural analysis are not clearly
interpretable, or because a large number of polymorphic
differences are present which mask the true mutation that
causes the disease. To solve this, it is essential to
correlate structure with function.

β-Thalassaemia is an excellent model system for this
type of study. It is by far the commonest form of thalass-
aemia in the β-related globin gene family and it is generally
divided into two classes. In the first, β^+-thalassaemia, a
low but detectable level of β-globin protein is found; this
low level of β-globin protein reflects a low level of β-globin
mRNA. It seems likely, <u>a priori</u>, that this disease would be
the result of a defect in the transcription or processing of
β-globin mRNA precursors. In β^0-thalassaemia no β-globin
protein is detected. In this case, therefore, a number of
causes of the genetic lesion can be envisaged. In some cases,
globin mRNA is readily detectable (14), whereas in others,
β-globin mRNA could not be detected using cDNA hybridization
techniques (e.g. see reference 15).   It is likely,
therefore, that there are several different molecular forms
of β^0-thalassaemia.

The molecular defect in β^+-thalassaemia has been
studied by structural analysis of cloned thalassaemic β-globin
genes (16,17). Interestingly, both genes show a single base
substitution when compared with the normal β-globin gene --
TTGG -> TTAG at a site 21 nucleotides to the 5' side of the
3' splice junction of the small intervening sequence of the
β-globin gene. Among the possible phenotypes that could be

postulated to explain this result, the most interesting is
that this generates a new splice acceptor site in the small
intervening sequence.   We have recently shown this using the
SV40-HeLa cell system (18). S_1 mapping of the β-globin RNA
produced in HeLa cells from a   β$^+$thalassaemic β-globin gene
shows that about 95% of the β-globin mRNA produced
is spliced to this abnormal splice site (not shown).
To establish the nature of the aberrant RNA produced, we have
used a 5'-labelled DNA fragment as a primer to synthesize, $_+$
using reverse transcriptase, a cDNA complementary to this β-
globin RNA. This cDNA has been purified and its sequence determined (Fig. 3).   As a control, cDNA synthesized from globin
mRNA from a patient with sickle cell haemoglobin was also
sequenced.   It can be seen that the abnormal splicing of the
small intron generates a β-globin mRNA with an additional 19
nucleotides which derive from the small intervening sequence;
since this is not a multiple of 3, this insertion results in
the alteration of the translational reading frame and
generates an in-phase stop codon 5 amino acids after the
splice.
        From the above, it should be clear that a combination
of structural and functional analyses will make it possible
in the future to elucidate the molecular nature of defects in
a large number of human diseases.   In the examples provided
here, we show that defects at the level of protein synthesis
and RNA processing can be elucidated using these systems.
Since the globin mRNA produced in these cells is spliced,
polyadenylated and exported to the cytoplasm (18), it is
likely that defects at the level of RNA transport can also be
studied.   It is to be anticipated that the combination of
structural and functional analyses in the next few years will
bring some interesting discoveries in the field of the molecular biology of human diseases.

SPECIFIC EXPRESSION OF THE HUMAN
β-GLOBIN GENE IN ERYTHROID CELLS

        In the above mentioned experiments, globin gene expres-
sion has been analysed in cells (such as HeLa cells) which do
not express their endogenous globin genes.   Although this
permissive behaviour is useful, this system is not suited to
a study of those factors which determine the specific expression of globin genes in erythroid cells.   To examine this,
we have introduced various segments of the human β-related
globin gene locus as cosmid recombinants into an erythroid
cell line, the well known Friend cell (murine erythroleukaemia
cells or MEL) and then determined the level of human β-globin
mRNA in the transformed MEL cells before and after the

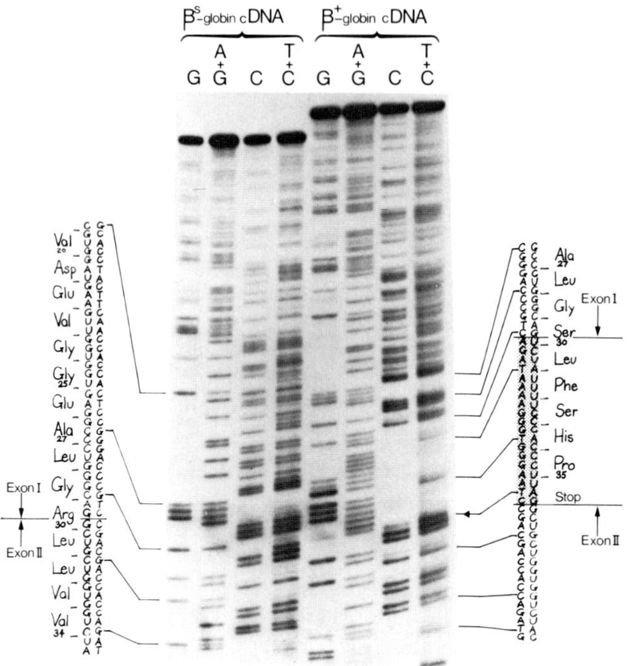

FIGURE 3.   cDNA sequence over the junction of Exon I
and Exon II of β⁺-globin mRNA.
We synthesized cDNA by hybridizing an end-labelled, 87
nucleotide CvnI-Cvn I DNA primer to β⁺-globin mRNA obtained
from transformed HeLa cells and by subsequent primer extension
with reverse transcriptase. Full-length cDNA synthesized up
to the 5' end of β-globin mRNA was sequenced, and then
separated on a 10% polyacrylamide sequencing gel.   βˢ-globin
mRNA isolated from reticulocytes of a patient with sickle cell
anaemia was used in the control experiment.   The cDNA
sequences and the complementary mRNA sequences, together with
the deduced amino acid sequences, are shown to the right and
left.   Amino acids are numbered starting at the N terminus
of the β-globin protein.   Shaded area: the additional 19
nucleotides of the β⁺-globin mRNA that are inserted between
Exon I and Exon II of the normal β-globin mRNA.

induction of erythroid differentiation.   The DNA segments
that we have introduced into these cells are shown in Fig. 4.
In the majority (15 of 24 clones examined) of clones that
contain the human β-globin gene, the level of β-globin mRNA
is stimulated by from three to greater than ten-fold by
inducers of erythroid differentiation (Fig. 5 shows an
example).   This result is found for all the relevant cosmid
clones;   since the cosmid clones used have from 2.5kb to about
 35kb of the DNA sequences flanking the β-globin gene on the
5' side and the same amount of 3'-flanking DNA, it follows
that DNA sequences far from the human β-globin gene are not
required for the induction phenomenon.   Of those clones that
contain the γ-globin genes, the majority are not inducible;
instead, constitutive synthesis is usually seen or a decrease
in the level of γ-globin mRNA after induction (see the MEL
clone in,e.g.,Fig. 6 which is inducible for β-globin mRNA but
not γ-globin mRNA).
     It is likely that the induction of human β-globin mRNA
is regulated at the level of transcription, although we have
not specifically addressed this question.   In all clones
examined, the mouse α-globin mRNA induction was monitored and
was greater than ten fold in all cases.   Secondly, in clones
where the β- and γ-globin genes (such as Fig. 5) were present
γ-globin mRNA, but not β-globin mRNA, was present before
induction;   upon induction β-globin mRNA is found.   It is
therefore unlikely that the human β-globin mRNA, but not the
γ-globin mRNA is being synthesized but selectively degraded
before induction.

MOLECULAR ANALYSIS OF A COMPLEX GENE LOCUS;
THE MAJOR HISTOCOMPATIBILITY COMPLEX

     The major histocompatibility complex (MHC) plays an
important role in the regulation of the immune response in
vertebrates.   There are three classes of MHC proteins, the
classical transplantation antigens (class I: in man HLA-A B
and C;   in mouse the H-2 K L and D proteins), immune response-
associated antigens (class II: in man HLA DR;   in mouse Ia
antigens) and the complement products (class III).   The genes
for the three classes of antigens are closely linked on
chromosome 17 of the mouse.   Adjacent to the H-2 complex in
mouse is the TL complex which contains several genes encoding
lymphoid differentiation antigens, the Qa and Tla antigens.
See Fig. 6 for a schematic genetic map of the MHC complex in
mouse.
     Both the H-2 antigens (K, D, L,and R) and the Qa and
Tla are integral membrane proteins of 40-45000 molecular
weight found associated with β_2-microglobulin.   It is,
therefore possible that these antigens are evolutionarily
related (see e.g. 19,20).

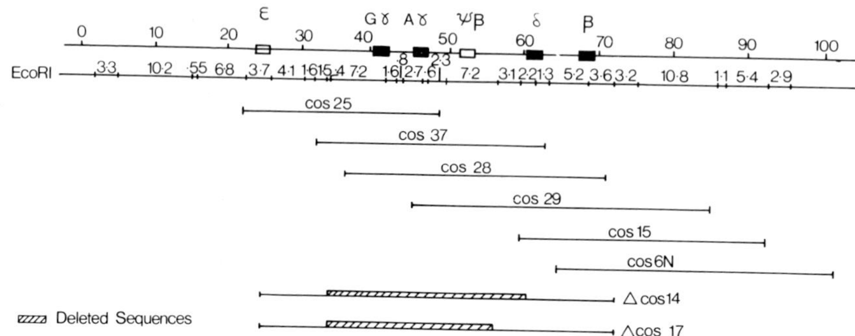

FIGURE 4. Cosmid clones used to transform MEL cells. A map of the human β-globin is shown; the coordinates of the cosmids are indicated by the solid lines.

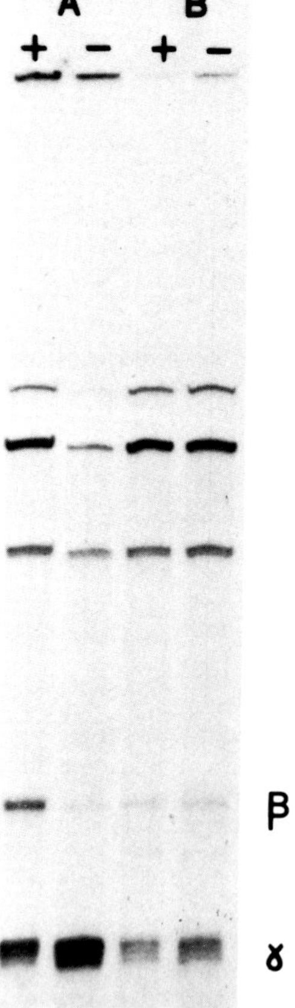

FIGURE 5. S_1-mapping of RNA from MEL cells before (−) and after (+) induction of erythroid differentiation. 3'-labelled probes were used to measure simultaneously γ- and β-globin mRNA. The position of the relevant bands is indicated. The bands seen in the upper part of the gel are input fragments.

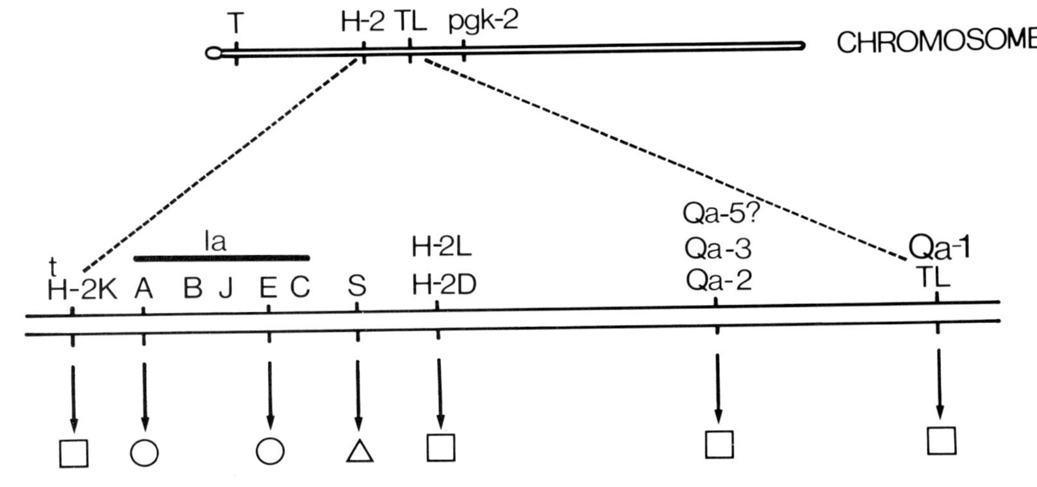

FIGURE 6.   Schematic genetic map of the H-2 and TL complexes of chromosome 17 of the mouse.

The relative order of the loci listed vertically is not known.   The distance from H-2K to H-2D is about 0.3 centimorgans (cm) and from H-2D to Tla is about 1 cm.

□ Class I antigens;   ○ Class II antigens;

△ Class III antigens.

     One of the most intriguing aspects of the H-2 antigens
is their extreme polymorphism. Allelic products from
different inbred mouse strains differ in most cases as shown
by both serological analyses and protein sequencing. In fact,
about 50 alleles at each of the H-2K and H-2D loci have been
observed (for a review see 21). H-2 haplotypes are defined
serologically and consist of a particular set of alleles found
at each H-2 locus (e.g. the C57 black/10, or B10, mouse is
H-2^b at each H-2 locus, the AKR strain H-2^k, DBA/2 H-2^d and
so on). Some inbred mouse strains appear to be natural recombinants between other strains (e.g. the A mouse strain with
the H-2^a haplotype is H-2K^k but H-2D^d). It is likely that
the analysis of the gene clusters by recombinant DNA technology will provide some clues to the nature of the polymorphic differences.
     A number of groups have isolated cDNA clones which
encode proteins whose sequences are homologous to known transplantation antigens (see e.g.,22 and 23). Southern blotting
analysis of genomic DNA probed with these cDNAs shows many
cross-hybridizing DNA fragments which reinforces the notion
that the MHC is a large multi-gene family. We have isolated
a large number of mouse genomic clones which contain sequences
complementary to H-2 cDNA from a cosmid library (24) using
spleen DNA from the B10 mouse (A. Mellor, L. Golden and
H. Bullman, unpublished). Analysis of these clones by
standard restriction enzyme digestion techniques establishes
that many of these clones fall into a number of gene clusters
(Table I); in total about 17 different H-2 like genes have
been isolated in these cosmids.
     Which H-2 genes are contained within these clusters?
The number of independent H-2-like DNA clones exceeds that
expected from the 4-5 known H-2 genes. This suggests that
other H-2 related sequences are present in the mouse genome.
A first step to identifying these genes is to determine where
they are located on the MHC genetic map. To do this we have
made use of polymorphic differences at restriction sites that
exist between different mouse strains in the DNA sequences
flanking a given H-2 gene. This is done by cutting DNA from,
say, B10 (H-2^b), AKR (H-2^k) and BALB/c (H-2^d) mice with a
number of restriction enzymes and blotting this onto nitrocellulose filters. The filters are then probed with a fragment from a given cosmid in order to detect polymorphic restriction site differences between the parental inbred strains.
For example, using a probe from a clone in cluster 4 in PstI
digests of these mouse DNAs shows a 2.5kb band in B10 DNA,
but a 3.8kb band in AKR DNA. Recombinant mice have been
produced which contain a given segment of the MHC map from one
haplotype (say AKR; H-2^k) and a second segment from a second

mouse (say C57 black/6 or B6; H-2^b).   The origin of genes which have been recombined in these mice has been shown by serological analysis of the recombinants.   Thus, in the recombinant mouse B6K2 the Tla and Qal loci are derived from the AKR (H-2^k) mouse, whereas the H-2 region comes from the B6 mouse.   Blotting _PstI_-cut DNA from the B6K2 DNA with the same probe shows that the AKR pattern (3.8kb band) and not the B6 pattern is obtained.   It follows, therefore, that gene cluster 4 is localized in the AKR segment of the B6K2 mouse; that is, in the Tla region.

We have used the same approach to localize DNA segments in the H-2K region, the H-2D region and the Qa2, 3 region. Thus, class I, H-2 genes, or pseudogenes are found spread over chromosome 17 between the H-2K and TL genetic loci; a region of 1.3 cM and perhaps as much as 10^6-10^7 base pairs of DNA. This gene cluster might then be two orders of magnitude larger than the β-globin cluster (Table I).

EXPRESSION OF AN H-2K^b GENE IN MOUSE L CELLS

Although cosmid clones can be mapped to a given locus by the mapping procedures described above, to establish the identity of any given gene, a more detailed analysis is required.   We have, therefore, introduced H-2 cosmids into mouse L cells to ask if these cells which contain a given cosmid express a known H-2 gene.   We have transformed L cells with cosmid DNA as a calcium phosphate coprecipitate and then selected for stable TK positive transformants.   Since the cosmid vector used for most of our experiments, pOPF1 (F. Grosveld and T. Lund, unpublished) contains the Herpes Simplex Virus (HSV) thymidine kinase (tk) gene, transformants are readily obtained.   Transformants were screened for the binding of monoclonal antibodies which react with the H-2K or H-2D molecules.   L cells transformed with the cosmid H8 bind the H-2K^b monoclonal specifically, while none of this group binds the H-2D^b monoclonal (not shown).Cosmid clones H8, H24 and H25 are all found in cluster 1 and map at the H-2K locus. This cluster (Fig. 7) contains two genes; since both H24 and H25 contain gene A and do not express H-2K^b we deduce that gene B is the H-2K^b gene.   Sequence data confirm this conclusion (E. Weiss, R. Zakut, unpublished).   L cells containing the B gene of cluster I (called gene IB here) bind a variety of other alloantisera directed against the H-2K^b molecule and immuno precipitations with anti H-2K^b antibodies reveal the expected 45K protein.   We are therefore confident that the cells express the H-2K^b protein.

TABLE I

Summary of H-2^b gene clusters isolated from B10 mouse library

Cluster	map location	number of genes	number of cosmids
1	H–2K	2	7
2	H–2D	1	4
3	Qa 2/3	3	25
4	Tla	4	39
5	?	3	11

H-2 cosmids were mapped using standard restriction enzyme digestion and blot-hybridization with probes for the 5' and 3' regions of H-2 cDNA, mouse genomic DNA and the vector used for the construction of the library,   nine cosmids have not yet been arranged into clusters.   Of these one maps to Qa 2/3 and one appears to map outside the entire locus.

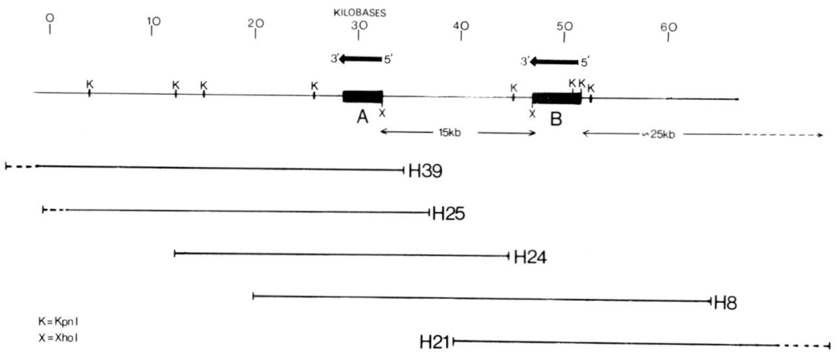

FIGURE 7.   A restriction enzyme map of the region cloned in the H8 cosmid cluster.   The top line shows the extent of the cloned region and the approximate positions of XhoI (X) and KpnI (K) restriction sites (see top scale calibrated in kilobases, kb).   Class I gene regions (A and B) are shown as thick lines and their 5' to 3' orientation is indicated by the thick arrows over the line.   The approximate extent of each cosmid is shown below the line beside the name assigned to each cosmid.   Dotted lines at the end of cosmids indicate that the end of the cosmid has not been determined precisely.

CYTOTOXIC T CELL MEDIATED LYSIS OF TRANSFORMED CELLS

The presence of the $H-2K^b$ molecule on a cell renders these cells as targets for allogeneic cytotoxic T-cell (T_C-cell) mediated killing by T_C cells raised against $H-2^b$ antigens (25). A number of indirect arguments suggest that the presence on the cell membrane of a given H-2 polypeptide, such as the $H-2K^b$ molecule, is sufficient to generate a target for T_C-cells. For example, this process can be blocked by monoclonal antibodies directed against a single H-2 polypeptide (26). To test this directly, and also to further characterize the $H-2K^b$ polypeptide detected by serological assays on the surface of L-cells transformed by cosmid H8, we generated T_C cells directed against $H-2^b$ or $H-2^k$ antigens and tested their ability to kill various target cells (not shown). Whereas anti-$H-2^k$ T_C cells show specific killing of all L-cell lines tested, anti-$H-2^b$ T_C-cells show specific killing of only control B10 target cells and two L-cell clones (LH8-1 and LH8-2) transformed by cosmid H8. Untransformed L-cells (Ltk⁻), a clone transformed by cosmid H25 (LH25-1) and other clones transformed by cosmids H24 and H39 (data not shown) are not seen as targets for killing by anti-$H-2^b$ T_C-cells. This data confirms that the $H-2K^b$ molecule detected by serological assays in cosmid H8 transformed cells, can act as a target for anti-$H-2^b$ T_C-cell killing.

We performed a similar experiment using B10.D2 ($H-2^d$) T_C-cells generated by immunisation with target spleen cells from the recombinant mouse B10.A (5R) (K^b,D^d). Such T_C-cells kill cells expressing the $H-2K^b$, but not the $H-2D^b$ molecule. These T_C-cells also kill L-cells transformed with cosmid H8, but not L-cells transformed with cosmid H39 (not shown).

CYTOTOXIC T CELL-MEDIATED LYSIS OF CELLS INFECTED WITH INFLUENZA VIRUS

T_C-cell-mediated lysis of cells expressing viral antigens on their surface can only take place when the viral antigen is presented to T_C-cells in association with an appropriate H-2 antigen; in this way H-2 antigens act as a restriction element for the cytotoxic T-cell response (25, 27). Thus, we tested the ability of the $H-2K^b$ molecule expressed on the surface of L-cells transformed by cosmid H8 to act as a restriction element for T_C-cell-mediated killing of influenza virus-infected cells.

Influenza-specific T_C clones restricted to one H-2 region can be selected and grown in the presence of T-cell growth factor (28). L-cells ($H-2^k$) and LH8-1 were infected with type A influenza virus (A/X31) and used as target cells

                                                  R. A. FLAVELL *et al.*

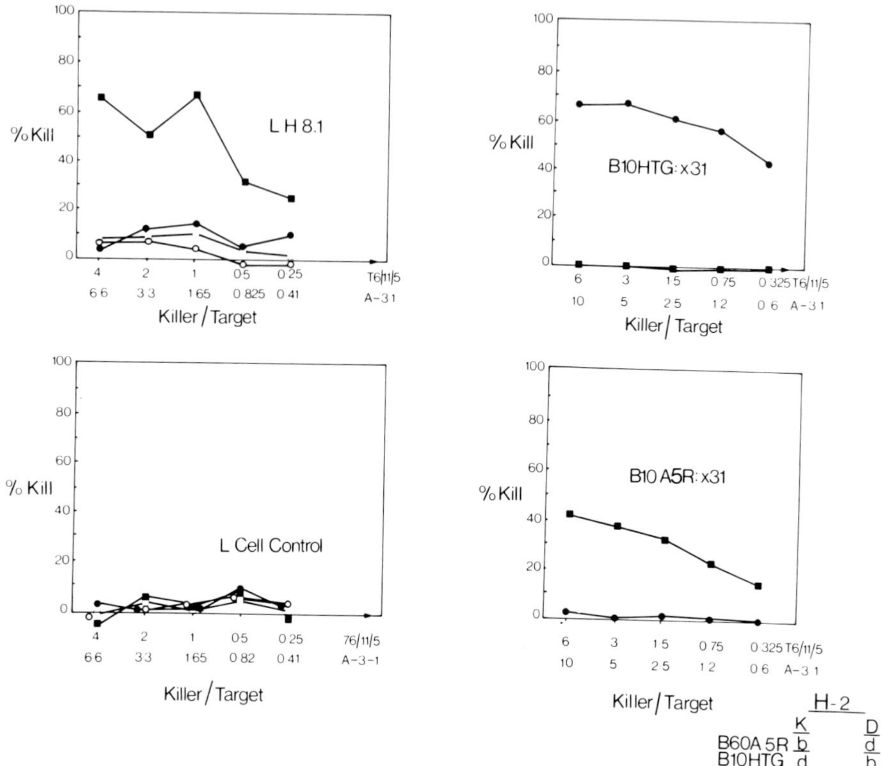

FIGURE 8.   Cytotoxic T-cell-mediated killing of transformed L-cells infected with influenza virus.

Two influenza specific T_C-clones, T6/11/5 (H-2K^b restricted; square symbols) or A3.1 (H-2D^b restricted; circular symbols) were incubated with influenza virus (strain X31) infected (closed symbols) or non-infected (open symbols) LH8-1 or Ltk$^-$ (L-cell control) target cells. The percent specific lysis estimated by [51]chromium release (29) at each killer (T_C-cell) to target (LH8.1 or Ltk$^-$) ratio was calculated using the formula $\frac{(E-C)}{(M-C)} \times 100\%$. Where E = cpm from target cells incubated with killer cells, C = cpm from target cells incubated with medium alone and M = cpm from target cells lysed with 5% triton. Controls using influenza virus infected macrophage* targets from B10HTG mice (H-2K^d, D^b) or B10A5R (H-2K^b, D^d) demonstrate the H-2K^b and H-2D^b restricted killing of T_C-clones T6/11/5 and A3.1 respectively. The isolation and characterization of these T_C-clones will be described in detail elsewhere (Townsend and Taylor, manuscript in preparation).

*3% thioglycollate induced P.E.C.

for two influenza specific $H-2^b$ T_C clones, one restricted to
$H-2K^b$ (T6/11/5) and the other (A/3.1) to $H-2D^b$ molecules
(A. Townsend and P. M. Taylor, unpublished data). Figure 8
shows that X31 infected, untransformed L-cells ($H-2^k$) are
unable to act as targets for killing by either the $H-2K^b$ or
$H-2D^b$-restricted T_C-cell clones. LH8-1 cells infected with
X31, however, act as targets for killing by the $H-2K^b$ restricted T_C cell clone, but not the $H-2D^b$ restricted clone.
This shows that the $H-2K^b$ molecule expressed on the surface
of LH8-1 cells is able to act as a restriction element for
T_C-cell killing of X31 virus infected cells. We have
recently carried out the same experiments with L cells transformed with an $H-2D^b$ cosmid. Again, using an $H-2D^b$-restricted,
influenza virus specific T_C cell clone, we have obtained
specific lysis. It is clear that this will be a general
phenomenon.

In summary, it is clear that the presence of the $H-2K^b$
gene in mouse L cells is sufficient to evoke essentially all
the immunological properties that have been associated with
this molecule by a number of more conventional approaches.
The fact that these processes can be studied by DNA mediated
gene transfer will make it possible to carry out a detailed
molecular analysis of the determinants recognised by antibodies and by the T cell receptor. Specifically, it should
now be possible to transform L-cells with hybrid H-2 genes,
constructed in vitro, which contain DNA segments from, say,
the $H-2K^b$ and $H-2D^b$ genes and to ask whether T_C-cell clones,
either allospecific or H-2 restricted, can kill such target
cells. In this way, we hope to map the determinants recognized by specific monoclonal antibodies and by the T-cell
receptor.

ACKNOWLEDGMENTS

M.B, L.G. and E. W. were the recipients of fellowships
from the Swiss National Fonds, the Helen Hay Whitney
Foundation and the Deutsche Forschungsgemeinschaft
respectively. Work in the laboratories of R.A.F., E.S. and
A.R.M.T. was supported by the British Medical Research
Council, work in the laboratory of H.F. by the Cancer Research
Campaign and work by R.J. by the Imperial Cancer Research
Fund. Cora O'Carroll is gratefully thanked for preparing
this manuscript.

REFERENCES

1. Goldberg, M. (1979). Sequence analysis of Drosophila histone genes. Ph.D. thesis, Stanford University, Stanford, California, U.S.A.
2. Proudfoot, N. J. (1979). Nature 279, 376.
3. Benoist, C., O'Hare, K., Breathnach, R. and Chambon, P. (1980). Nucl. Acids Res. 8, 127.
4. Grosschedl, R. and Birnstiel, M. L. (1980). Proc. Natl. Acad. Sci. U.S.A. 77, 1432.
5. Benoist, C. and Chambon, P. (1981). Nature 290, 304.
6. McKnight, S. L., Gavis, E. R. and Kingsbury, R. (1981). Cell 25, 385.
7. Grosveld, G. C., deBoer, E., Shewmaker, C. K. and Flavell, R. A. (1982). Nature 295, 120.
8. Dierks, P., van Ooyen, A., Mantei, M. and Weissmann, C. (1981). Proc. Natl. Acad. Sci. U.S.A. 78, 1411.
9. Mellon, P., Parker, V., Gluzman, Y. and Maniatis, T. (1981). Cell 27, 279.
10. Banerji, J., Rusconi, S. and Schaffner, W. (1981). Cell 27, 299.
11. Chang, J. C. and Kan, Y. W. (1979) Proc. Natl. Acad. Sci. U.S.A. 76, 2886.
12. Trecartin, R. F., Liebhaber, S. A., Chang, J. C., Lee, K. Y., Kan, Y. W., Furbetta, M., Angius, A. and Cao, A. (1981). J. Clin. Invest. 918, 1012.
13. Moschonas, N., deBoer, E., Grosveld, F. G., Dahl, H. H.M., Wright, S., Shewmaker, C. K. and Flavell, R. A. (1981). Nucl. Acids Res. 9, 4391.
14. Temple, G. F., Chang, J. C. and Kan, Y. W. (1977). Proc. Natl. Acad. Sci. U.S.A. 74, 3047.
15. Ottolenghi, S., Lanyon, W. G., Williamson, R., Weatherall, D. J., Clegg, J. B. and Pitcher, C. S. (1975). Proc. Natl. Acad. Sci. U.S.A. 72, 2294.
16. Spritz, R. A., Jagadeeswaran, P., Choudary, P. V., Biro, P. A., Elder, J. T., deRiel, J. K., Manley, J. L., Gefter, M. L., Forget, B. G. and Weissman, S. M. (1981). Proc. Natl. Acad. Sci. U.S.A. 78, 2455.
17. Westaway, D. and Williamson, R. (1981). Nucl. Acids Res. 9, 1777.
18. Busslinger, M., Moschonas, N. and Flavell, R. A. (1981). Cell 27, 289.
19. Michaelson, J., Flaherty, L., Vitetta, E. and Poulik, M. (1977). J. Exp. Med. 145, 1066.
20. Stanton, T. H. and Hood, L. (1980). Immunogenetics 11, 309.
21. Klein, J. (1979). Science 203, 516.

22. Kvist, S., Bregerere, F., Rask, L., Cami, B., Garoff, H.,
     Daniel, F., Wiman, K., Larhammar, D., Abastado, J. P.,
     Gachelin, G., Peterson, P. A., Dobberstein, B. and
     Kourilsky, P. (1981). Proc. Natl. Acad. Sci. 78, 2772.
23. Steinmetz, M., Frelinger, J. G., Fisher, D., Hunkapiller,
     T., Pereira, D., Weissman, S. M., Uehara, H., Nathenson,
     S. and Hood, L. (1981) Cell 24, 125.
24. Grosveld, F. G., Dahl, H. H. M., deBoer, E. and Flavell,
     R. A. (1981). Gene 13, 227.
25. Doherty, P. C. and Zinkernagel, R. M. (1975). J. Exp.
     Med. 141, 502.
26. Weyand, C., Hammerling, G. J. and Goronzy, J. (1981).
     Nature 292, 627.
27. Zinkernagel, R. M. and Doherty, P. C. (1979). Adv.
     Immunol. 27, 51.
28. Lin, Y-L. and Askonas, B. A. (1981). J. exp. Med., 154,
     225.
29. Sanderson, A. R. (1964). Nature 204, 250.

THE RELATIONSHIP OF GENE STRUCTURE TO PROTEIN STRUCTURE[1]

Charles S. Craik, Orgad Laub, Graeme I. Bell,
Stephen Sprang, Robert Fletterick & William J. Rutter

Department of Biochemistry & Biophysics
University of California
San Francisco, California   94143

ABSTRACT The cloning and sequencing of genes of proteins whose structure is known allows a comparison of
the localization of intervening sequences with the
structural features of the proteins. In some instances,
the intron positions seem to demarcate structural and/or
functional units, but in others, there is no obvious
relationship between the protein and gene structure.
Comparison with tertiary structures indicates that the
intron positions in the gene correspond to the external
boundary rather than the interior of the protein. This
localization is confirmed by a solvent accessible surface analysis which shows that the intron junctions are
almost always located at the protein-solvent interface.
While the surface rule is based on observations of a few
relatively small globular proteins with known tertiary
structure, it is further supported by the observation
that the consensus sequences at the 5' (but not the 3')
intron/exon splice junctions predominantly code for
hydrophilic and surface type amino acids. These results
suggest that intron placements may have played a role in
protein evolution. For example, intron junctions may
be sites of hypervariability and hence of structural
change. If so, such junctions would be better accommodated on the surface, rather than the interior where
changes would likely result in distortion or destruction
of the architecture of the molecule. Consistent with
these ideas, certain evolutionary variations in the serine protease family appear to correlate with intron
location. Experimental studies on the expression of
the human insulin gene in heterologous mammalian cells
using SV40 viral vectors show that the normal splicing
can occur, but that an alternative splice site, com-

[1]This work was supported by NIH grant AM21344 (to WJR)
and NSF grant PCM 8001950 (to RF).

pletely within the gene, is used in certain SV40 constructions. The 3'-extragenic region appears to be the
controlling factor for recruitment of the alternate site.
The transcript produced from the alternate site codes
for a protein that is identical to insulin at the N
terminus but is completely different at the C terminal
region because of a frame shift caused by the new splice.
This result may serve as a model in illustrating how
variations in gene splicing may generate genetic diversity as seen in the serine proteases.

INTRODUCTION

Our studies of gene expression in the rat pancreas have
led to the cloning and structural characterization of several
pancreatic cDNAs and genes (e.g.,the exocrine amylase,
carboxypeptidase, chymotrypsin, trypsin, elastase I, elastase
II, ribonuclease, and the endocrine genes, insulin, somatostatin and glucagon). Like most other eucaryotic genes,
these genes contain intervening sequences that interrupt the
coding regions. We have attempted to relate function of the
intron/exon mosaic to the structure/function of the resultant
protein. The introns may have arisen from the intrinsic
properties of''selfish'' DNA (1,2) or they may represent units
that provide phenotypic benefit to the organism. In consi-
dering a role for introns in the evolution of structure and
function of the gene product, a correspondence between an
exon and a structural or functional unit of the protein was
suggested (3-5). According to this concept the exons would
be the units of evolution in protein fabrication. Independent of their origin it is clear that the eucaryotic cell may
have exploited introns more generally in mechanisms involving
DNA replication, packaging, RNA maturation, gene regulation
or selection of the gene product. We present data in part I
for the intron positions for several pancreatic genes along
with a study in part II of the processing of the human insulin gene in a heterologous system to analyze this possible
role for introns.

I. INTRON/EXON SPLICE JUNCTIONS MAP AT THE PROTEIN SURFACE

The sequences of genomic DNAs have revealed certain
examples where location of the exons appear to articulate
well-defined domains or functional units, e.g., the immunoglobulins (6,7), proinsulin (8), somatostatin (L.P. Shen et
al., unpublished) and lysozyme (9). In other genes, the
location of the exons does not respect the usual definition

of domain structure or function in their segmentation of the protein. Thus, hemoglobin chains are made from three exons (10-12); the central exon product binds heme (13), but all three exon products are required for reversible oxygen binding (14). However, it has been shown that these three exons correspond not to domains but to "compact folding units" (15). The exons of collagen (16), α-fetoprotein (17), ovomucoid (18) and growth hormone (19) also correspond to subdomains and subfunction. In still other genes, the introns appear to be dispersed throughout the coding sequence without obvious regard to structure or function, e.g., carboxypeptidase (20), amylase (21), and exon 1 of alcohol dehydrogenase (22). Therefore the available data suggest a correlation between exons and structural or functional units in certain cases but in other cases the exons specify multiple units too small for a known function leaving the relationship between the splice junction and the protein structure unclear.

Using models or computer graphic displays of the protein tertiary structure we have observed that the amino acids at the splice junctions map at or near the protein subunit surface. These visual correlations are further supported by calculating the solvent accessible surface (23) of the boundary amino acids corresponding to the splice junction. The results of this calculation for seven proteins are in Table 1. Following Janin's definition of a buried amino acid (24), (20 to 25 $\mathring{A}^2$ or less solvent accessible surface, the approximate surface of a water molecule) we used 25 $\mathring{A}^2$ or less to indicate a buried amino acid and in every case but one, at least one atom of an amino acid side chain at the splice junction is exposed to the solvent. In this one exception, the first intron of lysozyme, the amino acid at the splice junction is located at the edge of the hydrophobic core.

There is a striking correspondence between intron/exon splice points and proteolytic cleavage sites of certain proteins. Activation of chymotrypsinogen, papain cleavage of the immunoglobulins and clostripain cleavage of hemoglobin all occur at exon splice junctions implying their surface location. Other surface splice junctions not included in Table 1 were found by visual inspection of the proteins encoded by the exons of the immunoglobulins and proinsulin. As expected many of the boundary amino acids at the splice junctions are also associated with turns in proteins which are usually at the surface (25).

<u>Boundary Amino Acids</u>. If splice junctions of all protein molecules map at the surface then the boundary amino acids should reflect this, i.e., they should be hydrophilic and/or contain the amino acids usually found at turns of

TABLE I

SOLVENT ACCESSIBLE SURFACE AREAS FOR JUNCTIONS
OF INTRONS AND EXONS IN CODING REGIONS

Protein (references)	Intron	Intron/Exon Splice position	Amino Acid at Splice Junction*	Amino Acid Solvent Accessibility (Å^2)	Interior Fraction[#] of Amino Acids <20 Å^2
Rat Cytochrome C		55	K G	44 66	0.20
Human Hemoglobin alpha	A	30	E R	48 87	0.23
	B	99	K L	143 18	
Human Hemoglobin beta	A	30	R L	89 11	0.20
	B	104	R L	146 20	
Chicken Lysozyme	A	28	N W	22 0	0.25
	B	81	S A	74 33	
	C	108	W V	11 86	
Trypsin	B	61	K S	55 82	0.35
	C	149	G T	57 85	
	D	192	Q G	120 20	
Rat Chymotrypsin	B	34	Q D	32 42	0.35
	C	62	K T	69 85	
	D	87	K V	117 27	
	E	148	N A	137 55	
	F	192	M G	106 30	

Chicken Dihydro-folate reductase	A	25	P	I	100	112	0.25
	B	37	A	G	48	47	
	C	67	L	K	8	102	
	D	105	E	L	115	18	
	E	140	H	E	96	59	
Rat Carboxypep-tidase**	C	17	D	E	103	70	0.39
	D	60	N	R	79	46	
	E	85	K	K	13	36	
	F	127	W	R	36	48	
	G	156	A	S	27	30	
	H	217	K	T	88	114	
	I	259	N	N	71	68	

*The precise position of the splice is sometimes equivocal due to the ambiguities of the DNA sequence around the intron/exon boundary so two amino acids are shown.

**Five of the seven splice junctions were determined by detailed restriction and heteroduplex mapping.

#Two or more consecutive amino acids with <20 $Å^2$ solvent accessible surface area (24) are counted as interior.

TABLE 2

AMINO ACIDS AT SPLICE JUNCTIONS

Protein		Boundary* Splice Amino Acids Junction	Mole Fraction**
Proinsulin	(human, rat)	1/1	0.75
Cytochrome C	(rat)	1/1	0.75
Somatostatin	(human, fish)	1/1	1.0
Large T antigen	(SV40)	1/1	0.75
Alcohol Dehydrogenase	(<u>Drosophila</u>)	2/2	0.75
Cytochrome в	(yeast mitochondrial)	2/2	0.88
Amylase	(rat)	2/2	0.75
α subunit of glyco- protein hormones	(human)	2/2	0.75
Actin	(<u>Drosophila</u>)	2/2	0.62
Lysozyme	(avian)	3/3	0.5
Trypsinogen	(rat)	3/3	0.75
Hemoglobin	(human)	4/4	0.5
Growth Hormone	(rat)	4/4	0.88
Chymotrypsinogen B	(rat)	5/5	0.7
Ovomucoid	(avian)	6/6	0.66
Ovalbumin	(avian)	6/6	0.79
Cytochrome Oxidase	(yeast mitochondrial)	5/7	0.64
Carboxypeptidase A	(rat)	7/7	0.89
α-Fetoprotein	(mouse)	12/12	0.77

*Number of cases for which at least two of the four amino acids at the splice point are (R,K,Q,E,D,H,N,S,T,P,G,and A)

**Mole fraction = fraction of boundary amino acids that are (R,K,Q,E,D,H,N,S,T,P,G,and A) per four amino acids at the splice junctions

References for the various sequences can be found in the text, in (26) and (27) or were personal communications; somatostatin, L.P. Shen: amylase, W. F. Swain.

proteins. We have defined the members of this set of boundary amino acids (R,K,E,D,Q,N,H,T,S,A,P and G); this group is
about 60-65% of all amino acids by natural abundance. Examination of the junctional sites of 20 proteins with a total of
87 junctions reveals that in all but two instances at least 2
of the 4 amino acids at the junctions are members of this
set (Table 2). Based on natural abundance of the boundary
amino acids the probability of observing one set of four
amino acids with two, three or four boundary amino acids
being members of this set is 0.86 and is expected. However
the probability that 85 or more of the 87 splice junctions
compiled in Table 2 should have this property is only 5×10^{-5};
this reveals the significance of this observation. Furthermore, the probability of finding clusters of members of the
boundary set within the interior of the molecule and particularly in the hydrophobic core is low. The two instances where
only one of four amino acids are boundary amino acids are in
yeast cytochrome oxidase (28) which is a membrane bound
protein and may therefore have a hydrophobic patch on its
surface.
     The basis of the result found in Table 2 in part lies in
the nucleotide sequence found at the intron/exon junction. A
compilation of the codons at known intron/exon junctions (26)
shows a marked preference for purines at the second base of
the codon at or immediately before the splice point (the GT
of Chambon's rule (29) was used to define the intron/exon
boundary). Analysis of the base frequencies at the three
reading frame positions of the codon reveals that 75% of the
amino acids are (in decreasing order of frequency) R,G,K,Q,S,
A,E,D or T at the 5' side of the boundary. In the proteins
sampled, there appears to be less preference for this type of
amino acid at the 3' side of the splice point. Several of
these 5' amino acids are hydrophilic and they are all frequently associated with turns in a protein and infrequently
found in its interior (24). Thus, the nucleotide sequences
at junctions to some extent predispose the protein sequence
to be at the surface of the molecule.

     <u>Surface Splice Junctions May Facilitate Protein Fabri-
cation</u>. Analyses of the sequence of homologous proteins from
different organisms have shown that the hydrophobic core is
more conserved than the surface amino acids (30,31). As a
consequence of the location of splice junctions at the surface, changes by additions or deletions at the junctions can
be made while maintaining the structural integrity of the
central core. Modifications in the interior hydrophobic core
of the molecule may lead to a disruption of the fundamental
molecular architecture of the molecule. Thus recombined

exons with interior junctions might be selected against and
only those genes having introns in unimportant places (i.e.
primarily on the surface) would survive. The junctional
nucleotides could be a purely accidental event, or a result
of this selection. Regardless of the mechanistic origin we
perceive certain salutary effects of the junctions correlating
with surface residues.

The localization of splice junctions at the surface of a
molecule can be related to the concept of a structural domain,
but it is perhaps too constraining to relate such domains to
a simple structural or functional motif and they need not be
compact folding units. It does however provide a mechanism
for understanding such genetic domains as evolutionary
instruments. If intervening sequences were present in primordial genes, they and their excision systems may have been
lost in the rapidly evolving simple genetic systems such as
prokaryotes. In contrast, intervening sequences may have
arisen with the development of increasingly complex genomes
and genetic processes in eukaryotic organisms. Regardless of
their evolutionary origin, once present, they may have been
recruited as agents of evolution. Additions or changes in the
nucleotide coding sequence could occur by recruiting the
relatively rapidly changing DNA from the intervening sequences
via variant splicing patterns or by losing sequences by a
similar process.

Alternatively, functional changes could be made between
homologous sequences by recombinational events. Novel amino
acid sequences that might result from the new splicing would
be isolated on the outside of the protein. This provides a
means for change in function without forfeiting the pre-
existing structure. This mechanism is not restricted to
exon-encoded structural or functional domains as has been
described and it also provides a rationale for the occurrence
of introns between and within folding units and active site
residues. For example, a new enzymatically active structure
could be formed by the addition of a polypeptide coding
sequence to the surface of a previously existing active site.
The intron/exon splice point separating the segments of the
functional domain might remain as evidence. A comparison of
gene structures of relatively simple gene families to their
more complex relatives might reveal plausible structural
relationships.

<u>Intron Positions in Serine Proteases are Conserved.</u>
The serine proteases have been suggested to be related to one
another on the basis of the homology that exists in primary
and resultant tertiary structure among the members of this
family (32-35). As shown in Figure 1 this homology exists

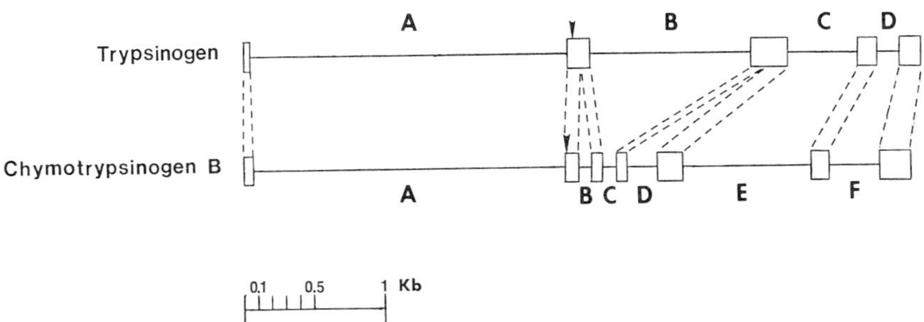

FIGURE 1. Homologous intron/exon structure of trypsinogen and chymotrypsinogen. Introns are denoted by capital
letters. "▼" reveals the position of amino acid number 1 of
the activated protein.

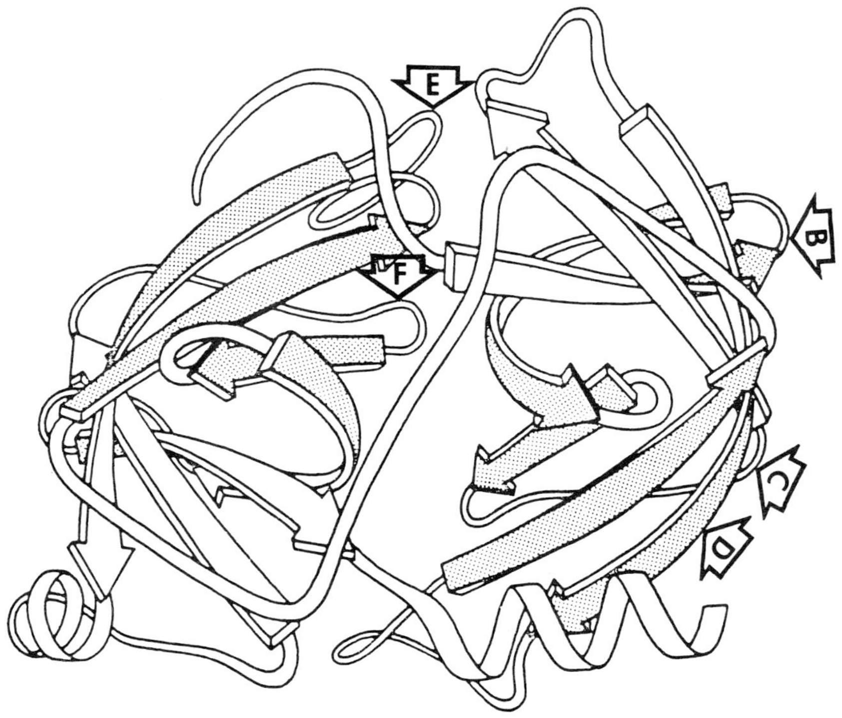

FIGURE 2. Serine protease domains I and II. Lettered
arrow heads refer to surface location of introns of the chymotrypsin gene. Stippled areas are the exterior of the β
barrel which comprise each domain. This figure was adapted
from Jane Richardson's drawing of elastase (37).

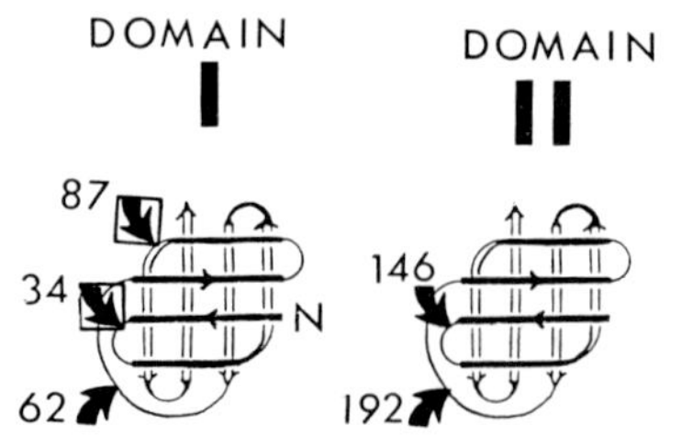

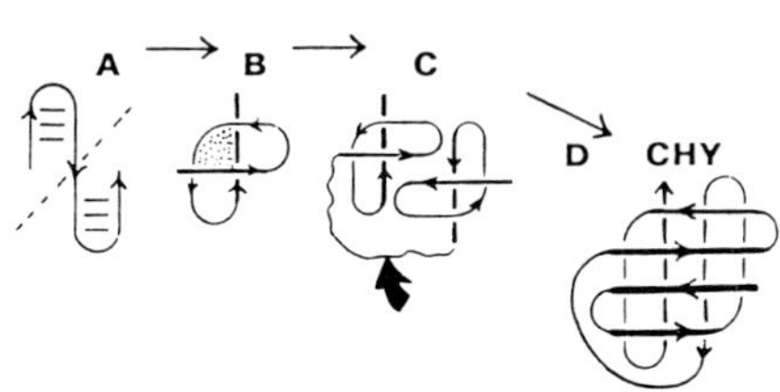

FIGURE 3. Topological representation of the domains and subdomains of chymotrypsin. The numbered arrows correspond to the amino acid sequence where an intron interrupts the coding sequence. Introns B,C, D and E correspond to amino acid number 34, 62, 87, 146 and 192, respectively. This figure was redrawn from ref. 39.

not only at the level of protein structure but also at the level of gene structure. Although the number of introns in the chymotrypsin gene is not the same as the number of introns in the trypsin gene the location of the introns present in trypsin is identical to the location of homologous introns in chymotrypsin. It has been suggested that trypsin is the primordial form of the serine proteases (36). If this is true then introns were either added to the chymotrypsin gene or deleted from the trypsin gene after divergence of the two genes occurred.

<u>Correspondence of Exons and Functional Domains in the Serine Proteases.</u> There is also evidence for internal homology in the serine proteases in that they are composed of two domains which show little primary sequence homology but striking tertiary structure homology (see Fig. 2). These domains may have arisen via gene duplication. On the basis of tertiary structure homology it has been suggested by McLachlan (38 and references therein) that the domains are composed of subdomains that were encoded by an ancestral genetic unit. Our structural characterization of the trypsin and chymotrypsin genes is consistent with this concept since the repeating structural motif (a three stranded β sheet, see Fig. 3) is localized within an exon. This exon could be involved as a functional folding domain in the construction of the gene for the serine proteases. The hydrophobic leader sequence which is required for extracellular transport of

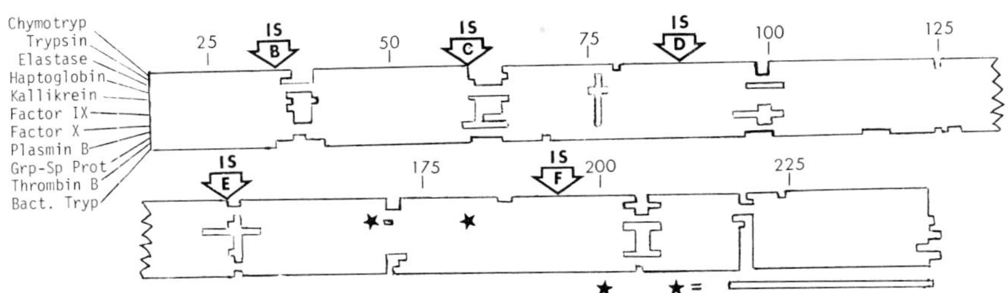

FIGURE 4. Aligned sequences for the serine protease family showing size variation of protein chain length at intron positions. Numbers refer to amino acid positions of the aligned sequences according to the numbering of chymotrypsinogen. Lettered arrow heads refer to the introns of the chymotrypsinogen gene. Data are from J. Greer (35 and his unpublished results).

trypsin and chymotrypsin may also be exon encoded. In both the trypsinogen and chymotrypsinogen genes, an intron interrupts the coding sequence between the prepeptide and the rest of the polypeptide chain. Therefore the functional unit involved in the secretory process is contained within a separate exon (see Fig. 1).

<u>Intron/Exon Junctions May Mark Regions of Change</u>. Given the observation of surface splice junctions and gene structure homology in gene families it might be expected that protein sequences contain additions or deletions in the region encompassing the splice junction. Alignment of the serine protease family (35) (Fig. 4) reveals that three of the five introns of chymotrypsin reside at areas of size variability. A similar alignment for prokaryotic and eukaryotic sequences of dihydrofolate reductase demonstrates that two of the five intervening sequences interrupt the coding sequences at points where the eukaryotic enzyme shows amino acid additions (D. Mathews, R. Fletterick, R. Schimke, unpublished). In certain cases size variation may be correlated with the splice junction, suggesting that introns serve to mediate this process. However, in other instances splice junctions fall within a highly conserved amino acid sequence; thus it is not inevitable that intron positions are sites of hypervariability.

The conservation of intron position in the serine protease genes is consistent with the view that the serine protease genes have arisen by gene multiplication from a primordial ancestor. Since the <u>Streptomyces</u> <u>griseus</u> protease has a similar tertiary structure to eukaryotic serine proteases, presum-

ably its gene is also related to this family. Since there
are no introns in bacterial genes, either the bacterial protease gene lost its introns during evolution or, as Hartley
suggested (33), horizontal gene transfer from eukaryotes to
prokaryotes accounts for this.

Intron/Exon Junctions May Predict Elements of Tertiary
Structure. If the surface rule for intron/exon junctions is
general, then the location of splice points in genomic DNA
sequences also provides useful tertiary structural information
about the gene product since certain regions of the primary
sequence can be reliably predicted to reside at the surface
(see Table 2). Thus these data can be used in part to predict
aspects of tertiary structure as well as provide a strategem
for the design of new proteins.

II. ALTERNATE SPLICING AS A MODEL FOR GENERATION OF GENETIC DIVERSITY

During studies of the expression of the human insulin
gene in a heterologous cell system using SV40 vectors as an
expression vehicle we discovered that an aberrant splicing of
the gene occurred. This alternate splice site resulted in the
formation of a messenger RNA coding for a protein with an N
terminal region (B chain) that is identical to insulin but due
to a codon phase change the C terminal moiety is completely
different. Further there is no dibasic amino acid sequence to
facilitate the removal of a connecting peptide. These results
illustrate the dramatic effects that alterations in the splicing patterns can have on the protein product. In these studies Simian Virus 40 (SV40) was employed as a vector to express the human insulin gene in permissive monkey kidney
cells. This system is particularly useful because of the high
multiplicity of virus that occurs and the resultant high
levels of net transcription that can be achieved. Figure 5
summarizes the procedure for constructing the SV40 insulin recombinants. In ESVins the insulin gene is placed under the
control of the late SV40 promoter. In LSVins2 the insulin gene
is adjacent to the early SV40 promoter. In LSVinsLP the late
SV40 leader sequences (KpnI/HpaI fragment) are inserted between the SV40 early promoter and the human insulin gene. In
LSVinsLP2 the insulin gene follows the early and late SV40
promoters cloned in tandem. In the LSVins construction the
insulin gene follows an SV40 DNA fragment (HpaI/HindIII) which
blocks the expression of the insulin gene. The human insulin
gene (8,39,40) including the putative signals in the 5' and 3'
flanking regions were isolated from pIns96 as a 1603 bp
HincII/BamHI fragment which was cloned between the SV40 HpaI

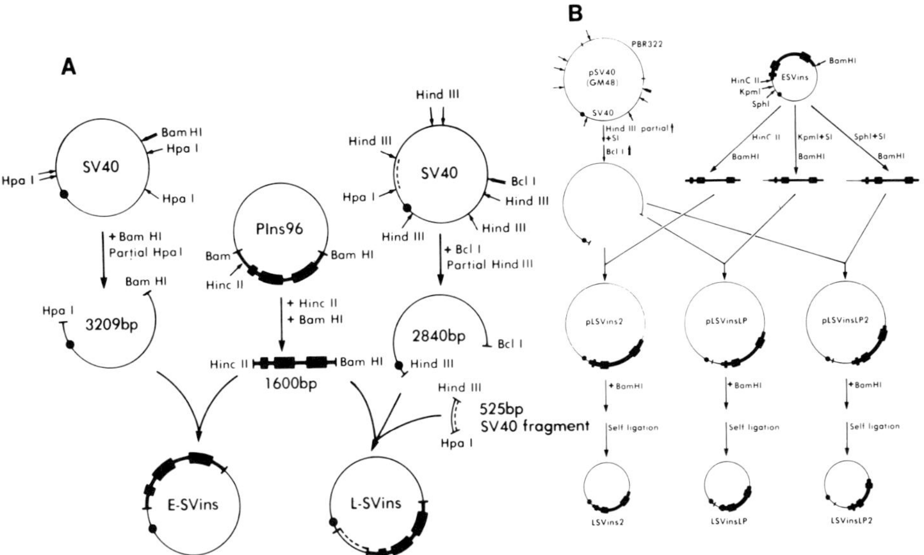

FIGURE 5.   SV40-human insulin recombinants designed to express the human insulin gene.   The human insulin coding sequences are indicated by three blocks.   The introns and insulin flanking sequences are indicated by a solid line. ESVins and LSVins recombinants are illustrated in 5A.   The details for the construction of the SV40 early replacement vectors LSVins2, LSVinsLP, LSVinsLP2 shown in 5B are summarized in the text.

and BamHI sites to construct the ESVins recombinant (Fig. 5A). The KpnI/BamHI (1825 bp) and SphI/BamHI (1907 bp) insulin containing fragments which were used to construct LSVinsLP and LSVinsLP2, respectively, were purified from ESVins.

The pLSV vector was generated from an SV40 genome cloned in the BamHI site of pBR322 (pSV40) and amplified in E. coli-GM48 cells.   DNA from the plasmid pSV40 was linearized by partial digestion with HindIII restriction endonuclease followed by S1 treatment to produce blunt ends.   The linear pSV40 fragment was digested with BclI restriction nuclease and the 7.2 kb pLSV vector was purified by preparative agarose gel electrophoresis.   All insulin DNA fragments had an S1 blunt 5' end and a BamHI 3' end.   These fragments were cloned between the HindIII/S1 blunt end and the BclI end within the pLSV vector (Fig. 5B).   Ampicillin-resistant colonies were screened by hybridization to a ^{32}P-labeled insulin probe and the positive colonies were analyzed by restriction endonuclease mapping.   The resulting plasmids contain a BamHI insert which includes an SV40 origin of replication, a functional late SV40 gene and a complete human insulin gene inserted in SV40 under

the control of the early SV40 signals. The BamHI fragments
containing the LSV-insulin hybrids were self-ligated to form
circular DNA and infected into COS cells (46). Because COS
cells produce SV40 T antigen and are permissive to the LSV
recombinants the viruses were propagated in these cells and
produced a high titer virus stock. The ESVins recombinant
was propagated in CV1 or COS cells. The complementing SV40
late gene functions were supplied by the non-insulin expressing LSVins recombinant.

S1 Mapping of RNA From COS Cells Infected with the
LSV-Insulin Recombinants. The method of Berk and Sharp (41)
was used to analyze the RNA present in the cytoplasm of COS
cells infected with each of the LSV-insulin recombinants. A
uniformly labeled insulin specific probe was prepared from COS
cells infected with the LSVins virus stock and labeled with
^{32}P-orthophosphate for 12-16 hr. Viral supercoiled ^{32}P-DNA
was purified (42,43) and the insulin probe was isolated as a
1.7 kb HincII fragment. The labeled probe contains the 1603
nucleotides of the human insulin insert and additional 103
bases derived from the 3' end of the early SV40 gene. The
^{32}P-probe was denatured and hybridized to polyA+ and polyA$^-$
RNA isolated from infected COS cells, under conditions favor-
ing RNA-DNA duplexes. The resulting hybrids were digested
with S1 nuclease and analyzed on a 5% acrylamide 8M urea
sequencing gel (44). As shown in Figure 6, lane A, polyA+
RNA from the ESVins recombinant produced five bands. The
bands 206 and 218 nucleotides in length represent the two
insulin coding exons which comigrate with human insulinoma
RNA analyzed under the same conditions (data not shown). Thus
it appears that the human insulin promoter, splicing and
polyadenylation signals are functioning in the ESVins construction. However, the 99 bp and 340 bp fragments represent
chimeric 5' and 3' insulin-SV40 exons indicating that most
insulin transcripts initiate and terminate at the late SV40
signals. LSVins (lane B) is a very poor template for insulin
transcription. From these results, it is apparent that the
SV40 DNA fragment (SV40 map units 0.86 and 0.76) exerts a
strong inhibitory effect on the transcription of downstream
sequences.
LSVins2, LSVinsLP and LSVinsLP2 (lanes C,D and E, respectively) are expressed efficiently and produce 5 to 10 fold
more RNA than the late SV40 replacement recombinant, ESVins.
The second exon of human insulin which is 206 nucleotides in
length is present in the polyA+ fraction of all LSVinsulin
constructions, however, exon 1 (42 bp) and exon 3 (218 bp) are
absent in all LSV recombinants. This result strongly suggests
that all RNA transcripts detectable initiate and terminate at

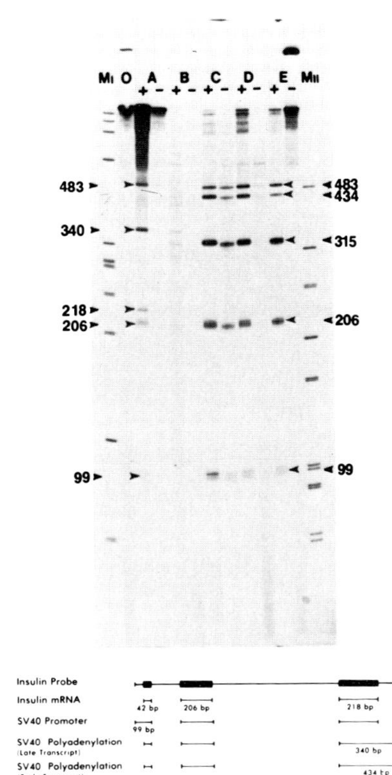

FIGURE 6. Nuclease S1 analysis of cytoplasmic RNA from the LSV-insulin recombinant. Purified non-polyA and polyA-containing cytoplasmic RNA from infected COS cells was used. RNA was hybridized to the insulin HincII/DNA fragment purified from uniformly labeled LSVins ^{32}P-DNA. Hybridization mixture in 80% formamide, 0.4M NaCl was denatured 2 min at 80° annealed at 52° for 3 hr, treated 1 hr with S1 nuclease and subjected to a 5% acrylamide 8M urea sequencing gel. Tracks MI and MII contain size markers; lane A, E/LSVins RNA; B, LSVins RNA; C, LSVins2 RNA; D, LSVinsLP RNA; E, LSVinsLP2 RNA; O, no RNA. The diagram below the figure represents the predicted S1 protected fragments.

SV40 early signals. RNA transcripts produced from insulin regulatory sequences must be less than 1% of the total. The band 99 nucleotides in length corresponds to RNA that initiates at the SV40 cap site and is spliced at the donor site of exon 1 of the insulin gene. The band 443 nucleotides long extends from the splice acceptor site of exon 3 of insulin to the 3' end of the insulin probe and therefore results from transcripts polyadenylated at the early SV40 sites. The largest protected band which is 483 nucleotides long was also observed with the ESVins recombinant (lane A). This fragment is most likely derived from a partial splicing event which does not remove the first intron of insulin. Thus this particular RNA class contains exon 1, exon 2 and the first intron of insulin. The band 315 nucleotides in length was not predicted and its origin was unknown.

S1 Endonuclease Mapping of the Cryptic Splice Junction. The 1603 bp HincII/BamHI insert was labeled using ^{32}P-γ-ATP and

                                                      CHARLES S. CRAIK *et al.*

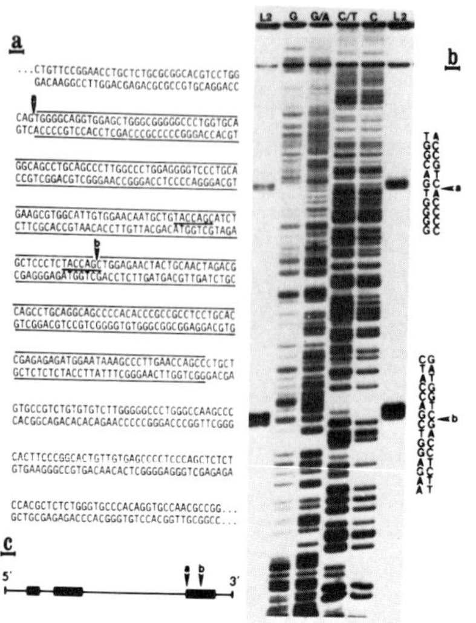

FIGURE 7. S1 mapping of the cryptic exon. The HincI/BamHI DNA fragment was labeled with T4 polynucleotide kinase using ^{32}P-γ-ATP. The labeled fragment was cut with SmaI and a 563 nucleotide DNA probe was purified. The DNA probe was annealed with LSVins2 RNA and analyzed by S1 nuclease as described in Figure 6. The protected fragments were sized on a sequencing gel alongside a sequencing ladder of chemical degradation products generated from the BamHI/SmaI probe (panel 7A). Panel 7B represents the location of the two alternative splice sites within the known sequence of exon 3 (8). The diagram in 7C represents the location of the two alternative splice junctions in the insulin gene.

polynucleotide kinase. The 5' end labeled fragment was digested with restriction endonuclease SmaI and a 563 bp labeled fragment was purified by gel electrophoresis and used as the probe for S1 mapping (45). This DNA probe extends from the 3' end of the insulin DNA insert into the second intron within the insulin gene; thus it includes exon 3. The DNA-^{32}P-probe was denatured and annealed to polyA+ RNA extracted from COS cells infected for 48 hr with LSVins2. The protected fragments were sized by electrophoresis on a gel alongside a sequencing ladder of chemical degradation products (44) generated from the same BamHI/SmaI probe. Figure 7 depicts the results. The position of the two protected fragments within the sequence can be seen clearly (7A), with the S1 bands comigrating with the predicted splice sites. The slight underdigestion by S1 nuclease is probably due to the partial sequence homology of the probe to the 3' end of the second exon of insulin. The 340 nucleotide band corresponds to the expected insulin transcript spliced at the acceptor site of exon 3. The second band which is 222 nucleotides long corresponds to the new insulin exon which uses a cryptic acceptor site located within the coding region of exon 3.

The Generation of Distinct Gene Products from a Single
Genetic Sequence. RNA mapping by S1 nuclease analysis demonstrated that two alternative splicing pathways are used in the
formation of stable specific mRNA derived from the insulin
gene. One processing pathway uses the authentic insulin
splicing sites to produce the mRNA coding for human proinsulin
however an alternative splice site was also used to produce a
novel mRNA. This should generate two insulin-related proteins
whose sequences are in part common and in part different. As
shown in this study the cryptic splice site is not so efficiently employed in the ESVins vector but is the predominant
form used in a series of LSVinsulin recombinants (estimated
5-10 fold difference). Since these vectors were constructed
independently the alternative splicing pathway is not a result
of a cloning rearrangement. Moreover, all these LSVinsulin
recombinants express insulin mRNA with different 5' untranslated sequences; thus these sequences appear to have no effect
on the splicing pattern.

The expressed cryptic splice site was accurately mapped
within the coding region of the third exon of insulin. This
splice site is within a sequence TACCAGC which is located in
front of a pyrimidine-rich DNA stretch and resembles a perfect
"consensus" sequence (26). The sequence TACCAGC is repeated
19 nucleotides upstream from the cryptic splice site, however,
it was not used in the SV40 expression system. Interestingly,
another similar sequence, ACCAGC, is found at the site of
polyadenylation.

Several aspects about the insulin gene expression in the
SV40 vectors are worthy of comment. The transcription and
processing from the insulin regulatory signals correspond to
those found in vivo; thus the expression system does not
appear artifactual. In these SV40 expression systems the late
promoter seems about 10-fold stronger than the insulin promoter. Further, the early promoter is still stronger (about 5
fold) than the late promoter and transcripts from the insulin
promoter cannot be detected. Intriguingly, the termination
signals in the three systems appear to correspond to the
strength of the promoter. The early terminator of SV40
appears stronger than the late terminator which in turn
appears stronger than the insulin terminator. Whether this
relationship is related to the position of the promoter in the
construction (the strongest promoter appears further downstream in all constructions) or whether it involves some kind
of interaction with the promoter region (5' leader) is not yet
clear. The recruitment of the aberrant splice site appears to
be partially influenced by extragenic sequences. The constructions employed in the early promoter regions produce
dominantly the aberrant transcript, while those in the late

promoter produce largely the normal insulin mRNA product. In
the described constructions the 5' promoters or the flanking
region exert no detectable influence on the splice selection
but the 3' region does. The manner of the interaction of the
3' region (polyadenylation site?) is unknown, but it could
involve either a question of distance alone or a specific
interaction dictated by sequence information.

There are precedents for the use of alternate splicing
sites in the regulation of expression of a single gene in two
different cells (e.g., liver and salivary amylase) (47) or in
the production of two separate proteins from the same gene
region. It was previously suggested (48) on the basis of an
analysis of RNA precursors and mature transcripts that differential RNA processing events may account for variation in the
expression of the gene coding for the small polypeptide hormone calcitonin. Rosenfeld et al. (49) have recently presented evidence suggesting that two structurally different mRNAs
encoding for different calcitonin-related protein products
arise as a consequence of alternative RNA splicing events from
the same gene. Our data on insulin RNA processing may provide
another illustration of the concept that a single gene can
encode more than one protein product. The alternative splicing pathways presumably generate two distinct proteins. The
two peptides share the amino terminal B chain of insulin but
differ in their carboxyl termini. We have attempted isolation
of this putative protein but if present it interacts poorly
with anti-insulin sera.

The experimental results illustrate the diversity of
transcripts that can arise from a single genetic sequence. It
seems likely that nucleotide environment external to the gene
can influence this choice. This suggests a evolutionary route
for the origin of gene families from constructed genetic
sequences. Whether such cogenic products will have related
three dimensional structures is unknown. However, the putative positioning of the intron/exon junctions at the surface
(as described in part I) would in principle allow the utilization of the common peptide element as a structural domain
within the hybrid molecule.

ACKNOWLEDGMENTS

We were aided by discussions with D.N. Standring. We
gratefully acknowledge Q.L. Choo, C. Quinto, R. Schimke, L.P.
Shen and W. Swain for communicating the gene sequences of
trypsin, chymotrypsin, dihydrofolate reductase and amylase,
respectively, prior to publication. We also thank Leslie
Spector for preparation of the manuscript.

REFERENCES

1. Doolittle, W.F., and Sapienza, C. (1980). Nature 284, 601.
2. Orgel, L.S., and Crick, F.H.C. (1980). Nature 284, 604.
3. Gilbert, W. (1978). Nature 271, 501.
4. Blake, C.C.F. (1978). Nature 273, 267.
5. Blake, C.C.F. (1979). Nature 277, 598.
6. Tonegawa, S., Maxam, A., Tizard, R., Bernard, O. and Gilbert, W. (1978). Proc. Nat. Acad. Sci. USA 75, 1485.
7. Sakano, H., Rogers, J.H., Huppi, K., Brack, C., Traunecker A., Maki, R., Wall, R., and Tonegawa, S. (1979). Nature 277, 627.
8. Bell, G., Pictet, R., Rutter, W.J., Cordell, B., Tischer, E., and Goodman, H.M. (1980). Nature 284, 26.
9. Jung, A., Sippel, A.E., Grez, M., and Schutz, G. (1980). Proc. Nat. Acad. Sci. USA 77, 5759.
10. Tilghman, S.M., Tiemeier, D.C., Seidman, J.G., Peterlin, B.M., Sullivan, M., Maizel, J.V., and Leder, P. (1978). Proc. Nat. Acad. Sci. USA 75, 725.
11. Jeffreys, A.J., and Flavell, R.A. (1977). Cell 12, 1097.
12. Leder, A., Miller, H.I., Hamer, D.H., Seidman, J.G., Norman, B., Sullivan, M., and Leder, P. (1978). Proc. Nat. Acad. Sci. USA 75, 6187.
13. Craik, C.S., Buchman, S.R., and Beychok, S. (1980). Proc. Nat. Acad. Sci. USA 77, 1384.
14. Craik, C.S., Buchman, S.R., and Beychok, S. (1981). Nature 291, 87.
15. Go, M. (1981). Nature 291, 91.
16. Wozney, J., Hanahan, D., Tate, V., Boedtker, H., and Doty, P. (1981). Nature 294, 129.
17. Eiferman, F.A., Young, P.R., Scott, R.W., and Tilghman, S.M. (1981). Nature 294, 713.
18. Stein, I.P., Catterall, J.F., Kristo, P., Means, A.R., and O'Malley, B.W. (1980). Cell 21, 681.
19. Barta, A., Richards, R.I., Baxter, J.D., and Shine, J. (1981). Proc. Nat. Acad. Sci. USA 78, 4867.
20. Quinto, C., Quiroga, M., Swain, W.F., Nikovits, W.C. Jr., Standring, D.N., Pictet, R., Valenzuela, P., and Rutter, W.J. (1982). Proc. Nat. Acad. Sci. USA 79, 31.
21. MacDonald, R.J., Crerar, M.M., Swain, W.F., Pictet, R.L., Thomas, G., and Rutter, W.J. (1980). Nature 287, 117.
22. Benyajati, C., Place, A.R., Powers, D.A., and Soefer, W. (1981). Proc. Nat. Acad. Sci. USA 78, 2717.
23. Lee, B., and Richards, F.M. (1981). J. Mol. Biol. 55, 379.
24. Janin, J. (1979). Nature 277, 491.
25. Chou, P., and Fasman, G. (1978). Ann. Rev. Biochem. 47, 251.

26. Breathnach, R., and Chambon, P. (1981). Ann. Rev. Biochem. 50, 349.
27. Sharp, P.A. (1981). Cell 23, 643.
28. Bonitz, S.G., Coruzzi, G., Thalenfeld, B.E., Macino, G., and Tzagoloff, A. (1980). J. Biol. Chem. 255, 11927.
29. Breathnach, R., Benoist, C., O'Hare, K., Gannon, F., and Chambon, P. (1978). Proc. Nat. Acad. Sci. USA 75, 4853.
30. Welling, G.W., Groen, G., and Beintema, J.J. (1975). Biochem. J. 147, 505.
31. Lenstra, J.A., Hofsteenge, J., and Beintema, J.J. (1977). J. Mol. Biol. 109, 185.
32. Hartley, B.S., Brown, J.R., Kauffman, D.L., and Smillie, L.B. (1965). Nature 207, 1157.
33. Hartley, B.S. (1970). Phil. Trans. Roy. Soc. Lond. B. 257, 77.
34. Stroud, R.M. (1974). Sci. Amer. 231, 74.
35. Greer, J. (1981). J. Mol. Biol. 153, 1027.
36. DeHaen, C., Neurath, H., and Teller, D.C. (1975). J. Mol. Biol. 92, 225.
37. Richardson, J. (1979). The Protein Structure Coloring Book, Little River Institute, Bahama, North Carolina.
38. McLachlan, A.D. (1980). In "Protein Folding" (R. Jaenicke, ed.), pp. 79-100. Elsevier, Amsterdam.
39. Bell, G.I., Swain, W.F., Pictet, R., Cordell, B., Goodman, H.M., and Rutter, W.J. (1979). Nature 282, 525.
40. Bell, G.I., Selby, M.J., and Rutter, W.J. (1982). Nature 295, 31.
41. Berk, A.J., and Sharp, P.A. (1977). Cell 12, 721.
42. Hirt, B. (1967). J. Mol. Biol. 26, 365.
43. Rudloff, R., Bauer, W., and Vinograd, J. (1967). Proc. Nat. Acad. Sci. USA 57, 1514.
44. Maxam, M.A., and Gilbert, W. (1980). Methods in Enzymol. 65, 499.
45. Weaver, R.F., and Weissman, C. (1979). Nucl. Acids Res. 6, 1175.
46. Gluzman, Y. (1981). Cell 23, 175.
47. Hagenbuchle, O., Tosi, M., Schibler, U., Bovey, R., Wellauer, P.K., and Young, R.A. (1981). Nature 289, 643.
48. Rosenfeld, M.G., Amara, S.G., Roos, B.A., Ong, E.S., and Evans, R.M. (1981). Nature 290, 63.
49. Rosenfeld, M.G., Lin, C.R., Amara, S.G., Stolarsky, L., Roos, B.A., Ong, E.S., and Evans, A.M. (1982). Proc. Nat. Acad. Sci. USA 79, 1717.

THE HUMAN α_1-ANTITRYPSIN GENE: ITS SEQUENCE HOMOLOGY

AND STRUCTURAL COMPARISON WITH THE CHICKEN OVALBUMIN GENE

Savio L.C. Woo, T. Chandra, Vincent J. Kidd,
George L. Long[*], Kotoku Kurachi[*] and Earl W. Davie[*].
Howard Hughes Medical Institute Laboratory
Department of Cell Biology, Baylor College of Medicine
Houston, Texas 77030
[*]Department of Biochemistry,
University of Washington
Seattle, Washington 98195

ABSTRACT. α1-Antitrypsin deficiency is a genetic disorder
that predisposes affected individuals to development of chronic obstructive pulmonary emphysema and/or infantile liver cir-
rhosis. The human chromosomal α1-antitrypsin gene has been
cloned. Electronmicroscopic analysis and restriction mapping
have indicated that this gene contains 3 intervening sequences
in the peptide-coding region. DNA sequences coding for the
amino and carboxyl-termini of α1-antitrypsin have been
identified. Human α1-antitrypsin and chicken ovalbumin
share significant sequence homology and belong to a common
protein super-family. Yet the number, position and size of
intervening sequences reveal that the two genes are dissimilar. The evolutionary origin of these two genes is discussed.

INTRODUCTION

α1-Antitrypsin is an important plasma protease inhibitor
which is capable of inhibiting a wide variety of serine proteases. It is a glycoprotein synthesized in the liver and
consists of a single polypeptide chain with a molecular weight
of 51,000 (1). It is transported by passive diffusion into
the lung and its primary physiological function is to protect
the pulmonary aveolar structure from destruction by polymor-
phonuclear leukocyte elastase (2,3). The balance between
elastase and α1-antitrypsin in the lung is perturbed geneti-
cally in individuals with an inborn deficiency of α1-anti-
trypsin. This imbalance can lead to hydrolysis of the elastic
fibers of the lung resulting in permanent damage to the alveo-
lar structure (4,5). Accordingly, individuals with α1-anti-
trypsin deficiency are predisposed to development of chronic
obstructive pulmonary emphysema with a risk 20-30 times that
of the general population (6,7). The genetic deficiency is
characterized by the presence of a variant protein in the
blood and is inherited as an autosomal recessive trait with a

prevalance of approximately 1 in 4,000 among Caucasians of
Northern European ancestry (8,9). Isolation of this human
gene would permit detailed characterization of the genetic
disorder at the nucleotide level and development of an analy-
tical method for prenatal diagnosis through DNA polymorphism
studies. The cloning of the human chromosomal α1-anti-
trypsin gene and its structural comparison with the chicken
gene coding for ovalbumin, a distantly related protein, are
reported in this communication.

RESULTS

Isolation of genomic α1-antitrypsin DNA clones from a
human gene library. A total of 16 independent phage isolates
were obtained when 2 x 10^6 plaques from the human genomic
DNA library (10) were screened using a baboon α1-antitrypsin
cDNA clone (11) as a hybridization probe. Subsequent analysis
of the 16 isolates indicated that they originated from four
independent clones. These clones, labeled αAT135, αAT35,
αAT80 and αAT101, were analyzed by restriction mapping and
Southern hybridization (Fig. 1) using as probes a Mbo II fragment of pBaαlal DNA, which contains the 3' terminal region
of the baboon cDNA (11), and a Hha I fragment of pBAαla2 DNA
which is a baboon cDNA clone lacking only about 100 nucleotides at the 5' end of the mRNA (12). These results have also
established the orientation of the human α1-antitrypsin gene.

Molecular structure of the human α1-antitrypsin gene.
The overall structure of the human α1-antitrypsin gene was
established by electron microscopic examination of hybrid
molecules formed between the cloned chromosomal DNA and baboon
α1-antitrypsin mRNA. The mature mRNA consists of approximately 1400 nucleotides. DNA was denatured thermally and hybrids were formed subsequently under conditions that favored
RNA/DNA hybridization but not DNA/DNA reassociation. It is
evident that there are three intervening DNA loops of various
sizes within the human α1-antitrypsin gene (Fig. 2). Nu-
monic measurements of the hybrid molecules have indicated that
the sizes of exon regions I, II, III, and IV are 0.71, 0.33,
0.13 and 0.27 Kb's in length, respectively. The sizes of introns A, B and C are 1.45, 1.15 and 0.8 Kb's, respectively,
and all three introns appear to be located within the 3' half
of the mRNA.

In order to characterize the human chromosomal α1-anti-
trypsin gene in greater detail, the 9.6 Kb Eco RI DNA fragment
containing the gene was sublconed into Eco RI site of pBR322.

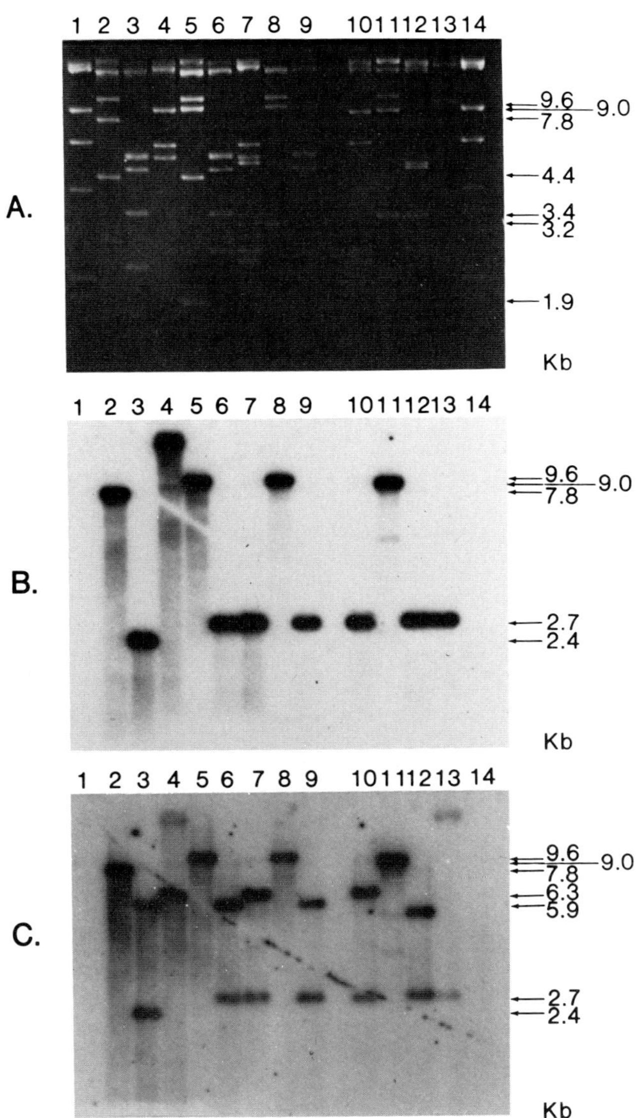

Fig. 1. Panel A, an EtBr stained agarose gel containing various recombinant phage DNA's. Panels B and C, Southern hybridization of the gel using a 3' specific probe and the entire
cDNA probe, respectively. Lanes 2-4, 5-7, 8-10 and 11-13 contained αAT 135, αAT35, αAT80 and αAT101 DNA, respectively.   Order of enzymes used for digestion of each DNA was
EcoRI, Hind III and EcoRI + Hind III.

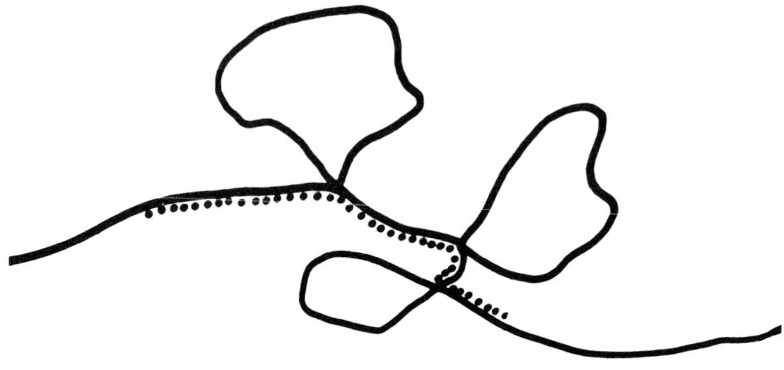

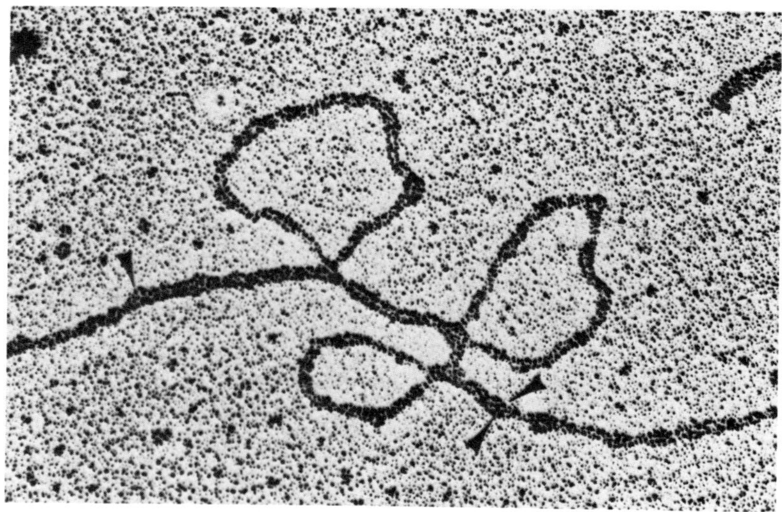

<u>Fig. 2.</u>   An electron micrograph and the corresponding line
drawing of a hybrid molecule formed between the human chromosomal αl-antitrypsin gene and baboon liver poly(A)-contain-
ing RNA. ——————————, single-stranded genomic human DNA; . .
. . . , baboon αl-antitrypsin mRNA.

The resulting clone, pAT9.6, was analyzed by restriction mapping and Southern hybridization. Four exon segments were
identified within the 9.6 Kb <u>Eco</u> RI DNA fragment using a combination of enzymes that do not cut the baboon αl-antitrypsin cDNA insert in pBaαla2 (12). These results, as summarized in Fig. 3, confirmed the existence of three introns in
the human αl-antitrypsin gene. Subsequent DNA sequencing
analysis has revealed the locations of the initiation and
termination codons on the 9.6 Kb <u>Eco</u> RI DNA fragment, and that
there are only 3 introns in the peptide-coding region of the
human chromosomal αl-antitrypsin gene (data not shown).

<u>Sequence homology and structural comparison between human
αl-antitrypsin and chicken ovalbumin</u>. Based on a partial
amino acid sequence of human αl-antitrypsin, an unexpectedly
high degree of amino acid sequence homology between human
αl-antitrypsin and chicken ovalbumin was originally observed
by Hunt and Dayhoff (13). Based on the assumption that the
rates of change for both proteins are similar to accepted
values for animal lysozyme and pancreatic secretory trypsin
inhibitor, it was concluded that the divergence time between
the two proteins could be over 500 million years ago. Consequently they proposed that the two proteins belong to a common
super-family and that they diverged before or during early
vertebrate evolution (13). We have subsequently determined
the complete amino acid sequence of baboon αl-antitrypsin
and shown that the two seemingly unrelated proteins have a 24%
amino acid sequence homology (12). Matrix plot computer analysis has revealed an apparent diagonal line between the two
proteins, indicating that homology exists at corresponding
positions throughout the two molecules (data not shown).
     Since human αl-antitrypsin and chicken ovalbumin ap-
parently belong to the same superfamily, it was predicted that
the two genes would contain similar exon-intron patterns
(13). The number and locations of the introns in the human
αl-antitrypsin gene were thus compared with those present in
the chicken ovalbumin gene (Figure 4). The chicken ovalbumin
gene contains seven introns and their exact locations within
the gene have been established by DNA sequencing (15-17). The
human αl-antitrypsin gene contains only three introns, which
are located within the 3' half of the corresponding mRNA,
whereas all seven introns in the chicken ovalbumin gene are
located within the 5' half of the mRNA. Based upon the

Molecular Structure of the Human Chromosomal Alpha-1-Antitrypsin Gene

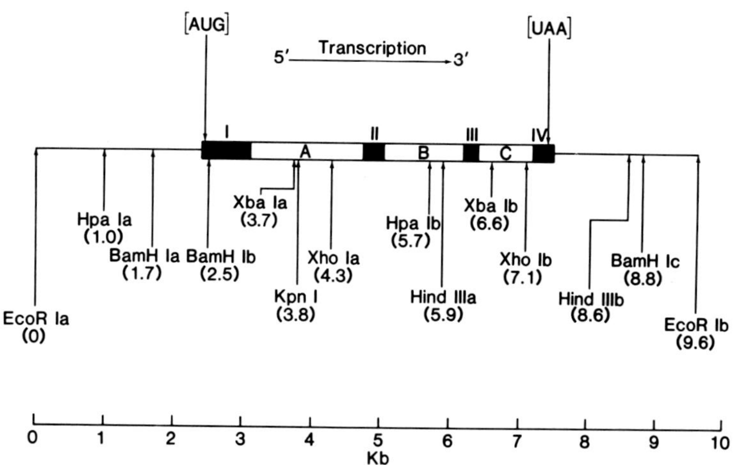

Fig. 3. Mosaic structure of the human chromosomal αl-antitrypsin gene. The various structural segments (▇) are indicated by Roman numerals and the various intervening sequences (▢) are indicated by capital letters. The transcriptional orientation of the αl-antitrypsin gene is shown from left to right.

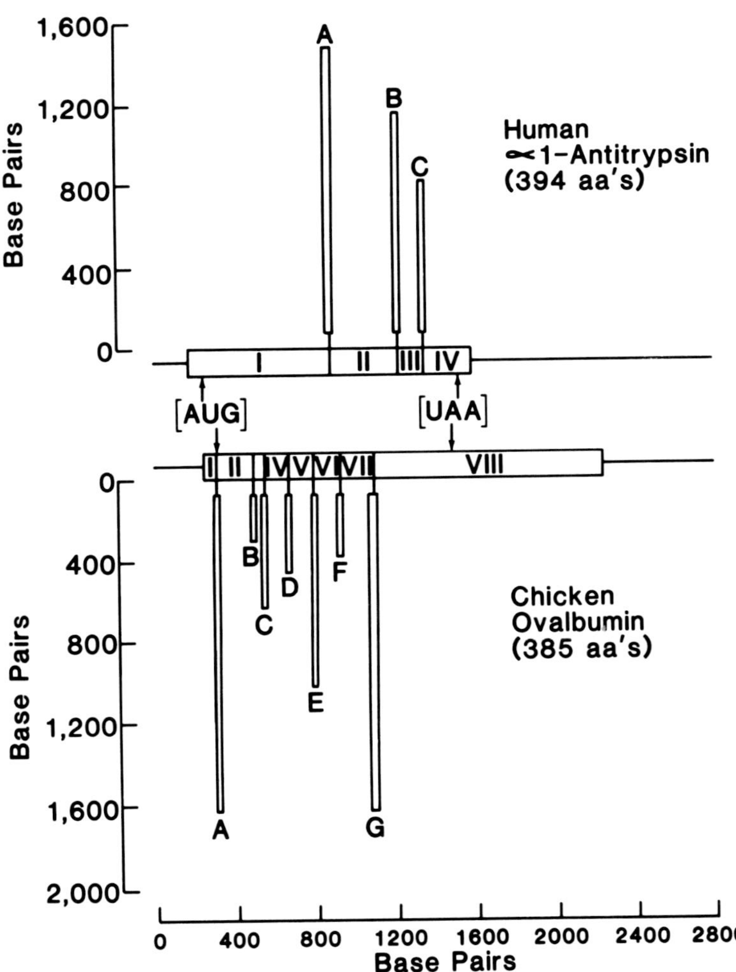

Fig. 4.   Comparison of the genes for αl-antitrypsin and ovalbumin. Coding regions for the two proteins are shown by open horizontal bars. The positions of the initiation and termination codons on the two genes are indicated by the arrows. Alignment of the two genes was based upon amino acid sequence homology for the two proteins (12). Intervening sequences are shown by the open vertical bars.

electron microscopic map, the only introns that might share a
common site in the two genes are intron A in the human
αl-antitrypsin gene and intron E or F in the chicken ovalbumin gene (Figure 4). Subsequent DNA sequencing analysis
however, has revealed that the precise location of intron A in
the human αl-antitrypsin gene is located 27 and 21 amino
acids away from ovalbumin introns E and F, respectively (data
not shown).

DISCUSSION

At the present time it is impossible to determine conclu-
sively whether human αl-antitrypsin and chicken ovalbumin
arose by convergent or divergent evolution. The lack of similarity in structure between the two genes is certainly not unexpected if convergent evolution were the pathway. However,
it would be very intriguing if these genes arose by divergent
evolution. In contrast to the other multi-gene families such
as the globins (18) and vitellogenins (19) where the number
and positions of introns are rigidly conserved, the αl-antitrypsin and ovalbumin genes contain different numbers of introns which are located at completely different sites in the
exonic sequences. This observation would thus suggest that
the ancestral gene to αl-antitrypsin and ovalbumin might
have undergone duplication <u>prior</u> to the addition or deletion
of intervening DNA sequences. Moreover, the ovalbumin gene is
linked to two other sequence-related genes (X and Y) in the
chicken genome (20, 21). The fact that the X-Y-ovalbumin gene
family (20,21) in the chicken have retained the number and positioning of the introns would suggest that there were subsequent gene duplication events involving the ancestral ovalbumin gene. In this regard, it is interesting to note that
the αl-antitrypsin gene is also linked to a sequence-related
gene in the human genome (data not shown).

The observation that the molecular structures of the
human αl-antitrypsin and chicken ovalbumin genes differ from
each other is similar to that made in the actin gene family.
There are 1, 0, 1, and 2 introns in the actin genes of yeast,
dictyostelium, <u>Drosophila</u>, and sea urchin, respectively, and
yet the positions of the introns are completely distinct within the exons which have retained extensive sequence homology
and resemble those of the mammalian cytoplasmic actins (22).
Based on these observations, it has been suggested that the
current model (23) for the functional and evolutionary conservation of locations of introns may not be valid for all
genes (24), and that some introns are vestiges of transpo-

son-like elements that have been inserted into genes, become
fixed and subsequently diverged in nucleotide sequence (22).
However, it has also been pointed out that owing to the limited numbers of introns present in the actin gene family, these
genes could also have evolved by random deletion of multiple
introns present in the primordial actin gene (22). Precise
deletion of an entire intron has been reported to occur in the
rat insulin gene, resulting in the existence of two genotypic
alleles (25). Similar observations have also been made in the
mouse α-globin gene (26). In the case of αl-antitrypsin
and ovalbumin however, if indeed the ancestral gene to these
proteins contained 10 or more introns and the two genes arose
by random intron deletion, it might be expected that a certan
number of introns would share common sites in the exons of the
two genes. The fact that the positions of the introns in the
two genes do not overlap at all makes this hypothesis less attractive. Consequently, it is possible that at least some of
the introns could be the vestiges of transposable elements
that had been inserted into preexisting exons of the two genes
after their divergence from an ancestral gene.

ACKNOWLEDGMENTS

The authors wish to thank Dr. Tom Maniatis for providing
the human gene library. This work was supported in part by a
NIH Grant HL 16919 to E.W.D. and a grant from the Cystic Fi-
brosis Foundation to S.L.C.W. G.L.L. is the recipient of a
NIH Senior Research Fellowship (HL 05962), and K.K. is the recipient of a NIH Research Career Development Award (HL
00404). S.L.C.W. is an Investigator of the Howard Hughes
Medical Institute.

REFERENCES

1. Heimburger, N., Haupt, H. & Schwick, H.G. (1970). In.
   Proc. Int. Res. Conf. on Proteinase Inhibitors (eds Fritz,
   H. & Tschesche, H.) W. deGruyter, erlin. p.1-22.
2. Laurell, C.B. & Erikson, S. (1963). Scand. J. Clin. Lab.
   Invest. 15, 132-140.
3. Beatty, K., Bieth, J. & Travis, J. (1980). J. Biol.
   Chem. 255, 3931-3934.
4. Starkey, P.M. & Barrett, A.J. (1976). Biochem. J. 155,
   265-271.
5. Kuhn, C. & Senior, R.M. (1978). Lung 155, 185-197.
6. Kueppers, F. (1978). In Lung Biology in Health and
   Disease (ed. Litwin, S.D.) p. 23 Marcel Dekker Press, New
   York.

7. Carrell, R.W. & Owen, M.C. (1980). Essays Med. Biochem. 4, 83-119.
8. Fagerhol, M.K. & Laurell, C.B. (1970). In Progress in Medical Genetics (eds. Steinberg, A.G. & Bearn, A.G.) Vol. VII, 96-111 (Bruen & Stratton, New York).
9. Allen, R.C., Harley, R.A. & Talamo, R.C. (1974) Am. J. Clin. Path. 62, 732-739.
10. Lawn, R.M., Fritsch, E.F., Parker, R.C., Blake G. & Maniatis, T. (1978). Cell 15, 1156-1174.
11. Chandra, T., Kurachi, K., Davie, E.W. and Woo, S.L.C. (1981). Biochem. Biophys. Res. Commun. 103, 751-758.
12. Kurachi, K., Chandra, T., Degen, S.J. Freizner, White, T.T., Marchioro, T.L., Woo, S.L.C. & Davie, E.W. (1981). Proc. Natl. Acad. Sci. U.S.A. 78, 6826-6830.
13. Hunt, L.T. & Dayhoff, M.O. (1980). Biochem. Biophys. Res. Commun. 95, 864-871.
14. Novotny, J. (1982). Nuc. Acids Res. 10 127-131.
15. McRaynolds, L., O'Malley, B.W., Nisbet, A.D., Fothergill, J.E., Givol, D.1 Fields, S., Robertson, M. & Brownless, G.G. (1978). Nature 273, 723-728.
16. Catterall, J.F., O'Malley, B.W., Robertson, M.A., Staden, R., Tanaka, Y. and Brownlee, G.G., (1978) Nature 257, 510-513.
17. Breathnach, R., Benoist, C., O'Hare, K., Gannon, F. and Chambon, P. (1978). Proc. Natl. Acad. Sci. U.S.A. 75, 4853-4857.
18. Maniatis, T., Fritsch, E.F., Lauer, J. and Laen, R.M. (1980). Ann. Rev. Genet. 14, 145-178.
19. Wahli, W., David, I.B., Wyler, T., Wever, R. and Ryffel, G.U. (1980). Cell 20, 107-117.
20. Colbert, D.A., Knoll, B.J., Woo, S.L.C., Mace, M.L., Tsai, M.J. and O'Malley, B.W. (1980). Biochemistry 19, 5586-5592.
21. Heilig, R., Perrin, F., Gannon, F., Mandel, J.L. and Chambon, P. (1980). Cell 20, 625-637.
22. Fyrberg, E.A., Bond, B.J., Hershey, N.D., Mixter, K.S. and Davidson, N. (1981). Cell 24, 107-116.
23. Gilbert, W. (1978). Nature 271, 501.
24. Firtel, R.A. (1980). Cell 24, 6-7.
25. Comedico, P., Rosenthal, N., Efstratiadis, A., Gilbert, W., Kolodner, R. & Tigard, R. (1979). Cell 18, 545-558.
26. Nishioka, Y., Leder, A. & Leder, P. (1980). Proc. Natl. Acad. Sci. U.S.A. 77, 2806-2809.

CRYPTIC SPLICE SITES IN THE RABBIT β-GLOBIN GENE
ARE REVEALED FOLLOWING INACTIVATION OF AN
AUTHENTIC 5' SPLICE SITE BY MUTAGENESIS <u>IN VITRO</u>

B. Wieringa, F. Meyer[+], J. Reiser,
and C. Weissmann

Institut für Molekularbiologie I
Universität Zürich
8093 Zürich, Switzerland

ABSTRACT.   The processing of transcripts of genes
altered by restructuring or site-directed mutagenesis was studied in a transient (acute) transformation system.
   Rabbit β-globin-specific splice products
were analyzed and quantitated by S_1 mapping. In
addition, cDNA primer-extension as well as cloning and sequencing of globin cDNAs was used for
precise characterization of the products.
   Deletion of the 5' splice site (SS) of the
large intron (from position +482 to +502) yielded
normal levels of shortened mRNAs, in which the
normal 3' SS was spliced to a cryptic SS at position +360 in the second exon.
   Out of six purine transitions around the 5'
SS only the GT⟶AT transition (position +495),
which breaks the GT-AG "rule" for intron ends,
affected splicing, giving rise to three abnormal
mRNAs. In each of these the normal 3' SS was
joined to a different cryptic (i.e., normally not
utilized) 5' junction. Two of these junctions, at
positions +360 and +499, are followed by the expected GT-sequence, the other however (at +443)
is followed by the sequence GC.
   Some of the cryptic 5' SS's show poor homology with the consensus sequence

$$\begin{smallmatrix}C\\A\end{smallmatrix}AG/GT\begin{smallmatrix}A\\G\end{smallmatrix}AGT$$

and a poor complementary fit with U1 RNA.

[+]Present address: CIBA-GEIGY AG, Basel, Switzerland

GENE REGULATION

65

The rule originally proposed by Breathnach and Chambon (1), which states that intron sequences invariably start with (5') GT and end with AG (3') has held true in all cases examined so far (2). More extensive consensus sequences have been derived for the exon-intron junctions, namely ^C_AAG/GT^G_AAGT around the 5' splice site and $\binom{C}{T}_n N^T_C$AG/G (with $\underline{n} \geqslant 11$) around the 3' splice site (2-6). In many 5' splice sites (5' SS) only 6 (and at least once only five) positions conform with the consensus sequence; in the case of the 3' splice site (3' SS), the consensus run of 11 pyrimidines may contain as many as six purines, but never the doublet AG (2). To examine the stringency of one particular 5' SS sequence, we generated single point mutations at six of the nine 5' SS consensus positions of the large intron of the rabbit β-globin gene, and determined the accuracy of splicing and the level of the globin transcripts in the cell. Of the mutations examined, only the one affecting the GT doublet abolished splicing at the correct 5' SS, and led to splicing at three cryptic 5' SS's, one in the large intron and two in the second exon; the most upstream of these sites had previously been found in a mutant from which the 5' SS and its flanking regions had been deleted (7). Surprisingly, in one of the cryptic splice sites, the "Chambon rule" GT was replaced by a GC.

RESULTS

Point mutations around the 5' SS were generated by the analog incorporation method described earlier (8,9), and deletions were prepared by exonuclease resection at the BamH1 site preceding the large intron (7,8; see also Fig. 3). Wild-type and modified β-globin genes, flanked by 425 bp of 5', and 358 of 3' neighboring sequences (10), were joined to an expression vector consisting of a 2.5 kb pBR327 (11) fragment (containing the origin of replication and the <u>bla</u> gene), the 3.05 kb <u>Hpa</u>II-<u>Bam</u>H1 fragment of SV40 (with the origin of replication, the 72 and 21 bp repeats (12,13) and part of the early region) and, as reference gene, a chromosomal mouse β-globin gene (including about 1300 bp of 5' and 400 bp of 3' flanking sequence) (14). The hybrid plasmid, shown in

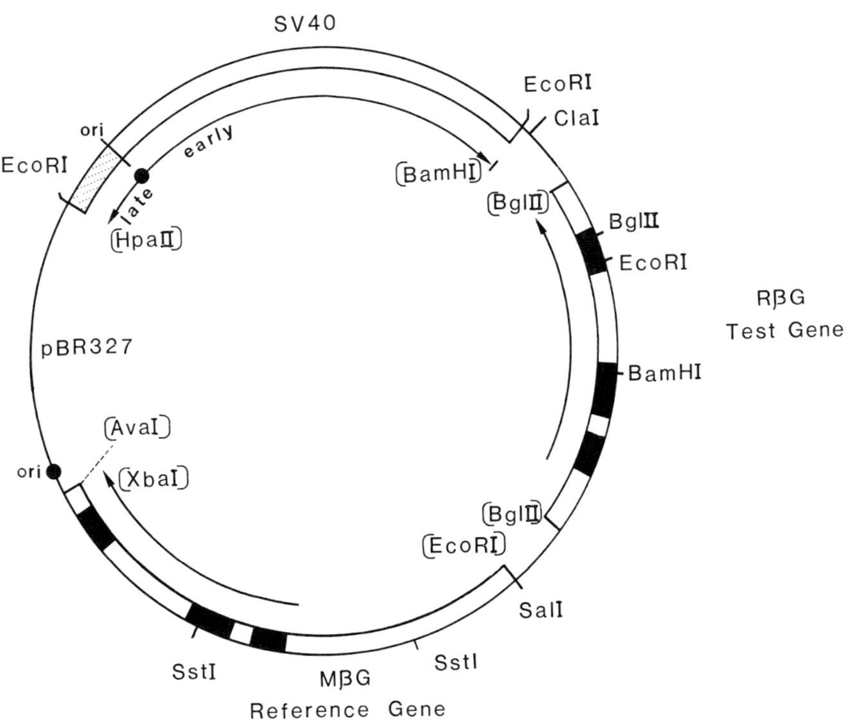

Fig. 1.   Expression plasmid containing a mouse **β**-globin (reference) gene and a rabbit **β**-globin (test) gene.

The expression plasmid consists of (i) a reference mouse β-globin major genomic DNA (3.1 kb EcoRI-XbaI) fragment (14, 15), (ii) a 3039 bp HpaII-BamHl fragment containing the whole early region and the beginning of the late region of SV40 (12), and (iii) a 2.1 kb partial BglII fragment of genomic rabbit **β**-globin DNA, inserted into pBR327 (11) lacking the region between SalI (650) and AvaI (1424).

Arrows indicate the 5'→ 3' orientation of the mouse and rabbit genes and of the SV40 late and early gene regions. Black boxes represent exon sequences of the globin genes. Obliterated restriction sites are placed in parentheses. The unique ClaI and SalI sites in the plasmid allow the easy introduction of rabbit **β**-globin genes modified in vitro and cloned in the BamHl site of pBR322 or pBR327.

a

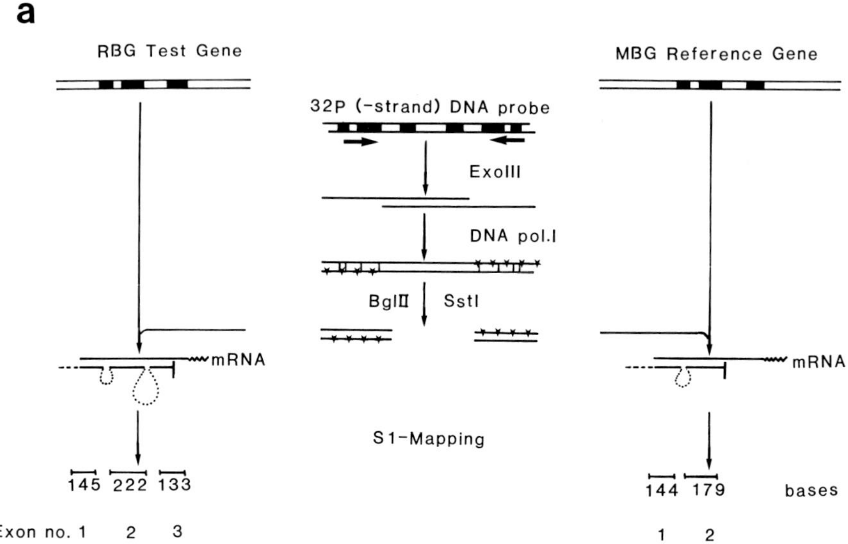

Fig. 2.   Simultaneous S_1 mapping of RNA transcripts from the mouse β-globin reference gene and a modified rabbit β-globin test gene.

a) Scheme. To prepare uniformly labeled probes, the expression plasmid (Fig. 1) was cleaved with ŚalI and the linear plasmid DNA (middle of figure) bearing the mouse β-globin and rabbit β-globin genes at its terminal regions (orientation indicated by arrows) was treated with exonuclease III to expose the (+) strands of both genes. The DNA was repaired with α-^{32}P-dNTPs using DNA polymerase I of E.coli and cut with BglII (rabbit-specific moiety) and SstI (mouse-specific moiety) to yield probes specific for (i) the first two exons and part of the third exon of rabbit β-globin RNA, and (ii) for exon 1 and the greater part of exon 2 of mouse β-globin RNA. Hybridization of the probe mixture to the RNA of transformed Hela cells under stringent conditions, S_1-nuclease digestion and analysis of the protected fragments on a denaturing polyacrylamide gel yielded fragments of 145, 222 and 133 nucleotides for rabbit β-globin mRNA and of 144 and 179 nucleotides for mouse β-globin mRNA.

b) S_1 analysis of β-globin transcripts from transfected Hela cells. Expression plasmids such as the one shown in Fig. 1, bearing a mutant rabbit β-globin gene were introduced into Hela cells by the calcium phosphate coprecipitation method (17,20) essentially as described by Wigler et al. (18).

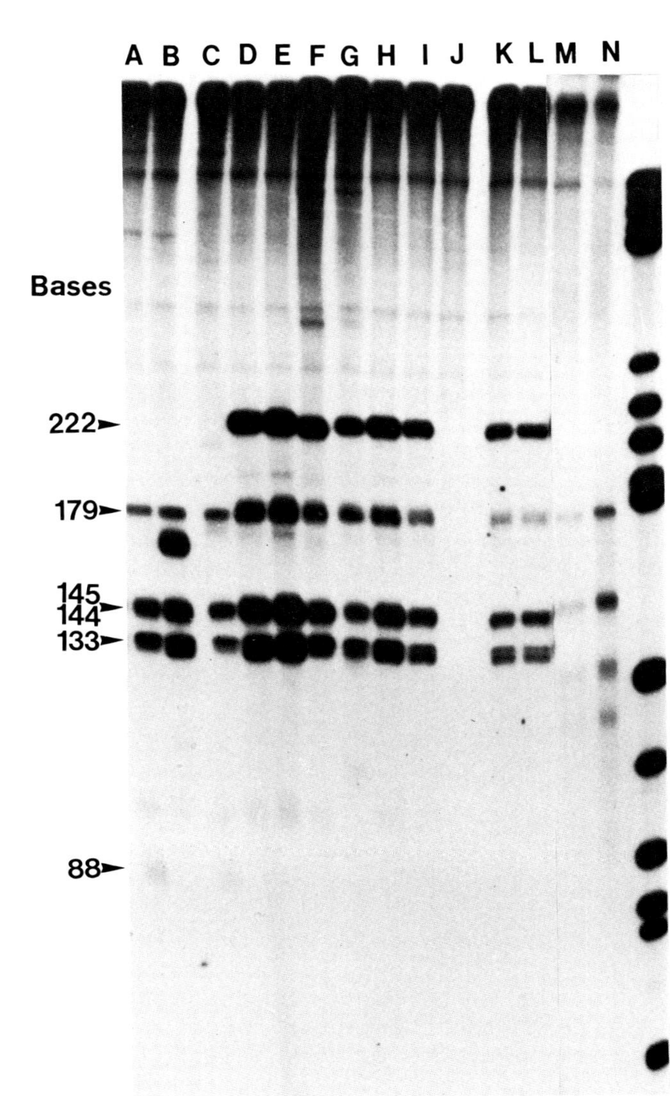

Total RNA was extracted after 48 h by the LiCl-urea method
(21) and the β-globin RNAs were mapped and quantitated. Lane
A: RβG Bam 59 (38 bp deleted upstream and 21 bp downstream
from the BamHl site (cf. Fig. 3-III). Lane B: RβG Bam 38 (38
bp deleted upstream of the BamHl site). Lane C: RβG Bam 21 (21
bp deleted downstream of the BamHl site (cf. Fig. 3-IV)).
Lanes D to H: rabbit β-globin gene with purine transitions at
positions 498, 497, 493, 494 and 492, respectively (cf. Fig.
3B). Lanes I and K: wild-type rabbit β-globin DNA, orientation

Fig. 1, was designed to have the following properties: (1) to allow the expression of the rabbit
(test) and the mouse (reference) gene from their
own promoters (in divergent directions, to preclude
mutual interference or interference from the SV40
promoters). (2) to provide the enhancer function
ascribed to the region containing the 72 bp and 21
bp repeats of SV40 and shown to be required for the
transcription of the β-globin gene in the transient
expression system (13), (3) to facilitate replication of the hybrid DNA in the host cell, by elimi-
nating the "poison sequences" identified in pBR322
(16).

      Hela cells were transformed with expression
plasmid by the Ca-phosphate method (17,18) and the
RNA was isolated after 48 h. Mouse and rabbit β-
globin-specific transcripts were analyzed by S_1
mapping (19), using mouse and rabbit chromosomal β-
globin probes uniformly labeled to the same specific
activity, as described in Fig. 2a. Since, under
stringent annealing conditions, there is no cross-
hybridization between the mouse and the rabbit sequences, the two transcripts could be assayed simultaneously. Both mouse and rabbit β-globin mRNA
gave a protected probe fragment of 144/145 nucleotides length, corresponding to the first exon in
each case; rabbit β-globin mRNA gave characteristic
fragments of 222 and 133 nucleotides corresponding
to the second and third exons, respectively, while
the mouse mRNA gave a fragment of 179 nucleotides,
corresponding to the second exon (Fig. 2a, 3). The
expression of the modified rabbit test genes was
monitored by determining the size of the protected
probe fragments, and relating the radioactivity in
the 222 (or 133) nucleotide fragment to that of the
179 nucleotide mouse specific fragment. The ratio
of wild type rabbit to mouse specific transcripts
was 2.5, irrespective of the orientation of the

(cont. Legend Fig. 2)
as indicated in Fig. 1. Lane J: no plasmid. Lane L: wild-type
rabbit *β*-globin DNA, orientation opposite to that in Fig. 1.
Lanes M and N: 50 and 200 pg mouse 9 S *β*-globin mRNA isolated
from Friend cells (22). Right outer lane: ^{32}P-labeled BspI
fragments of pBR322.

rabbit gene in the expression plasmid (cf. Fig. 3). The lower level of mouse transcripts may be due to a variety of causes, such as lower promoter activity, lesser efficiency of the enhancer, or greater lability of the pre-mRNA or mRNA.

Two deletion mutants, in which the junction between second exon and large intron had been removed, and which either lacked 51 bp of the second exon and 8 bp of the intron (Fig. 3, III) or 13 bp and 8 bp (Fig. 3, IV), respectively, gave almost normal transcript levels, however splicing of the large intron was aberrant, as we showed earlier using cells stably transformed with the restructured globin genes by the thymidine-kinase cotransfection method (7). Primer extension (23) and direct sequencing (24) of the extended ^{32}P-labeled primer showed that the normal 3' SS of the third exon had been joined to nucleotide 360 within the sequence ...AAG/GTGAAG..., second exon (J. Reiser and B. Wieringa, unpublished results). Splicing thus led to truncation of the middle exon, which gave an 88 nucleotide signal in the S$_1$ mapping experiment of Fig. 2 (lanes A and C). It is not clear why the 88 nucleotide band is so weak relative to the 145 nucleotide exon-1 and 133 nucleotide exon-3 signals, even if the fragment lengths are taken into account. A possible explanation is that a substantial fraction of splice products are due to a direct joining of the third to the first exon, as found in a similar natural mutant by Treisman et al. (25). However, in our case, primer extension showed no evidence for such products (unpublished results). Perhaps the recovery of short fragments by the S$_1$ method is lower than that of longer ones, or there is a specific problem with the S$_1$ mapping of certain fragments, as reported for an 0.5 kb adeno virus mRNA (26).

In the case of the mutant with the shorter deletion, a second cryptic splice at about position 440 (which is absent in the mutant with the longer deletion) was detected by S$_1$ mapping and primer elongation; it will be discussed below in more detail.

The six β-globin genes with single point mutations around the 5' SS of the large intron (Fig. 3b) were analyzed. Five of these, with transitions at positions 492, 493 and 494 (in the second exon, immediately preceding the 5' SS) and 497 and 498 (following the GT at the beginning of the large in-

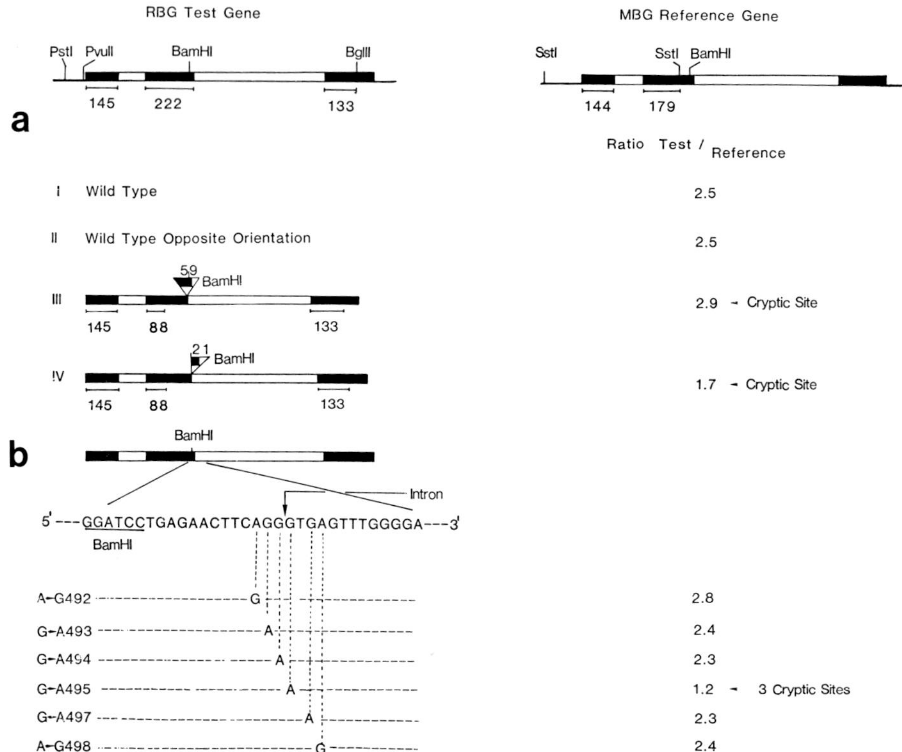

Fig. 3.   Summary of analyses of transcripts of rabbit β-globin genes modified at the 5' splice site region of the large intron.

Quantitation and analysis of transcripts were described in Fig. 2 legend. The ratio of test to reference transcripts (relative transcript level = RTL) was calculated from the radioactivities of the 222 nucleotide fragment (RβG specific) and the 179 nucleotide fragment (MβG specific) excised from the gel. In those cases (III and IV) where the second exon of RβG was shortened, the ratio was calculated from the radioactivity in the 133 nucleotide band (which is also unique for the rabbit β-globin transcripts) relative to the 179 nucleotide mouse fragment and multiplied by 1.7 to correct for the difference in length of the 222 and 133 nucleotide fragments. Numbers given are the average of two independent experiments.

a) Wild-type genes and genes with deletions. I: wild type, orientation as in Fig. 1. II: wild type, orientation

tron) had no effect on the structure or the quantity of the transcripts. The globin gene with a transition at position 495, which converts the GT at
the beginning of the intron into an AT, gave rise to
a somewhat reduced level of transcripts, all of
which were abnormal. As revealed by S_1 mapping with
a uniformly labeled probe the first and third exons
generated correct signals, however the second exon
gave rise to three fragments, of which one appeared
to have almost the correct length, and two were
shorter (data not shown). Primer extension analysis
(Fig. 4a), using a 5'-^{32}P-labeled minus strand primer, confirmed the existence of 3 distinct β-globin
RNA species, one slightly longer than wild type
mRNA (about 30% of the recognizable transcripts, as
judged by inspection of the autoradiographs), the
other two shorter by 50 and 130 nucleotides, respectively (about 20 and 50% of the total). This
suggested that one of the cryptic SS's was located
within the large intron, just a few nucleotides
downstream of the authentic 5' SS, and the other
two within the second exon, about 50 and 130 nucleotides upstream of the authentic 5' SS. To unequivocally locate the abnormal SS's, double-stranded
cDNA copies of the transcripts were cloned. One
"wild-type" clone, and one clone of each size class
of the aberrant transcripts was sequenced (Fig. 4b),
and the precise positions to which the third exon
had been joined was determined. As shown in Fig. 5a,

opposite to I. III: RβG Bam 59 (38 nucleotides upstream and 21
bp downstream of the BamHl site deleted). IV: RβG Bam 21 (21
bp downstream of the BamHl site deleted).

    b) Site-directed mutations. Rabbit β-globin DNAs with
point mutations were prepared as outlined by Weber et al. (8),
inserted into the expression vector and the analysis of transcripts carried out as described above. The positions of the
purine transitions in the test genes are indicated under the
sequence (which extends from position +476 to +507 (15). The
test/reference ratios can be compared to the ratios given for
the wild type transcripts under (a). The cryptic splice sites
referred to in mutant G⟶A_{495} are displayed in Fig. 5a. The
nucleotide sequence of the rabbit β-globin gene is given by
van Ooyen et al. (15).

a)
A B C m
1
wt
2
3
primer
b)
Wild
Type
1
498/499
2
442/443
3
359/360
G G C C A T

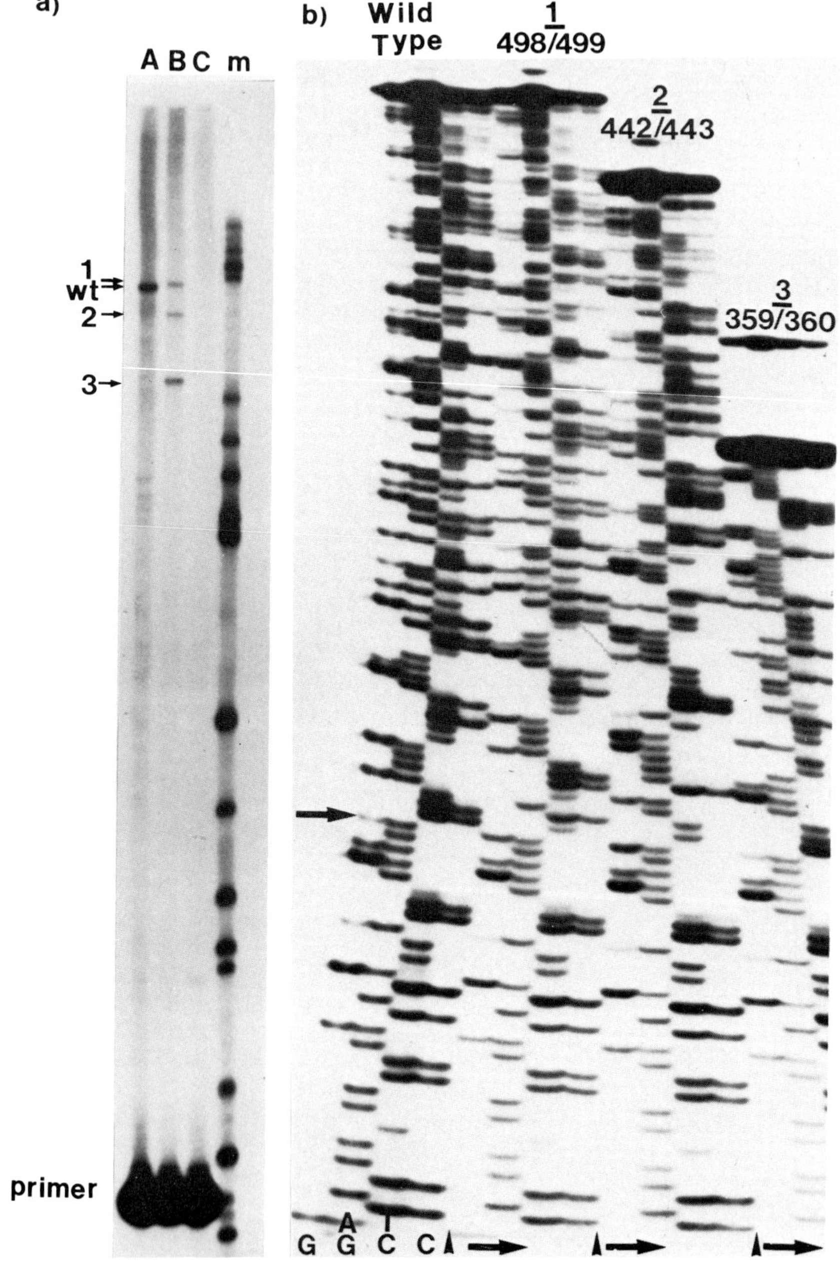

Fig. 4

Fig. 4. Characterization of incorrectly spliced tran-
scripts of rabbit **β**-globin DNA with the G$\longrightarrow$A$_{495}$ 5' splice
site mutation.
     a) Primer extension maps. A single-stranded AluI-EcoRI
primer fragment (minus strand, pos. +1068 to +1120 (15)) was
5'-^{32}P-labeled and hybridized to the poly(A$^+$) RNA prepared
from either 150 μg of total RNA of Hela cells transfected
with wild type plasmid DNA (lane A), from 60 μg of total RNA
obtained after transfection with the G$\longrightarrow$A$_{495}$ R**β**G mutant
DNA (lane B) or from 150 μg of total mock infected RNA (lane
C). Hybrids were elongated by reverse transcriptase, the
cDNAs resolved on a 6.5% acrylamide-urea sequencing gel and
autoradiographed. M, ^{32}P labeled BspI restriction fragments
of pBR322. The position of full-length cDNAs of wild-type
(a) and mutant (cryptic 1, 2 and 3) transcripts are indicated by arrows.
     b) Sequencing ladder of cloned **β**-globin cDNAs derived
from incorrectly spliced RNAs.
     Poly(A) RNA from Hela cells transformed with wild type
or with the G$\longrightarrow$A$_{495}$ mutant **β**-globin DNA was used as template
for oligo(dT)-primed synthesis of double-stranded (ds) cDNA
as described by Wickens et al. (27). Each ds cDNA preparation
was cut with TaqI and BglII, and electrophoresed on a 2% low
melting agarose gel. Fragments of 300-350, 220-300 and 150-
220 bp (the sizes expected for Taq-BglII fragments derived
from the incorrectly spliced **β**-globin transcripts characterized by S$_1$ mapping (not shown) and primer extension mapping,
see above) were ligated to a ClaI and BamHl digested pBR327
plasmid. The ligated DNAs were transfected into E.coli HB101
and the transformed colonies screened for globin-specific sequences with a ^{32}P-labeled **β**-globin DNA probe. DNAs of the
positive colonies were characterized by restriction analysis
and four different classes of cDNAs, corresponding to either
wild type or the expected three mutant transcripts were identified. One representative of each of the different types of
DNA was sequenced according to Maxam and Gilbert (24). In
each case the sequence reads upstream from the labeled EcoRI
end in the third exon into the second exon. The site where
splicing occurs in the various transcripts is indicated above
each panel; the 5' end of exon 3 is indicated by an arrow.

the abnormal transcripts resulted from splicing at
position 498/499 (...TGA/GTTTGG..., at the beginning
of the large intron), position 442/443 (...TAA/
GCTGAG...) and position 359/360 (...AAG/GTGAAG...),
the latter position being the same described above
in the case of the 5' SS deletion mutants. An asto-
nishing aspect of these results was that in the
case of the 5' SS in position 442/443, splicing
occurred 5' to a GC doublet, rather than to the GT
found in all other cases so far examined (2). It
could be argued that this conclusion is not defini-
tive, because the DNA used for transfection might
fortuitously have contained 20-30% of a species with
a GT (rather than CT) sequence at that position,
which could not have been detected by sequencing.
However, the DNA had been recloned three times prior
to sequencing and transfection, making this possibility quite unlikely. Moreover, the sequence
TAAGCTGAGT, but not TAAGGTGAGT, contains an AluI
site (underlined), and > 95% of the template DNA
was cleaved by AluI in this position (data not
shown). In addition, the cDNA sequence of the long-
est aberrant transcript allowed us to recheck the
sequence around position 443, which agreed with that
given above. Finally, a second cDNA clone corresponding to RNA spliced at the 442/443 position was analyzed by restriction mapping and found to be indistinguishable from the first, showing that we had
not fortuitously picked a non-representative cDNA
clone.

DISCUSSION

Normal splicing can be partially or completely
abolished by at least two types of events, namely
(a) deactivation of a natural SS and (b) generation
or activation of new SS.

(a) Deactivation of a natural splice site may
come about by mutations modifying or deleting a
splice site consensus sequence. As a consequence
thereof, splicing at that position may be abolished
and transcripts containing a complete or partial intron may appear in the cytoplasm, as shown for the
abnormal Ela mRNA generated by the adeno 5 dl502 deletion mutant in which a 5' SS and its flanking regions are deleted (28). A pair of point mutations
at one of the 5'SS's of the adeno 5 Ela gene converts
the sequence GAGG/GTGAGG to GAGG/GTGAAT, abolishes

splicing and leads to the accumulation of unspliced
precursor (26). The same 5' SS was also inactivated
by a site-directed T⟶G transversion which generated the sequence GAGG/GGGAGG (29). In other cases,
sequences within introns or exons, which show some
similarity to SS consensus sequences (cryptic splice
sites (7)) or authentic splice sites normally utilized in a different context may be used in lieu of
the deactivated natural SS, giving rise to incorrectly spliced RNAs, with all or part of an exon
missing, or with insertions due to a residual
intron sequence. The _in vitro_ GT⟶AT mutation we
describe in this paper inactivates the 5' SS of the
large intron of the rabbit β-globin gene and leads
to the utilization of three cryptic SS's. Similarly,
the natural GT⟶AT mutation at the 5' SS of the
large intron of a human β^+-thalassemic DNA abolishes
correct splicing and leads to cytoplasmic transcripts in which the third exon is joined either to
a cryptic 5' SS (..ATG/GTTAAG..) in the large intron, leading to RNA with a persisting partial intron, or to the normal 5' SS adjoining the first
exon (25), giving a product in which the middle
exon is missing. Orkin et al. (30) have described
a human mutant α2-globin gene in which the 5' SS of
the small intron was inactivated by a deletion of
five nucleotides at the beginning of the small intron; in the processed transcript, the second exon
was joined to a cryptic 5' SS in the first exon,
which had the sequence ...GG/GTAAG... (B.K. Felber,
personal communication).

A further example where deletion of a 5' SS
leads to utilization of an authentic 5' SS at a
neighboring exon-intron boundary has been described
for an aberrantly rearranged æ chain immunoglobulin gene (31,32).

In some cases abolition of a SS results in the
failure of any transcript, be it normal or aberrant,
to accumulate in the cytoplasm at anything close to
wild type levels (29,33,34), maybe due to retention
of incompletely spliced precursors within the nucleus, as is the case for an SS mutant in the rat
albumin gene (35,36), or to the instability of incorrectly spliced mRNA in the cytoplasm. It has
been shown that β-globin genes with mutations leading to premature termination codons accumulate at
decreased levels in the cytoplasm, perhaps because
untranslatable mRNAs are more susceptible to degra-

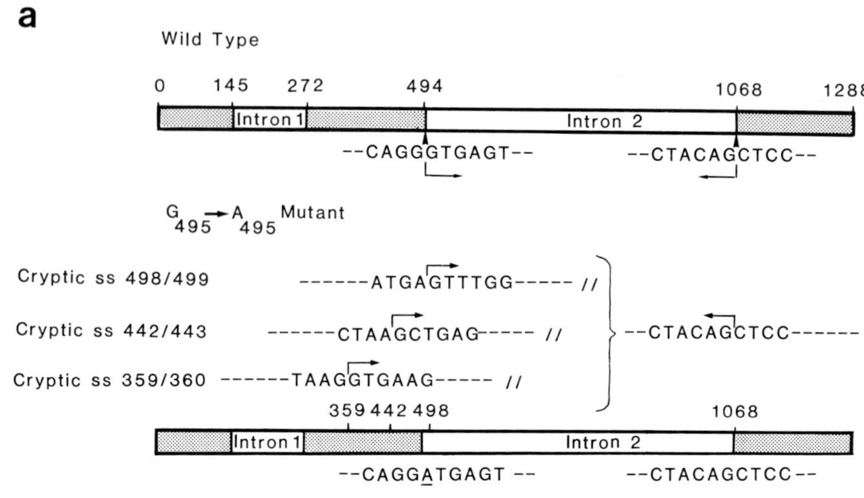

Fig. 5a and b

dation (37,38,39); incorrect splicing may give rise
to a frame-shift and therefore lead to premature
termination and incompletely translatable RNA.

It should also be mentioned that presence or
absence of an abnormal transcript may depend on
whether expression occurs from the endogenous gene
in vivo, or from its cloned counterpart in cultured
cells (25,38,40).

(b) Generation or activation of new splice-
sites. A form of β⁺-thalassemia has been described
in which a single mutation in the first intron, 21

c

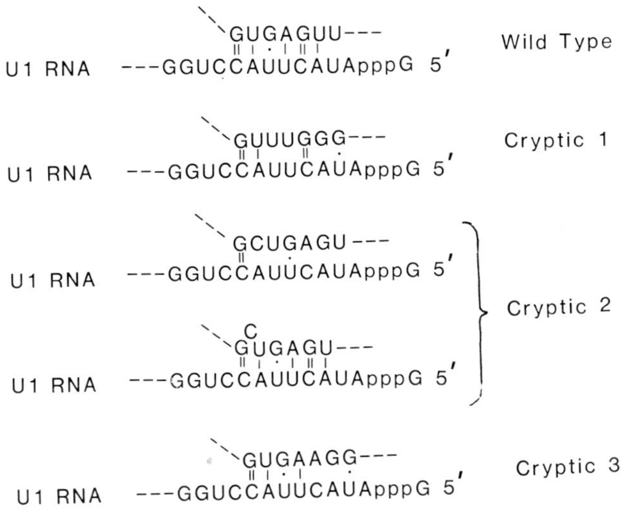

```
                       \
                        \GUGAGUU---                ,      Wild Type
                        || | · | || |
     U1 RNA    ---GGUCCAUUCAUApppG 5'


                        \
                         \GUUUGGG---                       Cryptic 1
                         || |   ||  ·
     U1 RNA    ---GGUCCAUUCAUApppG 5'


                        \
                         \GCUGAGU---              ⌉
                         ||   ·
     U1 RNA    ---GGUCCAUUCAUApppG 5'            |
                                                |
                                                 ⎬  Cryptic 2
                        \  C                     |
                         \GUGAGU---              |
                         || | · | || |           |
     U1 RNA    ---GGUCCAUUCAUApppG 5'            ⌋


                        \
                         \GUGAAGG---              ,      Cryptic 3
                         || | · |  ·
     U1 RNA    ---GGUCCAUUCAUApppG 5'
```

Fig. 5c

Fig. 5.   Cryptic splice sites revealed by mutational inactivation of the 5' splice site of the large β-globin intron.
  a) Location and sequence. Brackets with arrowheads indicate (at the DNA level) the regions that are spliced out from
wild type transcripts (above) or from transcripts of the gene
with the GT⟶AT mutations which abolishes correct splicing.
The numbers given indicate map positions (15).
  b) Comparison of wild type and cryptic 5' SS regions
with the 5' SS consensus sequence (2).
  c) Possible base-pairing of the 5' terminal region of
snU1 RNA with the sequence around the wild type or cryptic 5'
SS's. The 5' end of the snU1 RNA as well as 7 nucleotides at
the 5' ends of the correctly or aberrantly spliced large intervening sequences (starting at positions 499 (cryptic 1),
443 (cryptic 2), 360 (cryptic 3)) are paired in the register
proposed by Lerner et al. (5).

nucleotides from the 3' SS, converts the sequence
...CTCTGCCTATTGGT... to ...CTCTGCCTATTAG/T... (41,
42) which has a 3' SS consensus sequence. About 90%
of the transcripts were found to be spliced from
the new 3' SS to the authentic 5' SS, to yield a
product still containing part of the small intron,
which is transferred to the cytoplasm, along with
some correctly spliced mRNA (40,43). Concomitantly,
unspliced pre-mRNA (44,45) but not pre-mRNA with a
partially spliced small intron accumulated in the
nucleus, suggesting a slow-down of splicing (43), as
was found also for the β^0-thalassemia gene with the
inactive 5' SS reported by Treisman et al. (25). We
have found that if a 143 bp duplication of the β-
globin 3' SS is inserted downstream of the original
3' SS, then the splice occurs exclusively at the
more downstream copy (T. Kühne and B. W., unpublished results). This shows that factors other than
just the nucleotide sequences around the SS's determine the final splicing pattern, and that there
is no unique relationship between the normally utilized 5' and 3' SS as regards their relative position on the pre-mRNA.

It is conceivable that a third type of mutation, which does not affect the SS's proper may
alter the splicing pattern, but no well documented
examples are known to us at this time.

The compilation of SS regions (2) clearly shows
that a large variety of sequences are compatible
with splicing. Our results show that the consensus
sequence around one and the same splice site, the
5' SS of the large rabbit β-globin intron, can be
modified by point mutations at various positions
without detectably affecting splicing efficiency;
however, a G$\longrightarrow$A mutation converting the "Chambon
rule" GT to AT abolished splicing completely at this
site. It has been proposed that a specific base-
pairing (4,5,46) interaction between the small nuclear RNA and the 5' and 3' terminal regions of an
intron are part of the mechanism that mediates accurate splicing. The five base transitions which do
not affect splicing also do not disrupt the proposed
base-pairing. In that regard, our findings with the
point mutations do not contradict the model.

Fig. 5b shows a comparison of the regions comprising the wild type and the cryptic 5' SS's
described in this paper with the corresponding 9-
nucleotide consensus sequence of Mount (2). Whereas

the normal 5' SS of the large β-globin intron and
the cryptic SS #3 (pos. 359/360) conform in seven
and six positions, respectively, with the consensus
sequence, the cryptic SS's #1 (pos. 498/499) and
#2 (pos. 442/443) agree with it in only 3 and 2
positions, respectively, values that might be expected from random sequences. Moreover, when one
attempts to base pair the sequences downstream from
the cryptic SS's #1 and #2 with the 5' terminus
of U1 RNA as proposed by Lerner & Steitz (5), no
significant pairing is achieved, while that of SS
#3 forms 4 adjacent base pairs (as compared to 6
for the wild type SS) (Fig. 5c).

In the case of cryptic SS #2, where the GT
rule is formally broken, a better fit is obtained
if the C residue in the second position downstream
from the SS is looped out: this brings a U(T) residue next to the G, reconstituting, as it were, the
"Chambon rule" GT sequence, and allows five base
pairs to be formed. In the case of SS #1, two residues have to be looped out to allow similar base
pairing to occur. Alternatively, one might postulate that if base pairing of the SS region to a
"guide" RNA is required, there may be more than one
RNA species involved.

Inspection of the β-globin RNA sequence shows
that there are many potential splice sites, as defined by the Steitz-Lerner model, especially if
"looping-out" is permitted. For example, in the second β-globin exon, the sequence TCA/GUGAGGGU...
(bp 394 to 404 between cryptic SS #2 and #3)
could form five base pairs with U1 RNA (or seven if
a G is looped out) and thus represents as good a
candidate for a cryptic SS as any other; in fact,
no mRNAs spliced at that position are observed in
our experiment. Since the sequence is placed between two cryptic SS that are utilized, a scanning
model (6) cannot explain why it is not used while
the other two are. It could be that other sequence
elements are required: for example, the consensus
sequence shows the prevalence of the sequence
AAG preceding the 5' SS; this requirement is however also not met by cryptic SS's #1 or #2. Perhaps
some sequences which might potentially define a SS
are not used because they are masked by secondary
structure, as is thought to be the case for some
potential protein initiation sites in prokaryotes
(47,48). In addition, the tertiary structure of the

pre-mRNA may preclude the use of some combinations of splice sites, even though a specific tertiary structure is not required for splicing, as shown by the findings that a variety of hybrid RNAs, joined within an intron, are spliced normally (49) and that extensive deletions or modifications within introns or flanking exon regions (33,50,7) do not affect the accuracy of splicing.

In summary, it seems likely that a potential splice site and its affinity for the splicing system is primarly defined by one of a set of nucleotide sequences, but the actual use of a pair of sites is modulated by the secondary and tertiary structure of the RNA, and possibly by proteins associated with it.

ACKNOWLEDGMENTS

This work was supported by the Schweizerische Nationalfonds and the Kanton of Zürich; B.W. was recipient of a long-term EMBO Fellowship. We thank Hans Weber and Peter Dierks for helpful advice, and Ms. R.M. von Rotz for preparing the typescript with care and patience.

REFERENCES

1. Breathnach, R., Benoist, C., O'Hare, K., Gannon, F., and Chambon, P. (1978). Proc. Natl. Acad. Sci. USA 75, 4853-4857.
2. Mount, S.M. (1982). Nucl. Acids Res. 10, 459-471.
3. Seif, I., Khoury, G., and Dhar, R. (1979). Nucl. Acids Res. 6, 3387-3398.
4. Rogers, J., and Wall, R. (1980). Proc. Natl. Acad. Sci. USA 77, 1877-1879.
5. Lerner, M.R., Boyle, J.A.,Mount, S.M., Wolin, S. L., and Steitz, J.A. (1980). Nature 283, 220-224.
6. Sharp, P.A. (1981). Cell 23, 643-646.
7. Dierks, P., Wieringa, B., Marti, D., Reiser, J., van Ooyen, A., Meyer, F., Weber, H., and Weiss-mann, C. (1981). In "Developmental Biology using Purified Genes". ICN-UCLA Symp. on Mol. and Cell. Biol. vol. XXIII, (D.D. Brown and C.F. Fox, eds.), pp. 347-366. Academic Press, N.Y.
8. Weber, H., Dierks, P., Meyer, F., van Ooyen, A., Dobkin, C., Abrescia, P., Kappeler, M., Mey-hack, B., Zeltner, A., Mullen, E.E. and Weiss-mann, C. (1981). In "Developmental Biology Using

    Purified Genes".ICN-UCLA Symp. on Mol. and
Cell. Biol. vol. XXIII (D.D. Brown and C.F. Fox,
eds.), pp. 367-385. Academic Press, N.Y.

9. Müller, W., Weber, H., Meyer, F. and Weissmann,
C. (1978). J. Mol. Biol. 124, 343-358.

10. Dierks, P., van Ooyen, A., Mantei, N., and
Weissmann, C. (1981). Proc. Natl. Acad. Sci.
USA 78, 1411-1415.

11. Soberon, X., Covarrubias, L., and Bolivar, F.
(1980). Gene 9, 287-305.

12. Tooze, J. (1980). In "DNA Tumor Viruses" (J.
Tooze, ed.), 2nd edition, Cold Spring Harbor
Laboratory.

13. Banerji, J., Rusconi, S., and Schaffner, W.
(1981). Cell 27, 299-308.

14. Tilghman, S.M., Tiemeier, D.C., Polsky, F., Ed-
gell, M.H., Seidman, J.G., Leder, A., Enquist,
L.W., Norman, B., and Leder, P. (1977). Proc.
Natl. Acad. Sci. USA 74, 4406-4410.

15. van Ooyen, A., van den Berg, J., Mantei, N.,
and Weissmann, C. (1979). Science 206, 337-344.

16. Lusky, M., and Botchan, M. (1981). Nature 293,
79-81.

17. Graham, F.L., and van der Eb, A.J. (1973). Virology 52, 456-467.

18. Wigler, M., Pellicer, A., Silverstein, S., and
Axel, R. (1978). Cell 14. 725-731.

19. Berk, A.J., and Sharp, P.A. (1977). Cell 12,
721-732.

20. Lewis, W.H., Srinivasan, P.R., Stokoe, N., Si-
minovitch, L. (1980). Somatic Cell Gen. 6,
333-348.

21. Auffrey, C., and Rougeon, F. (1980). Eur. J.
Biochem. 107, 303-314.

22. Curtis, P.J., Mantei, N., and Weissmann, C.
(1977). Cold Spring Harbor Symp. Quant. Biol.
42, 971-984.

23. Nagata, S., Mantei, N., and Weissmann, C.
(1980). Nature 287, 401-408.

24. Maxam, A.M., and Gilbert, W. (1977). Proc.
Natl. Acad. Sci. USA 74, 560-564.

25. Treisman, R., Proudfoot, N.J., Shander, M., and
Maniatis, T. (1982). Cell, in press.

26. Solnick, D. (1981). Nature 291, 508-510.

27. Wickens, M.P., Buell, G.N., and Schimke, R.T.
(1978). J. Biol. Chem. 253, 2483-2495.

28. Carlock, L., and Jones, N.C. (1981). Nature
294, 572-574.

29. Montell, C., Fisher, E.F., Caruthers, M.H., and Berk, A.J. (1982). Nature 295, 380-384.
30. Orkin, S.H., Kazazian, H.H., Jr., Antonarakis, S.E., Goff, S.C., Boehm, C.D., Sexton, J.P., Waber, P.G., and Giardina, P.J.V., (1982) Nature 296, 627-631.
31. Choi, E., Kuehl, M., and Wall, R. (1980). Nature 286, 776-779.
32. Seidman, J.G., and Leder, P. (1980). Nature 286, 779-783.
33. Khoury, G., Gruss, P., Dhar, R., and Lai, C.-J. (1979). Cell 18, 85-92.
34. Lai, C.J., and Khoury, G. (1979). Proc. Natl. Acad. Sci. USA, 76, 71-75.
35. Esumi, H., Takahashi, Y., Sekiya, T., Sato, S., Nagase, S., and Sugimura, T. (1982). Proc. Natl. Acad. Sci. USA 79, 734-738.
36. Esumi, H., Takahashi, Y., Makino, R., Sato, S., Nagase, S., and Sugimura, T. (1982). Cold Spring Harbor Meeting on RNA Processing, abstract p. 132.
37. Chang, J.C., and Kan, Y.W. (1979). Proc. Natl. Sci. USA 76, 2886-2889.
38. Moschonas, N., de Boer, E., Grosveld, F.G., Dahl, H.H.M., Wright, S., Shewmaker, C.K., and Flavell, R.A. (1981). Nucl. Acids Res. 9, 4391-4401.
39. Trecartin, R.F., Liebhaber, S.A., Chang, J.C., Lee, K.Y., Kan, Y.W., Furbetta, M., Angius, A., and Cao, A. (1981). J. Clin. Invest. 68, 1012-1017.
40. Fukumaki, Y., Ghosh, P.K., Benz, E.J., Reddy, V.B., Lebowitz, P., Forget, B.G., and Weissman, S.M. (1982). Cell 28, 585-593.
41. Spritz, R.A., Jagadeeswaran, P., Choudary, P.V., Biro, P.A., Elder, J.T., deRiel, J.K., Manley, J.L., Gefter, M.L., Forget, B.G., and Weissman, S.M. (1981). Proc. Natl. Acad. Sci. USA 78, 2455-2459.
42. Westaway, D., and Williamson, R. (1981). Nucl. Acids Res. 9, 1777-1788.
43. Busslinger, M., Moschonas, N., and Flavell, R. A. (1981). Cell 27, 289-298.
44. Maquat, L.E., Kinniburgh, A.J., Beach, L.R., Honig, G.R., Lazerson, J., Ershler, W.B., and Ross, J. (1980). Proc. Natl. Acad. Sci. USA 77, 4287-4291.

45. Kantor, J.A., Turner, P.H., and Nienhuis, A.W. (1980). Cell, 21, 149-157.
46. Ohshima, Y., Itoh, M., Okada, N., and Miyata, T. (1981). Proc. Natl. Acad. Sci. USA 78, 4471-4474.
47. Lodish, H.F. (1970). J. Mol. Biol. 50, 689-702.
48. Gheysen, D., Iserentant, D., Derom, C., and Fiers, W. (1982). Gene 17, 55-63.
49. Chu, G., and Sharp,P.A. (1981). Nature 289, 378-382.
50. Gruss, P., and Khoury, G. (1980). Nature 286, 634-637.

CHROMATIN STRUCTURE OF THE OVALBUMIN GENE DOMAIN

William E. Stumph, Melvyn Baez, George M. Lawson[2], Ming-Jer Tsai, and Bert W. O'Malley

Department of Cell Biology, Baylor College of Medicine
Houston, Texas 77030

ABSTRACT The ovalbumin gene and the ovalbumin-related X and Y genes are expressed in chicken oviduct in response to steroid hormones. These three genes are linked within a 100 Kb domain of DNA which is preferentially sensitive to DNase I digestion in oviduct cell nuclei. No such preferential sensitivity to DNase is observed in nuclei isolated from other chicken tissues in which these genes are not transcribed. Thus, the DNase I sensitivity observed is correlated with the capacity for these genes to be expressed in oviduct. We have asked the question: Are there specific signals in the DNA which are responsible for defining this domain or for conferring upon it the active, DNase I sensitive, conformation? We have located DNA sequences belonging to a single repetitive DNA family, termed CR1, which are preferentially located in or near the boundary regions of the 100 Kb domain. Therefore, these CR1 sequences are possible candidates for such a function. We have also searched for, but have not observed, any tissue specific rearrangements of the DNA in the boundary regions of the domain. It is therefore unlikely that DNA rearrangements are involved in the establishment of the DNase I-sensitive domain in oviduct cells. However, we do note that a region at the far 3' end of the domain exhibits a cytidine methylation pattern which is highly variable among different chicken tissues. In particular, this region which is approximately 30 Kb downstream from the ovalbumin gene is undermethylated in oviduct as compared to other hen tissues, and thus could be a control region involved in domain activation.

[1]This work was supported by grants from the National Institutes of Health HD8188, and the Baylor Center for Population Research and Reproductive Biology HD7495.

[2]Present address: Clinical Chemistry Section, Hilton Building, Mayo Clinic, Rochester, Minnesota.

INTRODUCTION

During the past several years the nuclease DNase I has proven to be a useful tool for studying the chromatin structure of active genes. Numerous studies have indicated that in the chromatin of cells where a gene is able to be expressed, that gene is preferentially sensitive to digestion by DNase I. In other tissues where the gene is never expressed, it is digested no more rapidly than the bulk of the chromatin DNA sequences. This has been found to be true for genes as diverse as globin (1-3), ovalbumin (4-8), ovomucoid (6), heat shock proteins (9), insulin (10), immunoglobulin (11), alcohol dehydrogenase (12), protamine (13), ribosomal RNAs (14,15), and a variety of integrated viral genes (16-18). Presumably, this tissue specific susceptibility to digestion by nucleases reflects a difference in the nucleosomal structure or the chromosomal packaging of active genes as compared to the majority of the DNA sequences in the nuclear chromatin.

Depending upon the extent of digestion with DNase I and the chosen method of assay, it is possible to observe and distinguish between at least two or three different levels of DNase sensitivity. When low concentrations of DNase I are used such that only a limited number of nicks are introduced into the chromatin DNA, and the products are subsequently analyzed by the Southern procedure (19), the presence of DNase I "hot spots" or hypersensitive sites is apparent. The nearly universal and tissue-specific existence of hypersensitive sites 50-500 base pairs upstream from active genes suggests that they reflect a chromatin substructure important for gene expression (see ref. 20 for a review and additional references). In addition, the kinetics of complete disappearance of certain bands on Southern blots reveal that sequences throughout active genes are considerably more sensitive to DNase I than are inactive gene sequences (3). The preferential sensitivity measured in this way appears to extend into the flanking DNA sequences as well (7,21).

Alternatively, chromatin can be extensively digested using high concentrations of DNase I until approximately 20% of the chromatin DNA is rendered acid soluble. By using solution hybridization kinetics, the concentration of active gene sequences in this DNA can then be compared to control DNA from undigested nuclei. When this is done, the DNA from digested nuclei is observed to be specifically depleted in active gene sequences. This was first demonstrated a number of years ago by Weintraub and Groudine for the β-globin gene in erythroid cells (1) and by Garel and Axel for the ovalbumin gene in oviduct (4). Depletion of the specific gene sequence in question was not observed if the digested chromatin was from a tissue in which the gene was not expressed. These and subsequent studies have indicated that the establishment of a DNase I-sensitive chromosomal configuration appears to be a necessary prerequisite for gene expression.

Our laboratory has concentrated upon studying the DNase I sensitivity of genes which are expressed in the chicken oviduct, in particular the genes coding for the proteins ovalbumin and ovomucoid (6,22). The expression of these genes is specifically induced in oviduct tissue by steroid hormones. In this chapter, we will summarize our data concerning the nuclease sensitivity of the ovalbumin gene and its flanking DNA sequences following extensive digestion of nuclei with DNase I. We have localized the regions both 5' and 3' of the gene where the oviduct chromatin undergoes a transition from DNase I sensitivity to DNase I resistance. These regions define the boundaries of a DNase I-sensitive chromosomal domain in the oviduct which includes the ovalbumin gene and its closely related X and Y genes. Having defined this domain, we have turned our attention toward examining specific DNA sequences in and near the transition regions where the chromatin undergoes the change from an active to an inactive configuration. These data have provided clues as to possible signals in the DNA which may play a role in defining the ovalbumin gene domain. In addition, potential general mechanisms are suggested which may be involved in opening up domains of active gene expression during cellular differentiation and commitment.

METHODS

Nuclei prepared from oviduct or other hen tissues were digested with DNase I until 15-20% of the DNA was rendered soluble in cold perchloric acid. DNA was extracted from nuclei and used as driver in solution hybridization reactions. Restriction fragments were isolated from clones throughout the ovalbumin domain region and were used as probes after labelling to a high specific activity by nick translation. The entire procedure has been described in detail elsewhere (6,22). In order to facilitate quantitative comparisons, the solution hybridization data were transformed to linear plots of second order kinetic data. This was done by plotting $H/(1-H)$ versus C_0t where H represents the fraction of tracer driven into hybrid relative to the maximum hybridization observed. In such an analysis, the concentration of tracer sequence in a driver preparation is proportional to the slope of the straight line generated (4,23).

Southern filter hybridizations were performed overnight at 68°C in 6 X SSC and the filters were then washed at 68° in 1 X SSC over a period of several hours prior to autoradiography.

RESULTS

<u>The Ovalbumin Gene Exists in a 100 Kb DNase I Sensitive Domain</u>. At the top of Figure 1 is shown a diagram of the ovalbumin gene region of the chicken genome. The ovalbumin gene (OV) is preceded upstream by two closely related genes, called X and Y (24), which are transcribed in the oviduct but whose function still remains obscure. In oviducts from animals fully stimulated by estrogen, the X and Y genes are transcribed at rates approximately 3% and 10% respectively of the transcriptional rate of the ovalbumin gene (25). In the middle of the figure are shown a number of overlapping genomic DNA clones isolated from a Charon 4A gene library as described earlier (22). Taken together, these clones cover a region of over 130 Kb of contiguous DNA in the chicken genome containing and surrounding the X, Y, and ovalbumin genes. Various fragments from these different clones were used for the studies to be presented here.

At the bottom of the figure are shown data in which DNA from DNase I digested oviduct nuclei (open circles) or control total genomic DNA (closed circles) were hybridized to various cloned probes. It can be seen that probes from the central part of the region under study hybridized more slowly to the DNase I digested DNA than to the control DNA. This is shown by the $\sim$ 2.5-fold reduction in the slope of the line for the DNase I digested DNA compared to the slope of the line produced by the control driver DNA. These results indicate that DNA in this region is more susceptible to DNase I digestion. However, when probes located further 5' and 3' were used, the sequences gradually became more and more resistant to DNase I until the rate was almost identical to that of the control DNA. This indicates a loss of the preferential sensitivity associated with the active genes as one moves further along the chromosome in either direction.

Data accumulated from a large number of experiments using various probes from throughout the ovalbumin domain are summarized in Figure 2. The relative DNase I sensitivity plotted along the ordinate represents the relative reduction in the hybridization rate of the probe to DNA from DNase I treated nuclei compared to control DNA. (A relative sensitivity of 1.0 is indicative of no preferential sensitivity.) The results indicate that in oviduct chromatin there

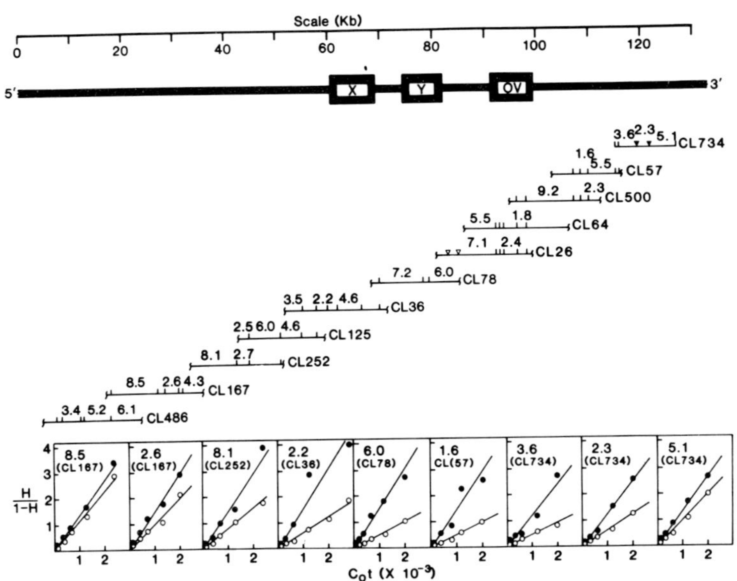

Figure 1. A series of overlapping clones are depicted
which contain genomic DNA from the ovalbumin region of the
chicken genome. The fragments are bounded by artificial Eco
RI sites used for cloning into Charon 4A vectors. Vertical
bars indicate natural Eco RI sites. Sites for Bam HI (open
triangles) and Xho I (closed triangles) are shown in selected
clones only. Probes isolated from these clones were used as
radiolabeled tracers in solution hybridization reactions.
Hybridization rates for some of these probes are shown at
the bottom of the figure. Open circles: Driver DNA isolated
from DNase I digested oviduct nuclei. Closed circles: Control DNA driver (total nuclear DNA sheared to the same size
as the DNase I-treated DNA).

is a nearly constant level (~ 2.5-fold) of preferential
DNase I sensitivity over approximately 80 Kb of DNA containing
and flanking the three genes in the domain. However, at
each terminus the transition to DNase I resistance appears
to occur in a gradient fashion over a distance of about 10
Kb. This results in a domain of preferential sensitivity
extending over approximately 100 Kb of DNA.
    We have also measured the DNase sensitivity of this
chromosomal domain in oviduct tissue from immature chicks

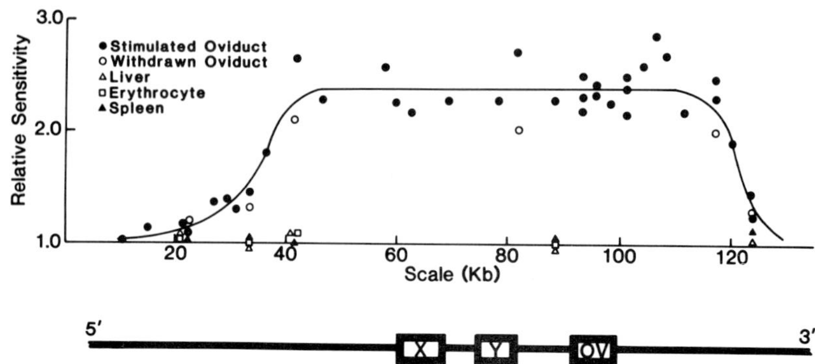

Figure 2.   DNase I sensitivity data of the ovalbumin
domain in stimulated oviduct, withdrawn oviduct, liver,
spleen, and erythrocytes. Relative DNase I sensitivity
was calculated by the ratio of the slopes of the straight
lines obtained for control DNA compared to DNA from DNase I
treated nuclei when plotted as shown at the bottom of Figure
1.

which had been given an initial primary exposure to estrogen but then subsequently withdrawn from hormonal treatment. Under such conditions of withdrawal, ongoing transcriptional activity of the ovalbumin gene ceases (26).
Nevertheless, the open circles in Figure 2 indicate that the
preferential DNase I sensitivity of the domain is still
maintained in the withdrawn oviduct. In contrast, this same
region of the chromatin exhibits no preferential sensitivity
to DNase I in the other tissues examined (liver, erythrocyte,
and spleen). Thus, the observed DNase sensitivity of the
domain is specific to oviduct tissue, but is apparently not
correlated with ongoing transcriptional activity per se since
it persists in hormonally withdrawn oviduct. Rather, it is
more likely a result of the developmental history of the
cell types involved and probably reflects the precommitment
of these cells to express the ovalbumin gene upon proper
hormonal stimulation.

Members of a Specific Repetitive DNA Sequence Family Are
Located Near the Boundaries of the Domain. Once the boundaries of the ovalbumin domain had been defined by the DNase I
sensitivity data, it was of interest to examine the transition regions at the 5' and 3' ends of the domain in more

detail. In particular, we were interested in finding out
if there might be specific signals present in the DNA which
could play a role in defining the domain or establishing
the DNase I sensitive chromatin configuration during the
process of cellular differentiation. Presumably, all genes
that are expressed in a cell type must be found in DNase I
sensitive domains. It is not unreasonable to assume that
the signals involved in setting up these independent domains
of gene activity may be similar from one domain to another.
One possible set of candidates which meet this criterion
would be the dispersed repetitive DNA sequences which appear
to be universally present in metazoan organisms (27).

In a previous report, we identified the Eco RI fragments in the ovalbumin domain which contain repetitive DNA
sequences (22). At that time no information was available
addressing the question of whether these repeats were members
of the same or different repetitive sequence families. However, the results shown in Figure 3 indicate that a specific
subset of these sequences are members of a certain family of
repetitive DNA sequences which we have termed the CR1 family.
We have recently reported the initial characterization of
this CR1 repetitive DNA family together with the nucleotide
sequence of two family members (28). One of the family
members sequenced had been discovered in a clone which contained a gene for U1 snRNA (28,29). The CR1 sequence from
that clone was used as a hybridization probe to look for
related CR1 sequences in Southern blots of cloned sequences
from the ovalbumin gene domain. As shown in Figure 3, the
CR1 probe hybridizes to specific fragments whose locations
in the ovalbumin domain are known (Figure 1). The hybridization pattern indicates the presence of CR1 sequences at
three locations in the region covered by these clones. The
six bands which hybridize represent three different pairs
of fragments in which the members of each pair overlap with
each other. For example, the 5.5 (CL57) and the 2.3 (CL500)
fragments overlap each other; thus, the same sequence is
presumably being observed in each case. The 2.7 (CL252)
and the 2.5 (CL125) fragments overlap, and the 4.3 (CL167)
and the 8.1 (CL252) fragments overlap. (In the latter case,
in many other experiments the difference in intensity between
the 4.3 and 8.1 fragments has not been observed. In the
particular experiment shown in Figure 3, it is evident
from the ethidium bromide stained gel that the CL252 lane
was loaded with more DNA than the CL167 lane; in addition, a
partial digestion product of the 2.7 (CL252) band migrates
very near to the 8.1 band and is contributing to the signal
observed.)

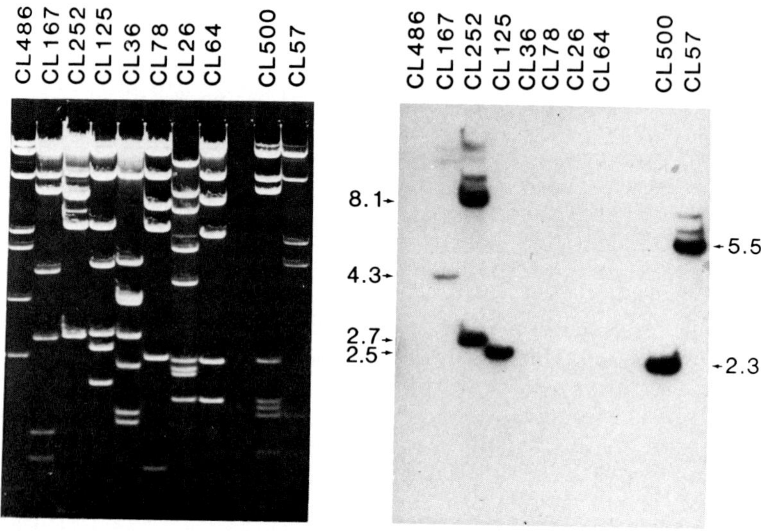

Figure 3.   Southern hybridization of a CR1 family (28) repetitive DNA sequence probe to clones from the ovalbumin gene domain.   Clones were digested with <u>Eco</u> RI (plus <u>Bam</u> HI for CL26), electrophoresed, and transferred to nitrocellulose. The left panel shows the ethidium bromide stained gel.   The right panel is an autoradiogram of the Southern filter after hybridization to a CR1 sequence probe which was originally isolated from a clone containinng a gene for U1 RNA (28,29).

In summary, the experiment in Figure 3 indicates that CR1 sequences are present in three distinct locations in the ovalbumin gene domain.   Remarkably, and perhaps not coincidentally these CR1 sequences are located at positions within or very near the transition regions from DNase I sensitivity to DNase I resistance.   This correlation leads us to speculate that the CR1 family repetitive DNA sequences may play a role in establishing or maintaining domains of gene expression in chicken cell chromatin.

<u>No Evidence for Gross DNA Rearrangements Correlated with Expression of the Ovalbumin Gene.</u> We have previously shown that CR1   sequences   exhibit   some   sequence   and   structural homology to mammalian ubiquitous repetitive DNA sequences, such as the human Alu family.   Many Alu-type sequences exhibit structural   analogies   to   known   transposable   elements

(30-32). This has led to the suggestion that Alu equivalent
sequences may be moveable genetic elements and that they
could play a role in promoting DNA rearrangements (31,33,34).
       It is conceivable that tissue-specific DNA rearrange-
ments could occur in the ovalbumin gene domain during the
process of oviduct cellular differentiation and commitment.
If so, this process could be responsible for the maintenance
of the DNase I sensitive state. It is also conceivable that
the CR1 sequences in particular could be responsible for
promoting such DNA rearrangements. In order to test this
hypothesis, the experiments shown in Figure 4 were performed.

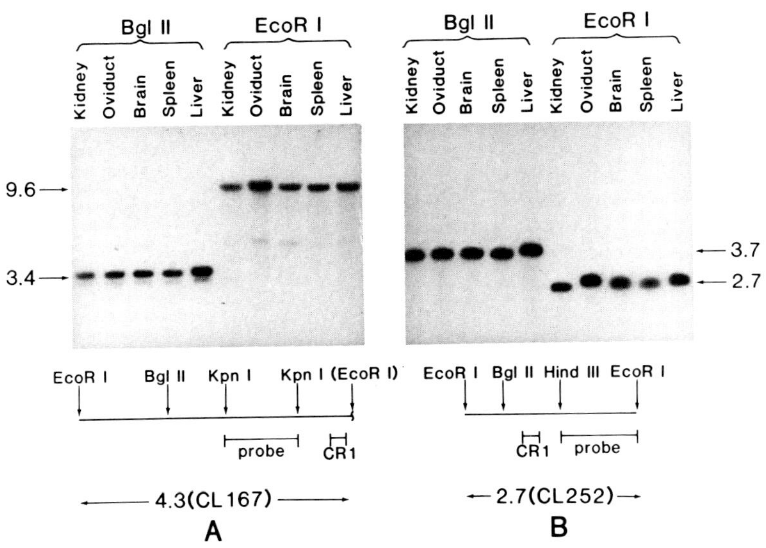

Figure 4. Tissue specific gross rearrangements of the
DNA surrounding CR1 sequences are not observed. DNA preparations were made from the oviduct, spleen, kidney, erythrocytes, and brain of a single hen. They were digested with
Eco RI or Bgl II, and a Southern transfer performed. Results
using two different hybridization probes are presented. The
probes were subfragments taken from Eco RI fragments that
had previously been subcloned. These are shown at the bottom
of the figure and are also shown in Figure 1. The locations
of the CR1 sequences in these subclones are also depicted.
The Eco RI site in parentheses is an artificial site not
found in genomic DNA.

Nuclear DNAs were prepared from kidney, brain, spleen, and liver, as well as oviduct tissue, digested with either Bgl II or Eco RI, and blotted onto nitrocellulose after agarose gel electrophoresis. In work to be reported elsewhere (Stumph et al., manuscript in preparation), we have precisely mapped and determined the nucleotide sequences of the three CRls indicated by the hybridization patterns in Figure 3. Unique sequence DNA probes from regions of the genome very close to, but not including, the CRl sequences were hybridized to the genomic DNA blots (Figure 4). Neither of the enzymes Bgl II or Eco RI cut between the CRl sequences and the probes, nor within the CRl sequences. Therefore, the probes utilized should be able to detect any tissue specific DNA rearrangements in the immediate vicinity of the CRl sequences as a variation of band size in the autoradiogram. As shown in Figure 4, however, the pattern obtained in all tissues is essentially identical, including the oviduct. When similar experiments were done using a probe from the region between the two CRl sequences at the 5' end of the domain, or using probes from the 3' end of the domain, no tissue-specific differences were noted. Although these studies are not exhaustive, the results obtained using a variety of probes indicate that DNA rearrangements in oviduct tissue are probably not a mechanism involved in the specific expression of the ovalbumin gene.

   Tissue-Specific Methylation at the 3' End of the Ovalbumin Domain. A characteristic often associated with active genes is the undermethylation of cytidine residues at the sequence CpG. Genes often appear to be undermethylated in tissues where they are expressed relative to tissues in which they are inactive (35–38). We have extended such methylation studies by examining patterns of methylation throughout the entire ovalbumin gene domain, particularly in the transition regions of DNase sensitivity. Some of these data are shown in Figure 5. DNA preparations from various tissues were digested with either of the enzymes Hpa II or Msp I, and Southern transfers performed as in Figure 4. Both of these enzymes recognize the same sequence (CCGG); however, Hpa II will not cut at this sequence if the internal C is methylated, whereas Msp I will. Figure 5 compares the extent of methylation at these sites in regions at the 5' and 3' borders of the domain. The results indicate that CCGG sequences at the 5' end of the domain (Panel A) are highly methylated in all of the tissues examined, including the oviduct. Other probes from upstream of the X gene also have not revealed any tissue specific methylation at Hpa II/Msp I sites in the 5' portion of the domain (data not

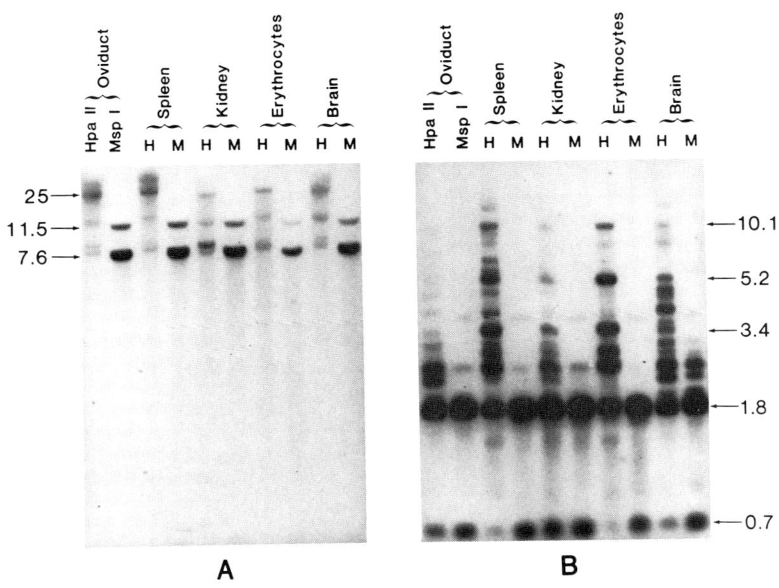

Figure 5.   Methylation at Hpa II/Msp I sites in the ovalbumin gene domain.   DNA's were prepared from various tissues of a single hen, exhaustively digested with either Hpa II or Msp I, electrophoresed in an agarose gel, and Southern hybridizations carried out.   In Panel A, the hybridization probe was the 8.5 fragment from CL167.   In Panel B, the probe was a 1.7 kb subfragment from 5.1 (CL734).

shown).   In definite contrast, a probe from the 3' end of the domain yields a very complex tissue specific pattern of methylation (Figure 5, Panel B).   Notably, Hpa II is able to digest oviduct DNA from this region of the genome to smaller fragments than what is observed for the same DNA sequences in other tissues.   This result indicates that oviduct DNA is undermethylated at Hpa II/Msp I sites at the 3' border of the ovalbumin domain compared to the same DNA region in other tissues.   In addition, these data and other direct mapping experiments indicate that there is about a 10-fold greater density of CCGG sequences in this 3' transition region of the domain than what is normally found in chicken DNA in general.   This increased occurrence of the rare di-

nucleotide CpG, taken together with the observed tissue
specific methylation of these sites (undermethylation in the
oviduct), could indicate that this 3' transition region of
the domain may be an important control region for ovalbumin
gene expression.

DISCUSSION

Many of the significant aspects of our data on the ovalbumin gene domain are summarized in Figure 6. Approximately
100 Kb of DNA containing and surrounding the X, Y, and ovalbumin genes exhibit a preferential sensitivity to DNase I
in the chromatin of oviduct cells. In other tissues where
these genes are not expressed, no such preferential sensitivity is observed. We interpret this to mean that in cells
where these genes are to be expressed the nucleosomes in
this 100 Kb domain are packaged somewhat differently from
the majority of the nucleosomes in the bulk of the chromatin.
Presumably, all active genes share a similar altered structure
in regard to their nucleosomal configuration.

However, the fact that a gene is in a DNase I sensitive
state is not sufficient for it to be actively transcribed.
This is demonstrated by the fact that the DNase I sensitivity
of the entire domain persists in nuclei isolated from hormonally withdrawn chicks in spite of the shut down of ovalbumin gene transcription. An analogous result is obtained
for the globin gene in transcriptionally inactive erythrocytes (1). Therefore, the DNase I sensitivity appears to
reflect the developmental capacity of a cell to express
the gene in question. It can be viewed as a necessary step
in the prior commitment of a cell to allow a certain gene
to be expressed. Such a mechanism would make it possible
for distinct cell types to respond to a single inducer
each in its own individual and distinctive manner.

Thus, it is important to understand what mechanisms are
involved in the establishment of DNase I–sensitive domains
during cellular differentiation. Initially we have chosen
to look for possible signals in the DNA itself. Sequences
either within or near to the regions where the chromatin
undergoes the change in configuration seem like a logical
place to begin the search. Gross rearrangements of the DNA
do not appear to be responsible for establishment of the
domain, since we have not detected any such rearrangements
in oviduct tissue as compared to a number of other tissues.

Logically, it seems quite possible that repetitive DNA
sequences could play a role in establishing domains. However,
there is no good correlation between the locations of all
repetitive DNA sequences in the domain in general and the

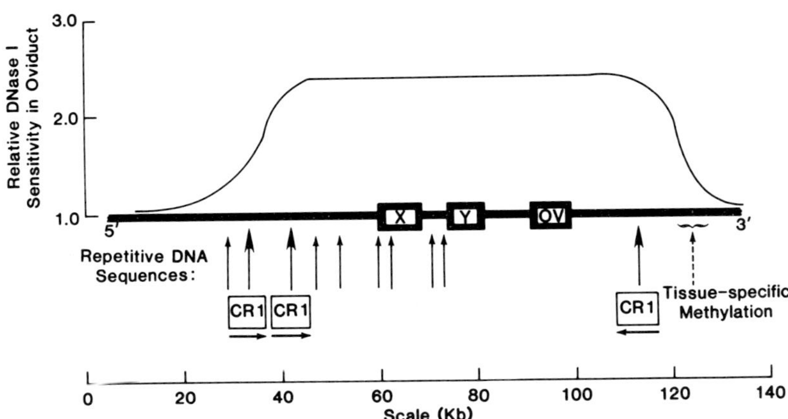

Figure 6. Summary of the ovalbumin gene domain. Solid
vertical arrows denote the locations of repetitive DNA sequences within the domain (22,54). The bold vertical arrows
represent the specific locations of CR1 family repetitive
DNA sequences. The orientations of the CR1 sequences are
denoted by the horizontal arrows. The dashed arrow indicates
a region where distinctive methylation patterns are noted
for different chicken tissues, and where oviduct DNA is
specifically undermethylated.

measured DNase I sensitivity (Figure 6). Nevertheless, we
have observed that a certain subset of these sequences belong
to a single family of repeats termed CR1 (28). These CR1
family members are preferentially located near the transition
regions at each end of the domain, as shown in Figures 3 and
6. We have also assigned a polarity to the CR1 sequences
based upon their previously noted homology to human Alu
sequences (28). (The Alu sequences have been assigned a
polarity based upon the direction in which they are transcribed in vitro by RNA polymerase III [30].) Interestingly,
the sequence data (not shown) indicate that the two CR1s at
the 5' end of the domain have a polarity the same as the
direction of transcription of the X, Y, and ovalbumin genes.
In contrast, the CR1 at the 3' end of the domain is in the
opposite orientation. Thus, the CR1s so far characterized
in the ovalbumin domain have inward orientations pointing
toward the central portion of the domain containing the
structural genes. This is shown by the horizontal arrows in

Figure 6. One additional CR1 sequence that we have characterized (28) is located about 2 Kb upstream from a chicken U1
RNA gene, and its orientation is also in agreement with
this observation. This inward orientation may be an important structural feature if the CR1s actually do play a role
in establishing gene domains. Obviously, it will be necessary
to investigate additional CR1s which flank other chicken
genes. Finally, we note that in the rat, a pair of inverted
repeats has been shown to flank a 12 Kb length of DNA which
contains a preproinsulin gene (39). Also, approximately 35
Kb of rabbit DNA which contains a cluster of four β-
globin genes is flanked by inverted repeats (40). It would
be interesting to know if the locations of these repeats are
correlated with a change in DNase I sensitivity in these
other systems.

If the CR1 sequences are potentially involved in establishing or defining the ovalbumin gene domain, as well as
other domains in the chicken cell chromatin, what kinds of
mechanisms might be involved? A number of possibilities can
be proposed. Some mammalian repetitive DNA sequences are
promoters for RNA polymerase III in in vitro systems (30,41).
Recently a class of small RNAs has been identified in Chinese
hamster cells which could represent in vivo transcripts of
such repetitive DNA sequences (41). By analogy, it is thus
possible that the chicken CR1 sequences may also act as
promoters for RNA polymerase in vivo. If so, perhaps just
the very act of them being transcribed might be sufficient
to result in the opening up of the domain either by affecting
nucleosome configuration or perhaps by relieving stress or
constraints in the DNA. This effect could then be propagated
throughout the entire domain.

A second possibility is that the CR1s may act as specific
binding sites for a class of proteins which induce conformational changes in the chromatin. Alternatively, the CR1s
could act to establish the boundaries of the domain by being
specific attachment sites of the chromatin to the nuclear
matrix, an architectural structure of the nucleus (42) which
continues to exist even after removal of most of the DNA,
RNA, and chromosomal proteins. Interestingly, several groups
have reported that most of the chromatin in cells exists as
supercoiled loops, most of which contain about 50-150 Kb of
DNA (43-46). The ends of each individual loop appear to be
anchored to adjacent points on the nuclear matrix. Perhaps
the CR1 sequences are found at these anchorage points.
Thus, the chromatin could be bound to the matrix by means of
specific interactions between CR1 sequences and functional
protein components of the matrix. In addition, evidence
has accumulated that transcriptionally active genes and

newly synthesized RNA are closely associated with the nuclear matrix (47-50). It is therefore possible that CRls
could be secondary binding sites which are utilized only
when the chromatin is to be closely associated with the
matrix in an active conformation.

It is also possible that complementary sequences (CRls)
at the 5' and 3' ends of the domain could specifically base
pair with each other. This would bring the termini of the
domain into close proximity with each other while permitting
the central part of the domain to loop out. An interesting
possibility is raised by the report that DNA containing
stretches of purines on one strand complemented by pyrimidines
on the opposite strand have the ability to form tetra-stranded
complexes (51). Such purine (pyrimidine) rich stretches
appear to be a characteristic of CRls when their nucleotide
sequences are examined (28). It is thus possible that CRls
at opposite ends of the domain could interact with each
other via such four-stranded structures.

In many instances, tissue specific undermethylation of
cytidine residues at the sequence CpG has been correlated
with active gene expression. Using DNA from different tissues, we have extensively studied methylation patterns at
CCGG sequences (Hpa II/Msp I sites) in the ovalbumin gene
domain. Hpa II/Msp I sites occur rarely throughout most of
the ovalbumin gene domain, thus reflecting the under-
representation of the CpG dinucleotide in eucaryotic cells.
Near to the ovalbumin gene itself, we observe a limited
degree of undermethylation at certain sites as described
earlier by other investigators (36). Using a variety of
probes from the 5' end of the domain we have not been able
to observe any tissue specific methylation patterns. However,
a region at the 3' boundary of the domain contains a highly
increased density of Hpa II/Msp I sites compared to the rest
of the domain. In addition, methylation at these sites is
highly variable from tissue to tissue, and oviduct DNA appears
to be specifically undermethylated (Figure 5). It is thus
attractive to postulate that this area is an important
control region for determining the state of activation of
the ovalbumin gene domain. We are currently examining
whether DNA sequences exist in this region which could
undergo a B $\rightleftarrows$ Z conformational transition as a possible
result of tissue-specific (de-)methylation (52). Since
transitions of this type would markedly alter the supercoil-
ing of the chromatin, such a mechanism seems plausible for
the activation of large domains of chromatin structure (53).

In conclusion, we have defined the 100 Kb ovalbumin, X,
and Y multi-gene domain based upon the relative sensitivity
of this region to DNase I in oviduct cells compared to other

tissues. We have also identified potential signals in the DNA sequence which may be involved in establishing this domain or forming its borders. Finally, this has allowed us to propose a number of experimentally testable mechanisms which could be utilized to activate domains of gene expression during cellular differentiation and development.

ACKNOWLEDGMENTS

We would like to thank Ms. Serlina Robinson and Ms. Valerie McMullian for exellent technical assistance.

REFERENCES

1. Weintraub, H., and Groudine, M. (1976). Science (Washington D.C.) 193, 848-856.
2. Young, N.S., Benz, E.J., Jr., Kantor, J.A., Kretschmer, P., and Neinhuis, A.W. (1978). Proc. Natl. Acad. Sci. USA 75, 5884-5888.
3. Stalder, J., Groudine, M., Dodgson, J.B., Engel, J.D., and Weintraub, H. (1980). Cell 19, 973-980.
4. Garel, A., and Axel, R. (1976). Proc. Natl. Acad. Sci. USA. 73, 3966-3970.
5. Bloom, K.S., and Anderson, J.N. (1979). J. Biol. Chem. 254, 10532-10539.
6. Lawson, G.M., Tsai, M.-J., and O'Malley, B.W. (1980). Biochemistry 19, 4403-4411.
7. Bellard, M., Kuo, M.T., Dretzen, G., and Chambon, P. (1980). Nucleic Acids Res. 8, 2737-2750.
8. Shepherd, J.H., Mulvihill, E.R., Thomas, P.S., and Palmiter, R.D. (1980). J. Cell Biol. 87, 142-151.
9. Wu, C., Wong, Y.-C., and Elgin, S.C.R. (1979). Cell 16, 807-814.
10. Wu, C., and Gilbert, W. (1981). Proc. Natl. Acad. Sci. USA 78, 1577-1580.
11. Storb, U., Wilson, R., Selsing, E., and Walfied, A. (1981). Biochemistry 20, 990-996.
12. Sledziewski, A., and Young, E.T. (1982). Proc. Natl. Acad. Sci. USA 79, 253-256.
13. Levy-Wilson, B., Kuehl, L., and Dixon, G.H. (1980). Nucleic Acids Res. 8, 2859-2869.
14. Stalder, J., Seebeck, T., and Braun, R. (1978). Eur. J. Biochem. 90, 391-395.
15. Giri, C.P., and Gorovsky, M.A. (1980). Nucleic Acids Res. 8, 197-214.
16. Panet, A., and Ceder, H. (1977). Cell 11, 933-940.
17. Flint, S.J., and Weintraub, H.M. (1977). Cell 12, 783-794.

18. Groudine, M., Das, S., Nieman, P., and Weintraub, H. (1978). Cell 14, 865–878.
19. Southern, E.M. (1975). J. Mol. Biol. 98, 503–517.
20. Elgin, S.C.R. (1981). Cell 27, 413–415.
21. Stalder, J., Larsen, A., Engel, J.D., Dolan, M., Groudine, M., and Weintraub, H. (1980). Cell 20, 451–460.
22. Lawson, G.M., Knoll, B.J., March, C.J., Woo, S.L.C., Tsai, M.-J., and O'Malley, B.W. (1982). J. Biol. Chem. 257, 1501–1507.
23. Britten, R.J., Graham, D.E., and Neufeld, B.R. (1974). Methods Enzymol. 29, 363–418.
24. Royal, A., Garapin, A., Cami, B., Perrin, F., Mandel, J.L., LeMeur, M., Bregegegre, F., Gannon, F., LePennec, J.P., Chambon, P., and Kourilsky, P. (1979). Nature (London) 279, 125–132.
25. Colbert, D.A., Knoll, B.J., Woo, S.L.C., Mace, M.L., Tsai, M.-J., and O'Malley, B.W. (1980). Biochemistry 19, 5586–5592.
26. Swaneck, G.E., Nordstrom, J.L., Kreuzaler, F., Tsai, M.-J., and O'Malley, B.W. (1979). Proc. Natl. Acad. Sci. USA 76, 1049–1053.
27. Davidson, E.H., Galau, G.A., Angerer, R.C., and Britten, R.J. (1975). Chromosoma 51, 253–259.
28. Stumph, W.E., Kristo, P., Tsai, M.-J., and O'Malley, B.W. (1981). Nucleic Acids Res. 8, 5383–5397.
29. Roop, D.R., Kristo, P., Stumph, W.E., Tsai, M.-J., and O'Malley, B.W. (1981). Cell 23, 671–680.
30. Duncan, C.H., Jagadeeswaran, P., Wang, R.R.C., and Weissman, S.M. (1981). Gene 13, 185–196.
31. Haynes, S.R., Toomey, T.P., Leinwand, L., and Jelinek, W.R. (1981). Mol. Cell. Biol. 1, 573–583.
32. Grimaldi, G., and Singer, M.F. (1982). Proc. Natl. Acad. Sci. USA. 79, 1497–1500.
33. Van Arsdell, S.W., Denison, R.A., Bernstein, L.B., Weiner, A.M., Manser, T., and Gesteland, R.F. (1981). Cell 26, 11–17.
34. Calabretta, B., Robberson, D.L., Barrera-Saldana, H.A., Lambrau, T.P., and Saunders, G.F. (1982). Nature (London) 296, 219–225.
35. McGhee, J., and Ginder, G.D. (1979). Nature (London) 280, 419–420.
36. Mandel, J., and Chambon, P. (1979). Nucleic Acids Res. 7, 2081–2103.
37. Van der Ploeg, L.H.T., and Flavell, R.A. (1980). Cell 19, 947–958.
38. Weintraub, H., Larsen, A., and Groudine, M. (1981). Cell 24, 333–344.

39. Lomedico, P., Rosenthal, N., Efstradiadis, A., Gilbert, W., Kolodner, R., and Tizard, R. (1979). Cell 18, 545-558.

40. Shen,C.-K.J., and Maniatis, T. (1980). Cell 19, 379-391.

41. Haynes, S.R., and Jelinek, W.R. (1981). Proc. Natl. Acad. Sci. USA 78, 6130-6134.

42. Berezney, R., and Coffey, D.S. (1974). Biochem. Biophys. Res. Commun. 60, 1410-1417.

43. Cook, P.R., and Brazell, I.A. (1975). J. Cell Sci. 19, 261-279.

44. Benyajati, C., and Worcel, A. (1976). Cell 9, 393-407.

45. Paulson, J.R., and Laemmli, U.K. (1977). Cell 12, 817-828.

46. Vogelstein, B., Pardoll, D.M., and Coffey, D.S. (1980). Cell 22, 79-85.

47. Miller, T.E., Huang, C.Y., and Pogo, A.O. (1978). J. Cell Biol. 76, 675-691.

48. Herman, R., Weymouth, L., and Penman, S. (1978). J. Cell Biol. 78, 663-674.

49. Nelkin, B.D., Pardoll, D.M., and Vogelstein, B. (1980). Nucleic Acids Res. 8, 5623-5633.

50. Robinson, S.I., Nelkin, B.D., and Vogelstein, B. (1982). Cell 28, 99-106.

51. Johnson, D., and Morgan, A.R. (1978). Proc. Natl. Acad. Sci. USA 75, 1637-1641.

52. Behe, M., and Felsenfeld, G. (1981). Proc. Natl. Acad. Sci. USA 78, 1619-1623.

53. Nordheim, A., Pardue, M.L., Lafer, E.M., Moller, A., Stollar, B.D., and Rich, A. (1981). Nature (London) 294, 417-422.

54. Heilig, R., Perrin, F., Gannon, F., Mandel, J.L., and Chambon, P. (1980). Cell 20, 625-637.

SUPERCOILED LOOPS AND TRANSCRIBING GENES[1]

Bert Vogelstein, Barry Nelkin, Drew Pardoll,
Sabina Robinson, and Don Small

Cell Structure and Function Laboratory
The Oncology Center
Johns Hopkins University School of Medicine
Baltimore, Maryland 21205

ABSTRACT   The DNA in a eucaryotic nucleus is arranged
into a series of supercoiled loops which are anchored at
their bases to a nuclear skeleton or matrix. It can be
shown that these supercoiled loops of DNA are mobile,
as they appear to be reeled through the matrix during
DNA replication. Using nuclease digestion, one can
progressively cleave DNA from the loops, thereby
isolating residual DNA that is close to the nuclear
matrix anchorage sites. Several transcribed genes
appear to be enriched in this residual DNA. The
sequences examined are the genes coding for ribosomal
RNA in rat liver cells, the SV40 genes in SV40
transformed mouse cells, the ovalbumin gene in hen
oviduct cells, and the <u>Alu</u> sequence family in human lung
cells. The results suggest that genes transcribed by
all three mammalian RNA polymerases are oriented
non-randomly with respect to supercoiled loop domains.

INTRODUCTION

There are at least three hierarchical levels involved in
the structure of chromatin. At a basic level, the DNA is
wound around histone cores to form nucleosomes (1,2). The
nucleosomes, in turn, are apparently packed in an orderly
fashion into solenoids or superbeads (3-6). At yet a higher
level of organization, these structures are arranged in the
nucleus as supercoiled loops, each containing approximately
10^5 base pairs of DNA (7-12).
Most of the evidence linking eucaryotic chromosome
structure to gene function has been adduced at the

[1]This work was supported by NIH grants and a gift from
the Bristol-Myers Co.

nucleosomal level. Active genes in chromatin are more sensitive to treatment with DNase I and micrococcal nuclease (13-15). This nuclease sensitivity has been shown to be in part due to the association of specific nonhistone proteins with nucleosomes containing active genes (16,17). In addition, DNA sequences surrounding active genes have been shown to contain nuclease hypersensitive regions(18,19).

Little information is as yet available on the role of the two supranucleosomal levels of DNA organization in nuclear function. We have been examining the arrangement of genes within supercoiled loops of DNA. These loops of DNA can be directly visualized in eucaryotic nuclei after extraction of the histones (and perhaps other proteins), which are responsible for compaction of the DNA <u>in situ</u> (12). In Figure 1, such an extracted nucleus is shown. There is a central residual nucleus surrounded by a halo of DNA. The central residual nucleus is similar in composition to the nuclear matrix that has been isolated from nuclei after treatment with nucleases and high salt (20,21). This matrix has been described as a general structural feature underlying the nucleus in a wide variety of eucaryotic cells (22-26). The residual nucleus (which will henceforth be referred to as the nuclear matrix) has the same size and shape as the nucleus from which it was made. The halo of DNA surrounding it is actually composed of a series of supercoiled loops; this has been shown using the intercalating agent ethidium bromide (12).

The diameter of the halo (approximately 12µ) is consistent with a loop size of approximately 75,000 base pairs. Similarly sized loops have been shown to be anchored to a central chromosome scaffold in metaphase chromosomes (27). By indirect means, the looped organization of DNA has been shown in numerous other cell types (7-11,28). Hence, the organization of DNA into a series of supercoiled loops seems to be a general feature of eucaryotic nuclear structure throughout the cell cycle.

<u>The loops are mobile</u>. These loops are not stationary organizational elements; rather, they have been shown to be mobile during DNA replication and perhaps during other nuclear processes. The most direct evidence for this is provided by autoradiography (12,29). When matrix-halo structures such as that shown in Figure 1 are prepared from cells pulse-labeled with ³H-thymidine, and subsequently autoradiographed, a distinct distribution of the newly labeled DNA is observed. With short pulse times most of the autoradiographic grains are located over the nuclear matrix; with longer pulse times, the grains (representing newly replicated DNA) gradually move from the matrix out into the

Figure 1. Visualization of loops of DNA anchored to a
central nuclear matrix. Nuclei from 3T3 cells were extracted
sequentially with nonionic detergent in isotonic solution and
then with a solution containing 2 M NaCl. The supercoiled
loops were then stained with DAPI, and subsequently relaxed
by exposure to 260 nm ultraviolet light. The central sphere
is the nuclear matrix, anchoring the loops of DNA.

halo region (Figure 2). Since these results are from a
logarithmically growing population of cells, the data
indicate that fixed replication complexes exist within the
nuclear matrix, and that the loops of DNA are reeled through
these complexes as they are replicated. The fact that the
size of the loops are the same as the sizes of the replicons
in diverse cells is consistent with this model of eucaryotic
DNA replication (Table I).

Supercoiled Loops and Gene Expression. Several
observations suggest that the loop anchorage sites might also
function in gene expression: (1) Miller et al. (30) and
Herman et al. (31) showed that essentially all the rapidly
labeled RNA remains associated with the nuclear matrix after
a salt extraction which removed over 90% of the nuclear
protein. It has been shown that the hnRNA is bound to the
nuclear matrix through a few specific proteins (32);(2) some
small molecular weight RNA's have shown to be quantitatively
associated with the nuclear matrix (33,34). These RNA's have
been postulated to be involved in RNA processing by base
pairing with sequences at the borders of introns (35,36);(3)
it has been shown that high affinity binding sites for

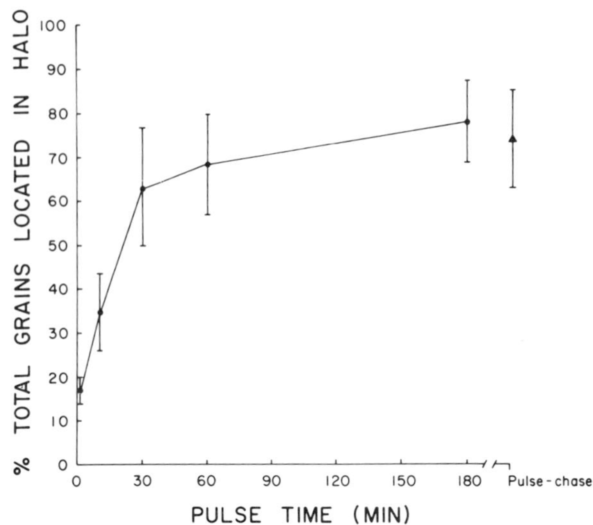

Figure 2. Autoradiographic demonstration of the
association of newly replicated DNA with the nuclear matrix.
3T3 cells were pulsed with ^{3}H-thymidine for various time
intervals, then extracted with buffers containing high salt.
Means and standard deviations are shown; 25 structures were
assessed for each point. (Reprinted with permission from
Cell <u>22</u>, 79. Copyright (c) MIT 1980.)

TABLE I

COMPARISON OF REPLICON LENGTH AND
SUPERCOILED LOOP LENGTH IN VARIOUS NUCLEI

Organism	Supercoiled Loop Length (Fluorescence Microscopy)	Replicon Length (Fiber Auto-radiography)
Human	75,000 bp	69,000 bp
<u>Drosophila</u>	36,000 bp	40,000 bp
Physarum	36,000 bp	37,500 bp

The supercoiled loop length was measured as the halo
diameter of matrix-halo structures in the presence of
ethidium bromide (12). The human cells (A549, a lung
carcinoma line) and <u>Drosophila</u> cells (Kc) used were grown in
tissue culture. Loops were relaxed by a 10-second treatment
vwith ultraviolet illumination in the fluorescence
microscope. The average replicon lengths as assessed by
fiber autoradiography were obtained from published data
(82-84).

steroid hormones are located on the nuclear matrix in target
tissues (37-40). In these tissues, steroid hormones modulate
gene expression at the level of transcription (41,42).

To investigate the involvement of the nuclear matrix in
the mechanics or control of gene expression, we have analyzed
the nature of the DNA that is closely associated with the
matrix anchorage points. We reasoned that if some of the
machinery involved in gene expression were located within the
nuclear matrix, then the DNA sequences to be expressed might
be bound to these elements.

To assess this possibility, the following approach has
been used. Nuclei are subjected to 2 M NaCl treatment to
remove histones and most other nuclear proteins. The
resultant nuclear matrices are then exposed to DNAase I to
randomly cleave the DNA; this treatment progressively
releases DNA from the loops. The extent of the DNAase I
cleavage can be monitored by fluorescence microscopy of the
nuclear matrices (Figure 1). The halo gradually shrinks in
size and then disappears completely with progressive DNAase I
treatment. As more DNA is released from the loops, the
residual DNA attached to the nuclear matrix is progressively
more enriched in sequences close to the loop attachment
points. This matrix-associated DNA can be recovered by low
speed centrifugation.

Using this approach, we initially isolated nuclear
matrix-associated DNA from rat liver nuclei (43). $C_o t$
hybridization analysis, using ^{125}I-rRNA (18S + 28S) as a
probe, demonstrated substantial enrichment of the genes
coding for ribosomal RNA in nuclear matrix-associated DNA
(Figure 3). Similar enrichments of genes coding for
ribosomal RNA have been found in mouse tissue culture cells
(data not shown). This experiment suggested that active
ribosomal genes might be associated with the nuclear matrix.
However, alternative explanations for the association of rDNA
with the nuclear matrix are possible. For example, it is
possible that the nucleolus has a different structure than
the rest of the genome. This structure could necessitate
more extensive nuclease treatment to release DNA from the
rDNA containing loops.

To further address this question, matrix-associated DNA
from SV40 transformed 3T3 cells (cell line SVB203) was
isolated (44). These cells have a partial duplication of the
SV40 genome integrated at one site per nucleus (45), and they
actively transcribe the SV40 sequences. Matrix preparations
from these cells were obtained containing 3.5% or 9.0% of the
total cellular DNA. These DNAs, along with reference total
cellular DNA, were extracted and cleaved with the restriction
endonuclease HindIII. Equal amounts of these DNA samples
were electrophoresed, transferred to nitrocellulose sheets

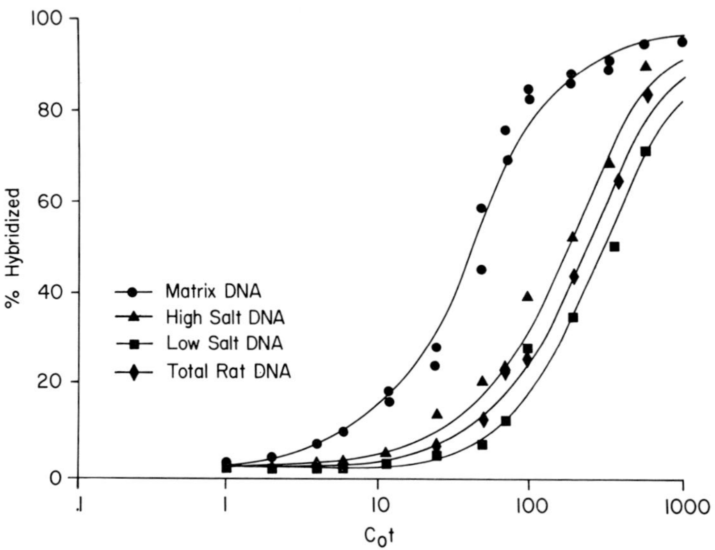

Figure 3. The genes coding for ribosomal RNA are associated with the nuclear matrix in rat liver cells. Rat liver nuclei were extracted with a low ionic strength buffer, treated with DNase I, and then extracted with 2 M NaCl. (Similar results were found if the 2 M NaCl extraction preceded nuclease digestion.) DNA was purified from matrix, low salt extract, high salt extract and total rat liver nuclei. Hybridization with ^{125}I-(18S+28S)rRNA was performed in 70% formamide, 0.2 M sodium phosphate (85) to the indicated C_0t values. (From reference 43, with permission.)

(46), and hybridized with SV40 DNA labeled (47) with ^{32}P ($5-10 \times 10^7$ dpm/μg). This provides a quantitative assay of the relative amount of SV40 DNA sequences in each DNA preparation (44). Figure 4 shows that the nuclear matrix DNA preparations are enriched in SV40 sequences relative to total DNA. Moreover, the preparation containing 3.5% of the total cellular DNA is more enriched in SV40 sequences than is the preparation containing 9.0% of the total cellular DNA. This finding implies that the transcribed SV40 sequences are close to the loop anchorage points. This experiment was repeated with two other SV40 transformed 3T3 cell lines with similar results (44). As a control, matrix and total cellular DNA samples from SVB203 cells were digested with HindIII,

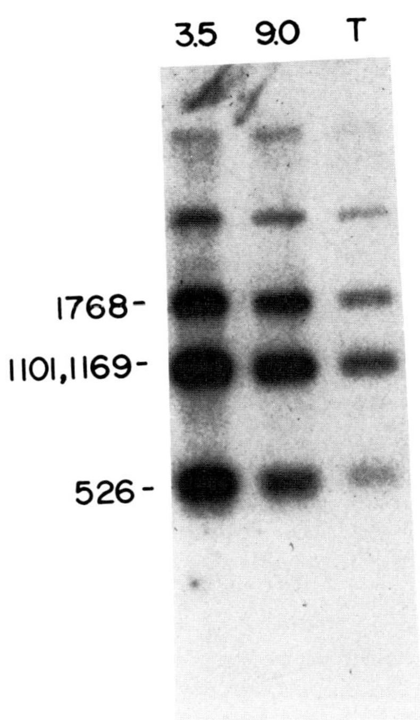

Figure 4. Nuclear matrix DNA from SV40 transformed cells
is enriched in SV40 sequences. Nuclei from SVB203, a 3T3
cell line transformed with SV40, were extracted with 2 M
NaCl, and then treated with DNase I. Equal amounts of total
cellular DNA and DNA from matrix preparations containing 9.0%
or 3.5% of the total cellular DNA were digested with
restriction endonuclease HindIII, electrophoresed,
transferred to nitrocellulose filters, hybridized with
^{32}P-SV40 DNA, and autoradiographed.(From reference 44, with
permission.)

electrophoresed, transferred and hybridized as before, except
that cloned mouse β-globin cDNA was used as a hybridization
probe. In this case, no enrichment of the globin sequence
was seen in matrix DNA preparations.
   The results with SV40 lend further credence to the
possibility of an association between active genes and the

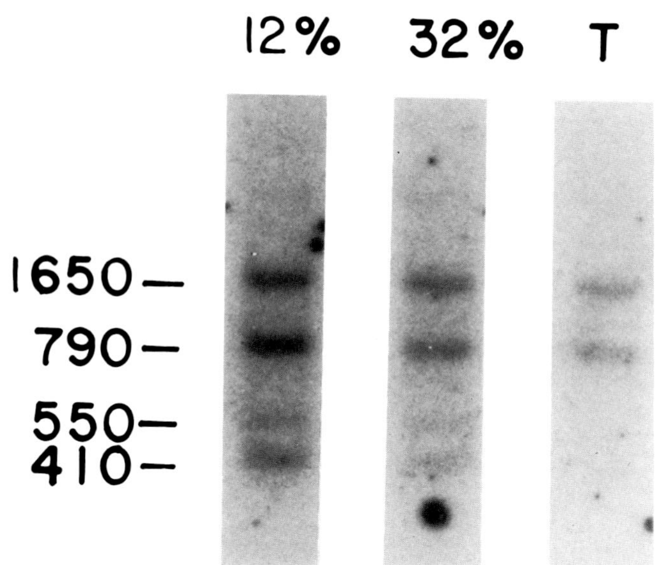

Figure 5.   The ovalbumin gene is preferentially
associated with the nuclear matrix of hen oviduct cells.
Nuclei were isolated from laying hen oviduct, extracted with
2 M NaCl, and then treated with DNase I.   Nuclear matrix
preparations containing 32% or 12% of the total cellular DNA
were recovered by centrifugation.   Equal amounts of DNA from
these preparations and from total cellular DNA were digested
with the restriction endonuclease HinfI, electrophoresed,
transferred to nitrocellulose filters, hybridized with
^{32}P-labeled, cloned ovalbumin cDNA, and autoradiographed.
(Reprinted with permission from Cell 28, 99.   Copyright (c) MIT
1982.)

nuclear matrix.   However, alternative explanations exist.   It
is possible that the SV40 genes' association with the matrix
is related to the genes' effects on cell transformation.   In
addition, sequences near the SV40 origin of replication bind
specifically to T antigens (48).   If SV40 T antigen is found
in the matrix of SV40 transformed cells, as is the case in
polyoma (49) and SV40 productively infected cells (50), then
preferential SV40 association with the nuclear matrix might
be due to its binding to T antigen.
       To clarify this question, we next asked whether a gene
transcribed in one cell type is associated with the nuclear

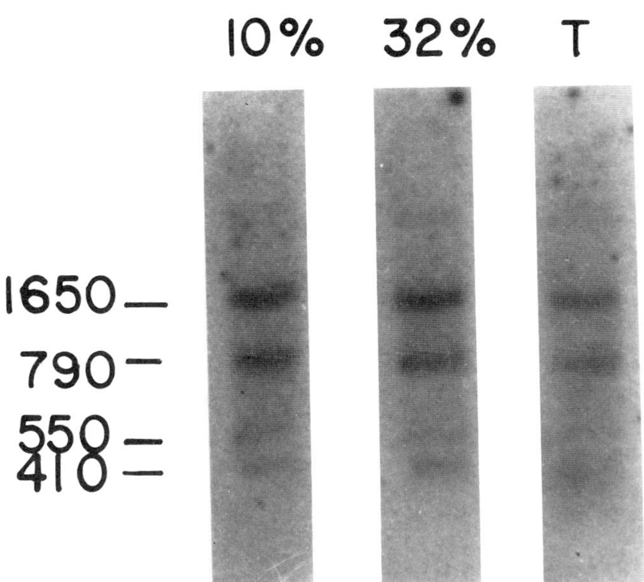

Figure 6. The ovalbumin gene is not preferentially
associated with the nuclear matrix of chicken liver cells.
DNA from nuclear matrices of hen liver cells was treated as
described in the legend to Figure 5.(Reprinted with permission
from Cell $\underline{28}$, 99. Copyright (c) MIT 1982.)

matrix in that cell type, but not in a cell type in which the
gene is not transcribed. Chicken ovalbumin is transcribed in
90% of the cells in the laying hen oviduct (51), but it is
not transcribed in chicken liver (52). Matrix preparations
were made from hen oviduct cells and hen liver cells. The
DNA from these preparations was purified and cleaved with the
restriction endonuclease HinfI. Equal amounts of total
nuclear DNA, or DNA from hen oviduct matrix or hen liver
matrix were then electrophoresed through agarose and
transferred to nitrocellulose. ^{32}P-labeled chicken globin or
ovalbumin cDNA sequences (cloned in plasmids) were used as
hybridization probes for these filters. In the filter
hybridization analyses with the ovalbumin probe, five
fragments could be detected (53). Four of these are the
fragments expected from the ovalbumin gene map (a 1650 bp
fragment from the 5' end, 550 and 410 bp fragments from the
middle, and a 790 bp fragment from the 3' end of the gene

[54]); the fifth fragment, 2,800 bp, is probably from the related X or Y genes, which are related to the ovalbumin gene and are also transcribed in hen oviduct cells (55,56). All of these fragments are preferentially associated with the nuclear matrix in oviduct, but not in liver cells (Figures 5 and 6). Furthermore, the enrichment is higher in DNA from the matrix preparation containing less of the total nuclear DNA (Figure 5). In several experiments of this type, it has been found that the majority of the ovalbumin gene sequences (up to 60%) are associated with nuclear matrix DNA fractions that contain 10-15% of the total nuclear DNA. These results provide substantial evidence that active genes are associated with the nuclear matrix.

As a technical control for these experiments, the content of chicken globin genes was also assessed in the matrix DNA fractions. The same matrix DNA preparations (cleaved with the same restriction enzyme) that contained elevated concentrations of the ovalbumin gene were <u>not</u> enriched for the inactive globin gene (53).

In addition to DNA sequences present in one or a few copies per genome, eucaryotic DNA contains sequences present in multiple (10^2-10^7) copies per genome (57). While some of these sequences have well characterized functions, the function of most of these sequences is unknown. Among the functions proposed for these sequences are roles in maintenance of the structure of chromosomes, in mitosis, and in transcriptional regulation (58-61).

One of the DNA sequences which has been proposed as a possible regulatory sequence is an interspersed, highly repeated sequence in human DNA (62). This sequence, known as the <u>Alu</u> family sequence, was found to be transcribed (63), and has since been found in the vicinity of numerous transcribed genes (64-66). It also serves as an RNA polymerase III promotor <u>in vitro</u> (67), and it appears to be transcribed by RNA polymerase III <u>in vivo</u> (68). Since we had found other transcribed sequences closely associated with the nuclear matrix, we also examined the proximity of the <u>Alu</u> family sequence to the nuclear matrix.

Matrix and total cellular DNAs were prepared from human A549 cells. Several complementary techniques were used to assay the concentration of <u>Alu</u> family sequence in each DNA preparation. First, aliquots of each DNA sample were nick translated and hybridized to excess DNA from an <u>Alu</u> family sequence cloned in pBR322 (BLUR8, kindly provided by Dr. P. Deininger) and immobilized on nitrocellulose filters. Table II shows that human nuclear matrix DNA is enriched in <u>Alu</u> family sequences. In a converse approach, equal aliquots of the matrix and total cellular DNA preparations were immobilized on nitrocellulose filters, and excess nick

TABLE II

QUANTITATION OF THE ALU FAMILY SEQUENCE CONTENT
OF HUMAN MATRIX DNA'S

A. Based on densitometry of gels containing MboI digests of
the indicated matrix preparations.

DNA	Relative Enrichment
Total nuclear	1.00
Matrix 31.9%	1.92
Matrix 12.2%	2.42
Matrix 7.1%	1.52

B. Based on hybridization of nick translated Alu family
sequence DNA (in BLUR8 plasmid) to total nuclear and matrix
DNA's immobilized on nitrocellulose filters. The mean and
range for three hybridizations per matrix preparation is
indicated.

DNA	Relative Enrichment	
	Mean	Range
Total nuclear	1.00	0.95-1.05
Matrix 31.9%	1.85	1.80-2.40
Matrix 12.2%	1.98	1.75-2.40
Matrix 7.1%	1.48	1.30-1.60
Matrix 1.3%	1.62	1.40-1.90

C. Based on hybridization of nick translated total and
matrix DNA's to Alu family sequence DNA immobilized on
nitrocellulose filters. The range of values for three
hybridizations per matrix preparation is indicated.

DNA	Relative Enrichment	
	Mean	Range
Total nuclear	1.00	0.96-1.08
Matrix 3.9%	1.67	1.55-1.83
Matrix 1.8%	2.20	2.15-2.25
Matrix 0.4%	1.91	1.85-1.98

translated BLUR8 DNA was used as a hybridization probe. This
experiment also demonstrated significant enrichment of Alu
family sequences in nuclear matrix DNA (Table II). Finally,
total cellular and matrix DNAs were digested with MboI, an
enzyme which cleaves the consensus Alu family sequence in two
places, leaving a 180 bp fragment (69) which can be
visualized in polyacrylamide gels. As shown in Table II,
this 180 bp band is more intense in matrix DNA samples,
indicating enrichment of Alu family sequences. Thus, by all
three methods it is found that some Alu family sequence
members are closely associated with the nuclear matrix.

It is clear from the above results that the organization

of DNA in supercoiled loops in eucaryotic nuclei is
strikingly nonrandom. From the results of the experiments
examining ribosomal RNA, SV40, ovalbumin and <u>Alu</u> family
sequence genes, it seems likely that sequences transcribed by
all three mammalian RNA polymerases are in close association
with the loop anchorage sites on the nuclear matrix.
Consistent with this possibility is the recent finding (74)
that DNA sequences complementary to whole nuclear RNA are
enriched in DNA associated with the nuclear cage (a structure
similar to the nuclear matrix). Further experiments will be
required to elucidate which elements are responsible for the
binding of transcribed genes to the nuclear matrix, and to
define the period during differentiation when this binding
occurs.

REFERENCES

1. Felsenfeld, G. (1978). Nature 271, 115.
2. Kornberg, R. D. (1977). Ann. Rev. Biochem. 46, 931.
3. Finch, J. T., and Klug, A. (1976). Proc. Natl. Acad.
    Sci. USA 73, 3966.
4. Worcel A., and Benjayati, C. (1977). Cell 12, 83.
5. Renz, M., Nehls, P., and Hozier, J. (1977). Proc. Natl.
    Acad. Sci. USA 74, 1879.
6. Stratling, W. H., Muller, V., and Zentgraf, H. (1978).
    Exp. Cell Res. 117, 301.
7. Cook, P. R., and Brazell, I. A. (1975). J. Cell Sci. 19,
    261.
8. Ide, T., Nakane, M., Anzai K., and Andoh, T. (1975).
    Nature 258, 445.
9. Benyajati, C., and Worcel, A. (1976). Cell 9, 393.
10. Pinon, R., and Salts, Y. (1977). Proc. Natl. Acad. Sci.
    USA 74, 2850.
11. Hartwig, M. (1978). Acta Biol. Med. Germ. 37, 421.
12. Vogelstein, B., Pardoll, D. M., and Coffey, D. S.
    (1980). Cell 22, 79.
13. Weintraub, H., and Groudine, M. (1976). Science 193,
    848.
14. Garel, A., and Axel, R. (1976). Proc. Natl. Acad. Sci.
    USA 73, 3966.
15. Senear, A. W., and Palmiter, R. D. (1981). J. Biol.
    Chem. 256, 1191.
16. Levy-Wilson, B. and Dixon, G. H. (1979). Proc. Natl.
    Acad. Sci. USA 76, 1682.
17. Weisbrod, S., and Weintraub, H. (1979). Proc. Natl.
    Acad. Sci. USA 76, 630.

18. Wu, C., Bingham, P. M., Livak, K. J., Holmgren, R., and
     Elgin, S. C. R. (1979). Cell 16, 797.
19. Stalder, J., Larson, A., Engel, J. D., Dolan, M.,
     Groudine, M., and Weintraub, H. (1980). Cell 20, 451.
20. Berezney, R., and Coffey, D. S. (1974). Biochem.
     Biophys. Res. Commun. 60, 1410.
21. Peters, K. E., and Comings, D. E. (1980). J. Cell Biol.
     86, 135.
22. Hodge, L. D., Mancini, P., Davis, F. M., and Heywood, P.
     J. (1977). Cell Biol. 72, 192.
23. Long, B. W., Huang, C.-Y., and Pogo, A. O. (1979). Cell
     18, 1079.
24. Herlan, G., and Wunderlich, F. (1976). Cytobiologie 13,
     291.
25. Mitchelson, K. R., Bekers, A. G. M., and Wanka, F. J.
     (1979). Cell Sci. 39, 247.
26. Poznanovic, G., and Sevaljevic, L. (1980). Cell Biol.
     Int. Rep. 4, 701.
27. Paulson, J. R., and Laemmli, U. K. (1977). Cell 12, 817.
28. Razin, S. V., Mantieva, V. L., and Georgiev, G. P.
     (1979). Nuc. Acids Res. 7, 1713.
29. McCready, S. J., Godwin, J., Mason, D. W., Brazell, I.
     A., and Cook, P. R. (1980). J. Cell Sci. 46, 365.
30. Miller, T. E., Huang, C.-Y., and Pogo, A. O. (1978). J.
     Cell Biol. 76, 675.
31. Herman, R., Weymouth, C., and Penman, S. (1978). J. Cell
     Biol. 78, 663.
32. Van Eekelen, C. A. G., and Van Venrooij, W. J. (1981).
     J. Cell Biol. 88, 554.
33. Zieve, G, and Penman, S. (1976). Cell 8, 19.
34. Miller, T. E., Huang, C.-Y., and Pogo, A. O. (1978). J.
     Cell Biol. 76, 692.
35. Lerner, M., Boyle, J., Mount, S., Wolin, S., and
     Steitz, J. A. (1980). Nature 283, 220.
36. Rogers, J., and Wall, R. (1980). Proc. Natl. Acad. Sci.
     USA 77, 1877.
37. Barrack, E. R., Hawkins, E. F., Allen, S. L., Hicks, L.
     L., and Coffey, D. S. (1977). Biochem. Biophys. Res.
     Commun. 79, 829.
38. Agutter, P. S., and Burchall, K. (1979). Exp. Cell Res.
     124, 45.
39. Hardin, J. W., and Clark, J. H. (1979). J. Cell Biol.
     83, 252a.
40. Barrack, E. R., and Coffey, D. S. (1980). J. Biol. Chem.
     255, 7265.
41. Gorski, J., and Gannon, F. (1976). Ann. Rev. Physiol.
     38, 425.

42. O'Malley, B. W., Roop, D. R., Lai, E. C., Nordstrom, J. L., Catterall, J. F., Swaneck, G. F., Colbert, D. A., Tsai, M.-J., Dugaiczyk A., and Woo, S. L. C. (1979). Rec. Prog. Horm. Res. 35, 1.

43. Pardoll, D., and Vogelstein, B. (1980). Exp. Cell Res. 128, 466.

44. Nelkin, B. D., Pardoll, D., and Vogelstein, B. (1980). Nuc. Acids Res. 8, 5623.

45. Ketner, G., and Kelly, T. J. (1980). Mol. Biol. 144, 163.

46. Southern, E. M. (1975). J. Mol. Biol. 98, 503.

47. Rigby, P. W. J., Dieckman, M., Rhodes, C., and Berg, P. J. (1977). Mol. Biol. 113, 237.

48. Reed, S. I., Ferguson, J., Davis, R. W., and Stark, G. R. (1975). Proc. Natl. Acad. Sci. USA 72, 1605.

49. Buckler-White, A. H., Humphrey, G. W., and Pigiet, V. (1980). Cell 22, 3.

50. Deppert, W. (1978). J. Virol. 26, 165.

51. Palmiter, R. D. (1973). J. Biol. Chem. 248, 8260.

52. Ono, T., and Getz, M. (1980). J. Dev. Biol. 75, 481.

53. Robinson, S., Nelkin, B., and Vogelstein, B. (1982). Cell 28, 99.

54. O'Hare, H., Breathnach, R., Benoist, C., and Chambon, P. (1979). Nuc. Acids Res. 7, 321.

55. Royal, A., Garapin, A., Cami, B., Perrin, F., Mandel, J. L., LeMeur, M., Bregegegre, F., Gannon, F., LePennec, J. P., Chambon, P., and Kourilsky, P. (1979). Nature 279, 125.

56. Lawson, G. M., Tsai, M.-J., and O'Malley, B. W. (1980). Biochem. 19, 4403.

57. Britten, R. J., and Kohne, D. E. (1968). Science 161, 529.

58. Appels, R., and Peacock, W. J. (1978). Int. Rev. Cytol., suppl. 8, p. 70.

59. Britten, R. J., and Davidson, E. H. (1969). Science 165, 349.

60. Walker, P. M. B. (1971). Prog. Biophys. Mol. Biol. 23, 145.

61. Jelinek, W. R., Toomey, T. P., Leinwand, L., Duncan, C. H., Biro, P. A., Choudary, P. V., Weissman, S. M., Rubin, C. M., Houck, C. M., Deininger, P. L., and Schmid, C. W. (1980). Proc. Natl. Acad. Sci. USA 77, 1398.

62. Houck, C. M., Rinehart, F. P., and Schmid, C. W. (1979). J. Mol. Biol. 132, 289.

63. Jelinek, W., Evans, R., Wilson, M., Salditt-Georgieff, M., and Darnell, J. E., (1978). Biochem. 17, 2776.

64. Duncan, C., Biro, P. A., Choudary, P. V., Elder, J. T.,
    Wang, R. R. C., Forget, B. G., deRiel, J. K., and
    Weissman, S. M. (1979). Proc. Natl. Acad. Sci. USA 76,
    5095.
65. Fritsch, E. F., Lawn, R. M., and Maniatis, T. (1980).
    Cell 19, 95.
66. Bell, G. I., Pictet, R., and Rutter, W. J. (1980). Nuc.
    Acids Res. 8, 4091.
67. Pan, J., Elder, J. T., Duncan, C. N., and Weissman, S.
    M. (1981). Nuc. Acids Res. 9, 117.
68. Haynes, S. R., and Jelinek, W. R. (1981). Proc. Natl.
    Acad. Sci. USA 78, 6130.
69. Rubin, C. M., Houck, C. M., Deininger, P. L., Friedmann,
    T., and Schmid, C. W. (1980). Nature 284, 372.
70. Southern, E. M. (1975). J. Mol. Biol. 94, 51.
71. Horz, W., Hess, I., and Zachau, H. G. (1974). Eur. J.
    Biochem. 45, 501.
72. Horz, W., and Zachau, H. G. (1977). Eur. J. Biochem. 73,
    383.
73. Waring, M., and Britten, R. J. (1966). Science 154, 791.
74. Jackson, D. A., McCready, S. J., and Cook, P. R. (1981).
    Nature 292, 552.
75. Flamm, W. G., Walker, P. M. B., and McCallum, M. (1969).
    J. Mol. Biol. 40, 423.
76. Hemminki, J., and Vainio, H. (1979). Cancer Letters 6,
    167.
77. Matsuura, H., Ueyama, H., Moni, S., and Ueda, K. (1979).
    J. Nutr. Sci. Vitaminol. 25, 495.
78. Tew, K. D., Wong, A. L., and Schein, P. S. Submitted for
    publication, 1981.
79. Hartwig, M., Matthes, E., and Arnold, W. (1981). Cancer
    Letters 13, 153.
80. Hunt, B. F., and Vogelstein, B. (1981). Nuc. Acids Res.
    9, 349.
81. Varshavsky, A. (1981). Cell 25, 561.
82. Hand, R. (1977). Human Genetics 37, 5.
83. Funderud, S., Andreassen, R., and Haugli, F. (1979).
    Nuc. Acids Res. 6, 1417.
84. Blumenthal, A. B., Kriegstein, H. J., and Hogness, D. S.
    (1973). Cold Spring Harbor Symp. Quant. Biol. 38, 205.
85. Vogelstein, B., and Gillespie, D. H. (1977). Biochem.
    Biophys. Res. Commun. 75, 1127.

CHROMATIN CONFORMATION AND GENE ACTIVITY

Gary Felsenfeld, James McGhee, Donald C. Rau, William Wood,
Joanne Nickol, and Michael Behe

National Institutes of Health
National Institute of Arthritis, Diabetes,
and Digestive and Kidney Diseases
Laboratory of Molecular Biology
Section on Physical Chemistry
Bethesda, Maryland   20205

ABSTRACT The physical properties of chromatin have
been explored at the level of individual nucleosomes,
extended polynucleosome filaments, and folded polynucleosome fibers. To compare the physical properties of
bulk chromatin with the properties of chromatin in
transcriptionally active genes, we have examined the
stability of the 30 nm chromatin fiber in the neighborhood of the adult β-globin gene in chicken red cells.
We find that the DNA in the neighborhood of this gene
is indistinguishable from bulk chromatin in its degree
of compaction and in its reaction with the HMG proteins
HMG 14/17.

The sensitivity to each of 3 nucleases (DNase I, DNase
II, and micrococcal nuclease) is remarkably uniform
throughout the region containing the adult β globin gene,
the 5′ end of the neighboring embryonic ε gene, and the
2.7 Kbp of DNA separating them. A notable exception is
the hypersensitive region just 5′ of the adult β-globin
gene. We have mapped this region in detail, using a
variety of nucleases, and show that it extends over
about 200 base pairs of DNA between about −60 and − 260
nucleotides from the starting point of transcription.
The hypersensitive region is highly enriched in the sequence CpG, the usual site of methylation. We have
used the restriction enzyme Msp I to liberate a 115 bp
fragment from 14-day embryo red cell nuclei. About one
third of the fragments are protein-free. The yield of
fragment is greater than 50%; its size and accessibility
suggest that the region is nucleosome-free in the nucleus.

The correlation of levels of methylation with transcriptional activity in this hypersensitive region and

in other active regions of the genome raises questions about the effects of methylation on DNA and chromatin structure. In order to detect such possible effects, we have examined the properties of synthetic polydeoxynucleotides with the sequence $(m^5dC-dG)_n$. Such sequences undergo transition from the B to Z form under physiological conditions of ionic strength. We find that the binding of histones stabilizes the B form, and that Z DNA does not form a well-defined nucleosome structure. The possible significance of these results for structure <u>in vivo</u> will be discussed.

INTRODUCTION

Chromatin structure has been the subject of a large number of physicochemical studies.(1) What is the relationship between the average physical properties revealed by these studies and the properties of chromatin in the neighborhood of expressed genes? In this paper we discuss some recent investigations of chromatin structure in solution, and show how these can be extended to the examination of defined regions of specific genes in the genome.

HIGHER ORDER STRUCTURE

Almost all DNA of the eukaryotic nucleus is complexed to histones to form nucleosomes. Under the ionic conditions prevailing in the nucleus, strings of nucleosomes are folded to form a chromatin fiber approximately 30 nm in diameter. Most studies of the 30 nm chromatin fiber support a model in

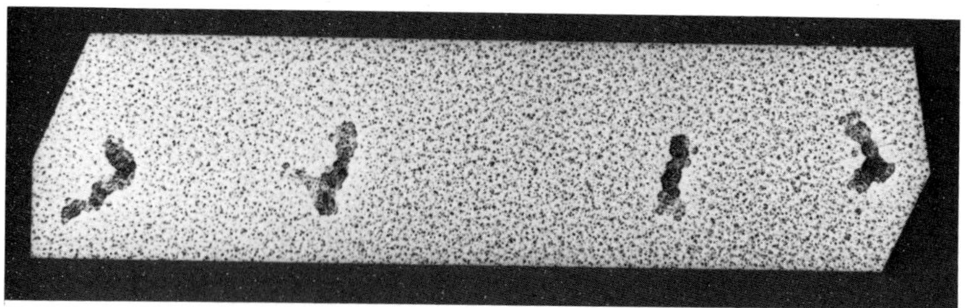

FIGURE 1. Electron micrograph of typical condensed chromatin particles obtained from nuclei by mild micrococcal nuclease digestion and sucrose gradient fractionation. Spread unfixed from 0.05 mM $MgCl_2$. The average width is ~30 nm. (From Ref. 1.)

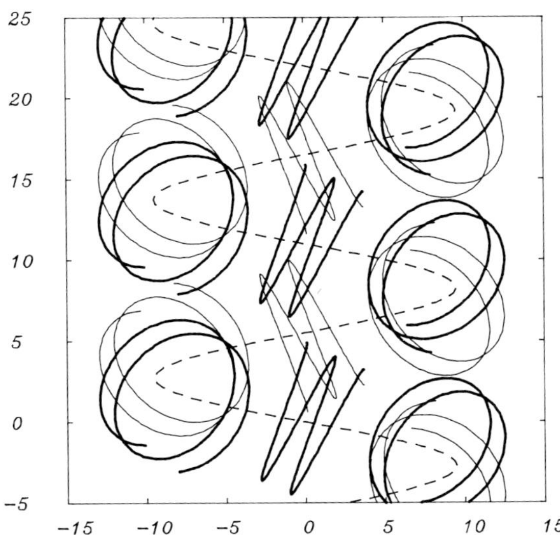

FIGURE 2. Schematic diagram of the proposed path (4) of
the DNA in polynucleosome solenoids of the kind shown in Fig.
1. This model is deduced from electric dichroism measurements (see text). Solid lines: DNA on chromatosomes. Dotted line: solenoidal path, drawn through center of chromatosomes. Linker DNA between chromatosome is omitted.

which the nucleosomes are arranged in a superhelical, solenoidal array, with about six nucleosomes per turn of super-
helix.(2-4)
     Such chromatin fibers can be released from nuclei by
mild digestion with micrococcal nuclease, and fractionated
on sucrose gradients. A typical solenoidal fragment obtained
in this way from chicken red cell nuclei is shown in Figure 1. We have recently carried out (4) electric dichroism
measurements to determine the orientation of the nucleosomes
packed within such fragments. The arrangement (Fig. 2) that
best fits our data (combined with data (2) from neutron
scattering studies) requires that the faces of the nucleosome
cores be oriented radially in the solenoid. The faces are
tilted 20-25° relative to the long solenoid axis. The estimated amount of tilt is dependent upon certain assumptions
about the path of the spacer DNA between cores, but the
angles within which the tilt can vary are severely limited by
geometric constraints. Further studies of chromatin isolated
from cells with different nucleosome repeat lengths show
that with the same assumptions the tilt of the nucleosome
cores is in all cases close (±10°) to the value observed in
chicken red cell nuclei (McGhee et al., manuscript in
preparation).

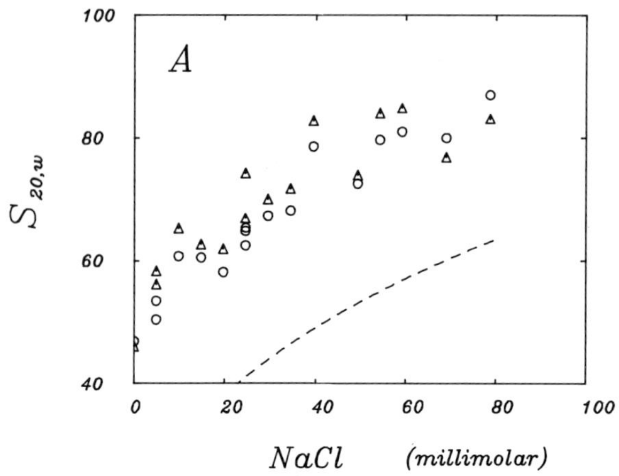

FIGURE 3. $s_{20,w}$ of chromatin as a function of NaCl concentration. Unfractionated chromatin (weight average size
26 nucleosomes) from 14-day chick red cells. All solutions contained 5 mM Tris HCl, pH 8.0, 0.1 mM EDTA, 1 mM
sodium butyrate. O = chromatin; Δ = chromatin with 2.0 ± 0.1
molecules of HMG 14/17 added per nucleosome. ----expected
sedimentation behavior if H1 and H5 were removed. (From
Ref. 6).

When the ionic strength of the solvent is lowered, the
30 nm fiber unfolds to form the familiar beaded-string 10-nm
filament, composed of alternating cores and spacers. The unfolding process is reversible upon addition of monovalent or
divalent cations; the conversion between 10-nm filament and
30 nm fiber can be monitored by measuring changes in physical
properties such as dichroism or sedimentation rate. For example, since the 30-nm fiber is more compact than the 10-nm
filament the folding process is accompanied by an increase
in the sedimentation coefficient   (Fig. 3).
These changes in physical properties can be used to
study the stability of higher order chromatin structure and
particularly to study if various biologically interesting
molecules that interact with chromatin affect the stability.
An example is provided by our recent investigation of the
interaction between chromatin and the high mobility group
proteins HMG 14 and HMG 17. These proteins are associated
<u>in vivo</u> with transcriptionally active chromatin (5), and
it has been suggested that they might assist in maintaining
chromatin in an "unfolded" conformation.

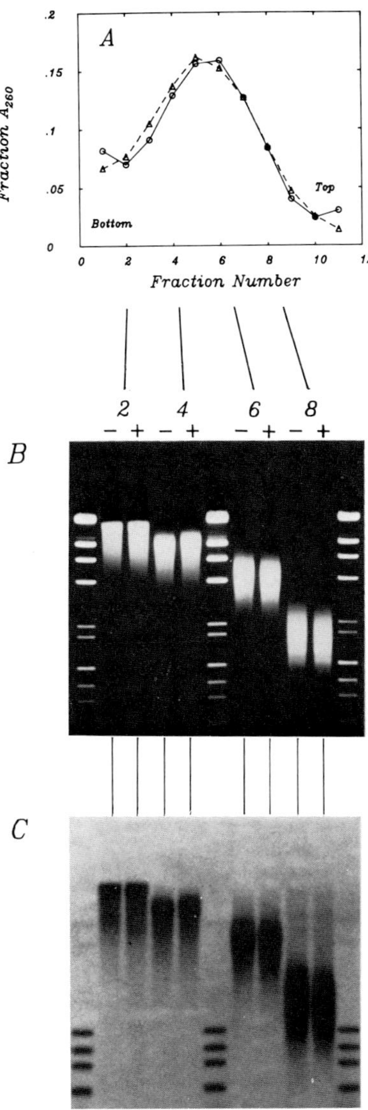

FIGURE 4.   Sucrose gradient sedimentation of 14-day chick red cell polynucleosome fraction. Top: Optical density profile. Middle: Ethidium stained gel of DNA isolated from fractions across the gradient. Bottom: Blot of same gel probed with 6.2 kb adult β-globin fragment.

To test this hypothesis, we carried out (6) a series of
experiments in which HMG 14/17 (an approximately equimolar
mixture of the two proteins) were added to an unfolded polynucleosome preparation, and the salt dependence of the re-
folding was measured. (The molar ratio of HMG proteins to
nucleosomes in this experiment was 2:1, corresponding to the
number of tight binding sites for HMG proteins on nucleosome
core particles (7)). The results of the experiment are shown
in Figure 3; when HMG 14/17 are bound, the sedimentation coefficient of polynucleosomes at a given salt concentration is
slightly higher than in their absence (presumably because of
increased mass) but the HMG proteins have no effect on the
formation of the 30-nm chromatin fiber. Similar studies of
the effect of divalent ions on the electric dichroism of
chromatin – HMG protein complexes lead to the same conclusion.

These physical studies reflect the average properties of
bulk chromatin, but it is possible to use simple extensions
of some of these techniques to measure the properties of defined, unique sequences within the genome. The data in Figure 4 summarize an experiment with chromatin and HMG proteins like that in Figure 3, but in addition the distribution of sequences coding for the chicken adult β globin gene
has been measured across the gradient. The results show
that the globin gene in chromatin isolated from 14-day-old
chick red cell nuclei reacts with HMG proteins in a manner
indistinguishable from bulk chromatin. Furthermore, the data
suggest (and more extensive studies in our laboratory confirm) that the region containing the β-globin gene is packaged in chromatin fibers with the same structure and stability as those described above. (McGhee and Felsenfeld, manuscript in preparation).

It thus seems unlikely that there are any large-scale
structural features at this level of organization that distinguish the region containing this gene from the surrounding bulk chromatin.

NUCLEASE SENSITIVITY AND STRUCTURE

Despite the above structural similarities between active
and inactive genes, there are differences that can be detected by enzymic probes. It is well known, for example, that
genes with a history of transcriptional activity are particularly sensitive to attack by pancreatic DNase (DNase I). (8)
It has also been reported that active genes are sensitive to
micrococcal nuclease (9), and that they are preferentially
excised from the nucleus (10,11) as large fragments by spleen

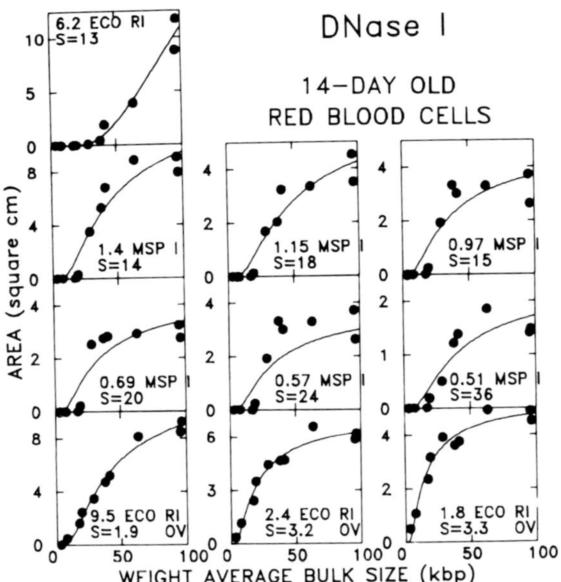

FIGURE 5.   DNase I sensitivity analysis of individual restriction fragments from the chicken genome.   Each graph shows the rate of disappearance of a given fragment as a function of bulk DNA size.   The data have been fit to an equation which describes the rate of disappearance in terms of a sensitivity parameter S.   (From Ref. 14)

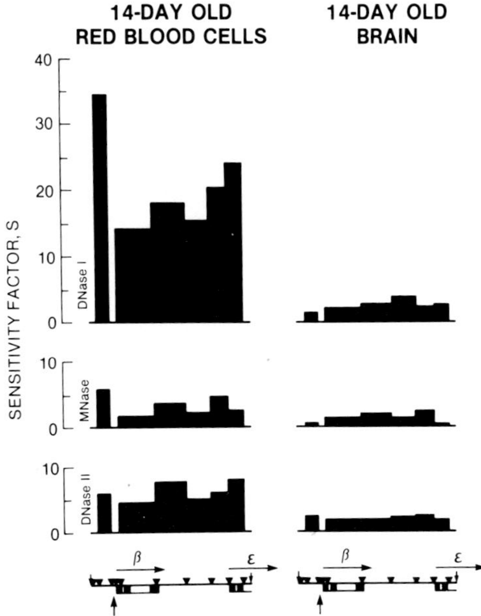

FIGURE 6.   Sensitivity to various nucleases in the chicken adult β globin gene.   (From Ref. 14)

acid DNase (DNase II). A convenient way of surveying the
domain of nuclease sensitivity in the neighborhood of a gene
is to digest nuclei and measure the rate of disappearance of
specific restriction fragments in that neighborhood (12,13).

We have made use of this method (14) to map in detail
the pattern of sensitivity near the chicken adult β globin
gene. Three enzymes (DNase I, DNase II and micrococcal nuclease) were used; the rates of disappearance of individual
bands were plotted as a function of the average molecular
weight of the bulk sample. The data were fit to a model
which assumes that cuts are random, but that chromatin containing a given restriction fragment has a nuclease sensitivity characterized by a length-independent factor S (S = 1
implies normal bulk sensitivity; S > 1 implies enhanced sensitivity per unit length). Typical data are shown in Fig. 5,
and the profile of sensitivity across the 6.2 kb region containing the adult β-globin gene is shown in Fig. 6 for red
cell and brain nuclei. For each of the three enzymes, S is
significantly higher in red cell nuclei than in brain, although the values of S are different for each enzyme. Within
an estimated two-fold error, S is uniform across the entire
globin gene region for any given enzyme. The results are
consistent with the presence of a uniformly perturbed structure within the adult β globin gene domain in chick red cells.

The same general approach can be used to determine the
relative accessibility of the adult β-globin gene to chemical probes. Figure 7 shows the results of an experiment in
which 14-day chick red cell nuclei were treated with increasing amounts of the alkylating agent dimethyl sulfate,
the DNA isolated, digested with MspI, cleaved at the methylated guanine residues, electrophoresed and blotted. The
MspI digest fragments disappear at approximately the same
rate as does genomic DNA of the same size, which shows that
the globin gene structure is not distinguished from bulk
chromatin by this probe.

A second class of sites with even greater sensitivity
to nucleases (one order of magnitude greater than that discussed above) was first described by Wu et al, (15). They
showed that very mild treatment of nuclei with DNase I or
micrococcal nuclease resulted in cleavage of the DNA at a
few rather widely separated specific sites. Subsequent studies have shown that these sites are often located near the
5′ ends of genes, and that the hypersensitivity is often
correlated with gene activity (16,17). We have found such a
"site" near the 5′ end of the chicken adult β-globin gene
(18). It is hypersensitive to a wide variety of nucleases,
including restriction endonucleases of appropriate specificity. The use of a variety of nucleases permits detailed

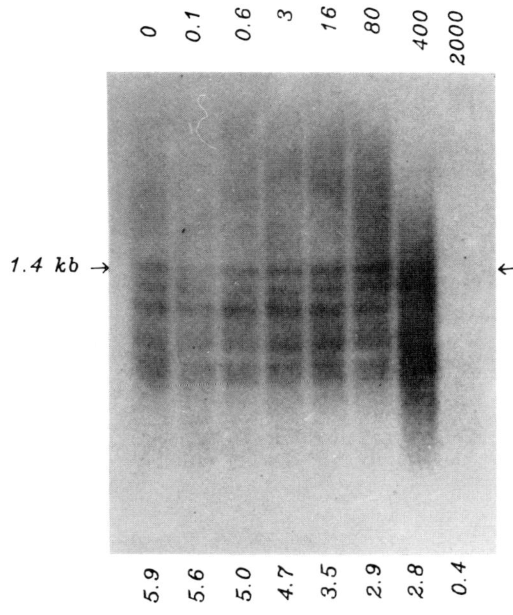

FIGURE 7. The sensitivity of the chicken adult β-globin gene to alkylation by dimethyl sulfate. Nuclei from 14-day chicken red cells were isolated as described in (14) and incubated for 20 hours at 0-4°C with the indicated concentrations of dimethylsulfate. The reaction was quenched with 0.5 M β-mercaphethanol and the reacted DNA was deproteinized, digested with the restriction endonuclease MspI, reisolated and finally cleaved at alkylated guanine residues by incubation for 15 min. at 90°C with 1 M piperidine. The cleaved DNA fragments were lyophilized, electrophoresed on an alkaline 1.2% agarose gel, transferred to DBM paper and globin specific sequences detected by hybridization (probe A of ref.18). The 1.4 kb MspI fragment containing the adult β-globin gene coding sequence is indicated by arrows. The average single standard sizes of the bulk DNA in the absence of Msp I digestion are indicated at the bottom of the figure and were determined from densitometer scans of ethidium bromide stained alkaline agarose gels.

mapping of the location of cuts, and shows that the "site" is in fact a region roughly 200 base pairs long, starting at a point ~70 base pairs 5´ of the mRNA cap site and extending ~200 base pairs in the 5´ direction (Fig. 8).

Treatment of 14 day chick red cell nuclei with the restriction endonuclease <u>MspI</u> results in cleavage at several sites in the above hypersensitive region, including two sites located 115 base pairs apart (Fig. 8). The fragment created in this way is released in greater than 50% yield from treated nuclei. About two thirds of all such released fragments probably have protein bound to them, but the complex does not behave like a nucleosome on gels or sucrose gradients (data not shown). In any case the fact that <u>MspI</u> can attack two sites 115 base pairs apart and that other nucleases cleave at different sites in the neighborhood strongly suggests that nucleosomes are not present in this region.

Another important feature of this hypersensitive region is the high frequency of occurrence of the sequence CpG, which is present about six times more frequently than elsewhere in the genome. Such sequences are potential sites of methylation (at the 5 position of cytosine), and in fact an earlier study showed that sites in this region, not known at that time to be hypersensitive, vary in their level of methylation inversely with adult β-globin gene expression (19, 20).

METHYLATION AND DNA STRUCTURE

The correlation noted above between DNA methylation and gene activity prompted us to ask how methylation affects the structure of DNA and chromatin. In earlier work, we studied the synthetic polynucleotide poly $(m^5dC-dG) \cdot poly(m^5dC-dG)$, and compared its properties to those of the corresponding unmethylated polymer. We found that the methylated polymer undergoes the transition to the left-handed Z form at low salt concentrations (21). In contrast, the unmethylated alternating C-G polymer is stable in the Z form only at high salt concentrations (22).

We have recently asked whether left-handed polymers of this kind can be accommodated in nucleosome-like structures (23). When in the B form poly $(m^5dC-dG) \cdot poly(m^5dC-dG)$ can be reacted with histones to form a complex that, when digested with micrococcal nuclease, yields a particle indis

FIGURE 8. Sequence of the hypersensitive region near the 5´ end of the chicken adult β-globin gene, with location of cutting sites for various enzymes. (From Ref. 18).

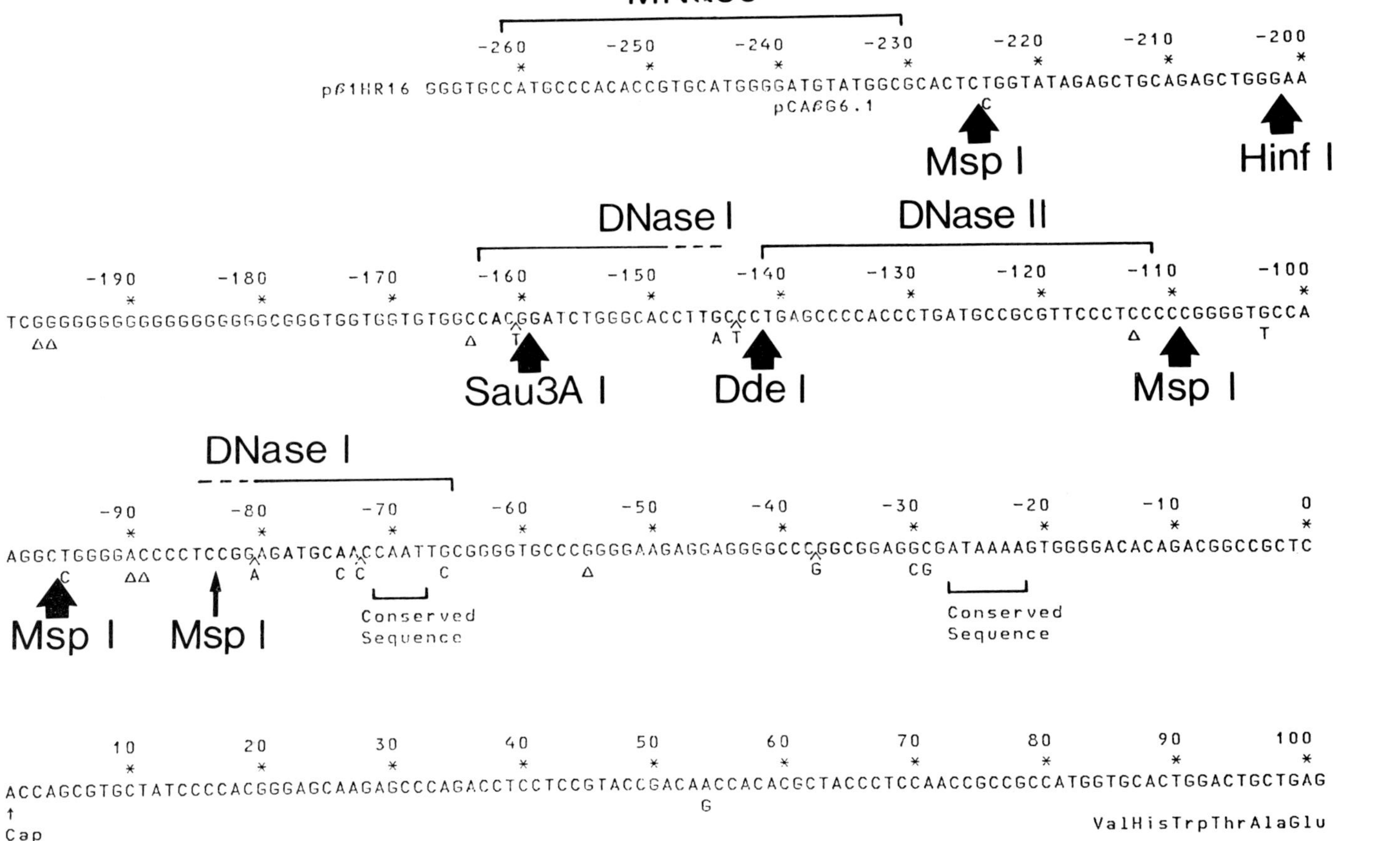

MNase
-260   -250   -240   -230   -220   -210   -200
pß1HR16 GGGTGCCATGCCCACACCGTGCATGGGGATGTATGGCGCACTCTGGTATAGAGCTGCAGAGCTGGGAA
pCAßG6.1
Msp I
Hinf I
DNase I   DNase II
-190   -180   -170   -160   -150   -140   -130   -120   -110   -100
TCGGGGGGGGGGGGGGGGGGGCGGGTGGTGGTGTGGCCACGGATCTGGGCACCTTGCCCTGAGCCCCACCCTGATGCCGCGTTCCCTCCCCCGGGGTGCCA
ΔΔ
Δ
T
A T
Δ
T
Sau3A I   Dde I   Msp I
DNase I
-90   -80   -70   -60   -50   -40   -30   -20   -10   0
AGGCTGGGGACCCCTCCGGAGATGCAACCAATTGCGGGGTGCCCGGGGAAGAGGAGGGGCCCGGCGGAGGCGATAAAAGTGGGGACACAGACGGCCGCTC
C
ΔΔ
A
C C
C
Δ
G
CG
Msp I   Msp I
Conserved
Sequence
Conserved
Sequence
10   20   30   40   50   60   70   80   90   100
ACCAGCGTGCTATCCCCACGGGAGCAAGAGCCCAGACCTCCTCCGTACCGACAACCACACGCTACCCTCCAACCGCCGCCATGGTGCACTGGACTGCTGAG
G
Cap
ValHisTrpThrAlaGlu

tinguishable in physical properties from a normal nucleosome
core particle. On the other hand, if the histones are complexed while the DNA is maintained in the Z form, we are
unable to detect discrete protected lengths of DNA as inter-
mediates in the digestion. Furthermore, when the B form of
the methylated polymer is complexed to histones to form a
core particle, it is stabilized against conversion to the Z
form. Although we cannot rule out the possibility that dis -
crete particles containing Z DNA exist under some conditions,
the results suggest that normal nucleosomes will not form on
regions of DNA that are already in the Z form.

Regions of DNA that react with antibodies to the Z form
have been identified in <u>Drosophila</u> polytene chromosomes (24);
our observations may have some relevance to chromatin structure in these regions. On the other hand,the Z form is not
likely to play an important role in the hypersensitive region
of the globin gene described earlier in this paper. Z DNA is
unusually resistant to nuclease cleavage (25), in contrast to
the behavior of the hypersensitive region. Furthermore, that
region has no extended alternating purine-pyrimidine sequences, which are a minimum requirement, so far as we presently
know, for the formation of left-handed structures. Alteration in levels of methylation near the β-globin gene is
likely to have subtler effects on DNA structure or capacity
to bind specific proteins.

DISCUSSION

These results extend earlier studies in our own and
other laboratories that suggest that chromatin in the
neighborhood of transcriptionally active genes shares many
many structural features with the bulk of chromatin in the
nucleus. The DNA in the region containing the adult β-glo-
bin gene appears to be packaged in nucleosomes. These
nucleosomes are, in turn, packed to form particles with
structures that are, by our methods, indistinguishable from
the usual 30-nm fiber in hydrodynamic properties, stability,
and reactivity with the high mobility group proteins HMG 14
and 17.

Despite these structural similarities, the active genes
can be distinguished by their sensitivity to nucleases. In
the case of the chicken adult β-globin gene the first kind
of nuclease sensitivity appears to be distributed almost
uniformly across the entire region containing the gene, for
each of three nucleases we have used. This suggests that the
structural modification giving rise to the enhanced sensitivity is also uniform in the region.

The nature of this modification is not known. It is certainly not related to preferential unfolding of the 30 nm fiber, since the experiments described earlier show no such difference in the neighborhood of the adult β globin gene. Weintraub and his collaborators (26) have shown that sensitivity to DNase I requires the binding of HMG 14/17, and it has also been demonstrated that these HMG proteins bind selectively to nucleosome core particles containing active gene sequences (7). Nonetheless, we have shown that HMG 14/17 do not selectively perturb the higher order structure of an active gene. So far, no quantitative differences have been detected in the composition of these particles that could explain the selectivity.

The second kind of sensitivity we have discussed is distinctly non-uniform in distribution along the genome. In the case of the adult β-globin gene this sensitivity is localized in a small domain at the 5′ end about ten-fold more sensitive to nucleases than the surrounding nuclease-sensitive region described above.

Our data show that this domain, about 200 base pairs long, is almost certainly free of nucleosomes, and thus distinctly different in structure from the surrounding chromatin. A part of this domain can be liberated from nuclei in high yield by digestion with MspI. About two thirds of those fragments released from 14 day chick red cell nuclei are apparently complexed to protein; the remainder are protein-free DNA. It remains to be determined whether protein is also bound to this DNA region within the nucleus. (There is also a hypersensitive domain near the 3′ end of the gene, not discussed here (18). Unlike the domain at the 5′ end, it does not vary in sensitivity with the level of adult β-globin gene expression).

The DNA of the 5′ hypersensitive region has an unusually high number of methylatable sites. The level of methylation at some of these sites is inversely correlated with expression of the gene (19). It is certainly reasonable to speculate that there could be some relationship between DNase I hypersensitivity and absence of methylation. However, given the nucleotide sequence of the region, it is unlikely in this case that methylation can have the effect of stabilizing the DNA Z form. If methylation does affect chromatin structure at the 5′ side of the globin gene, it probably does so by producing much smaller changes in DNA conformation, or by altering the strength of binding to specific proteins. This does not exclude the possibility that other regions of the genome, with appropriate nucleotide sequence, could undergo major changes in DNA structure when they are methylated. In any case, future studies of the details of chromatin structure must take into account not only the protein complement bound to the DNA, but also the levels and pattern of DNA methylation.

REFERENCES

1.   McGhee, J. D., and Felsenfeld, G. (1980). Ann. Rev. Biochem. 49, 1115.
2.   Suau, P., Bradbury, E. M., and Baldwin, J. P. (1979). Eur. J. Biochem. 97, 593.
3.   Thoma, F., Koller, T., and Klug, A. (1979). J. Cell. Biol. 83, 403.
4.   McGhee, J. D., Rau, D. C., Charney, E., and Felsenfeld, G. (1980). Cell 22, 87.
5   Levy, B., Wong, N. C. W., and Dixon, G. H. (1977). Proc. Natl. Acad. Sci. USA 74, 2810.
6.   McGhee, J. D., Rau, D. C., and Felsenfeld, G. Nucleic Acids Res., in press.
7.   Sandeen, G., Wood, W. I., and Felsenfeld, G. (1980). Nucleic Acids Res. 17, 3757.
8.   Weintraub, H., and Groudine, M. (1976). Science 193, 848.
9.   Bloom, K. S., and Anderson, J. N. (1978) . Cell 15, 141.
10.   Gottesfeld, J. M. and Partington, G. A. (1977). Cell 12, 953.
11.   Goldsmith, M. (1981). Nucleic Acids Res. 9, 6471.
12.   Zasloff, M., and Camerini-Oetero, R. D. (1980). Proc. Natl. Acad. Sci. U.S.A. 77, 1907.
13.   Stalder, J., Groudine, M., Dodgson, J. B., Engel, J. D., and Weintraub, H. (1980). Cell 19, 973.
14.   Wood, W. I., and Felsenfeld, G. J. Biol. Chem., in press.
15.   Wu, C., Bingham, P. M., Livak, K. J., Holmgren, R. and Elgin, S. C. R. (1979). Cell, 16, 797.
16.   Wu, C. (1980). Nature 286, 854.
17.   Stalder, J., Larsen, A., Engel, J. D., Dolan, M., Groudine, M., and Weintraub, J. (1980). Cell 20, 451.
18.   McGhee, J. D., Wood., W. I., Dolan, M., Engel, J. D., and Felsenfeld, G. (1981). Cell 27, 45.
19.   McGhee, J. D., and Ginder, G. D. (1979). Nature 280, 419.
20.   Ginder, G. D., and Kelley, K. K. (1982). Fed. Proc. 41, 1023.
21.   Behe, M., and Felsenfeld, G. (1981). Proc. Natl. Acad. Sci. U.S.A. 78, 1619.
22.   Pohl, F. M., and Jovin, T. M. (1972). J. Mol. Biol. 67, 375.

23. Nickol, J., Behe, M., and Felsenfeld, G. Proc. Natl.
     Acad. Sci. U.S.A., in press.
24. Nordheim, A., Pardue, M. L., Lafer, E. M., Moller, A.,
     Stollar, B. D., and Rich, A. (1981). Nature 294, 417.
25. Behe, M., Zimmerman, S., and Felsenfeld, G. (1981).
     Nature 293, 233.
26. Weisbrod, S., and Weintraub, H. (1980). Cell 19, 289.

THE RELATIONSHIP OF CHROMATIN STRUCTURE TO DNA SEQUENCE ORGANIZATION AT A HEAT SHOCK LOCUS OF <u>DROSOPHILA</u> <u>MELANOGASTER</u>[1]

Iain L. Cartwright and Sarah C.R. Elgin

Department of Biology, Washington University,
St. Louis, Missouri 63130

ABSTRACT   The patterns of digestion produced by micrococcal nuclease and the cleavage agent 1,10-phenanthroline-cuprous complex, in protein-free DNA from the region 67B of <u>Drosophila</u> <u>melanogaster</u> have been investigated. These patterns proved to be very similar and clearly related to the functional arrangement of the DNA. Relatively few preferential cuts were observed in the DNA across known genes, while many such cuts were seen in the spacers between genes. While both cleavage reagents show some sequence preferences, this is not sufficient to explain the patterns. It appears likely that secondary structural features along the DNA double helix are recognized in common by these two dissimilar reagents.

INTRODUCTION

Recent investigations have shown that position-specific structural features of chromatin exist, and have been detected as differential packaging of DNA by nuclease digestion studies. In particular, DNase I hypersensitive sites have been detected as small regions of DNA (ca. 50-200 bp) at specific positions along the chromatin fiber which display enhanced sensitivity to digestion by DNase I. Such sites have consistently been found at, or within a short distance of, the 5'-end of regions of the genome encoding transcripts, often in a tissue-specific or developmental stage-specific manner correlated with the pattern of gene expression. For a review of this data, see reference 1. In addition, there are several reports that the nucleosomal organization of chromatin (a general feature of eukaryotes) is not entirely random, and in some cases nucleosomes appear to be positioned quite specifically with respect to

[1]This work was supported by NIH grants GM-20779 and GM-30273 to SCRE.

the underlying DNA sequence. This has been observed both in
heterochromatin and in transcriptionally inactive
euchromatin (2,3; reviewed in 4). Work in our laboratory
using mild micrococcal nuclease digestion at the heat-shock
locus 67B of _Drosophila melanogaster_ has revealed an
unexpected feature at the level of DNA sequence organization
(5). The enzyme appears to cut _protein-free_ DNA with
preferences that leave the known transcription units (6,7)
relatively unscathed, while the intervening spacer regions
are cut with high frequency. Recently it has been shown
that micrococcal nuclease has marked sequence preferences in
its mode of cutting on DNA (8,9). Obviously it would be
very useful to have a sequence-neutral DNA cutting agent
that would allow us to distinguish between structure-
specific as opposed to sequence-specific features in the
organization of the genome. We describe here the use of the
1,10-phenanthroline-cuprous complex as a probe for such
features at locus 67B in _Drosophila_ DNA.

RESULTS AND DISCUSSION

Degradation of nucleic acids in the presence of trace
metal ions is a phenomenon that has been appreciated for
many years (10,11); it can be inhibited effectively by the
simple expedient of maintaining nucleic acids in solution
with low concentrations of EDTA. The deleterious effects of
hydrogen peroxide on DNA were also noticed (12) and could be
enhanced by the presence of certain metal ions (13). The
chemical basis of these degradative actions appears to
derive from production (and subsequent reaction with DNA) of
the hydroxyl radical, either from dissolved oxygen or from
added hydrogen peroxide, mediated by a redox cycle involving
a suitable metal ion (14). This series of reactions is
shown in Figure 1. Also shown in Figure 1 is the structure
of 1,10-phenanthroline, a planar tricyclic heteroaromatic
nucleus, that most probably interacts with DNA by an inter-
calative mechanism (15). Recently it was shown by two
groups that a complex of cuprous ion with 1,10-
phenanthroline at low concentration could degrade double-
stranded DNA rapidly in a reaction that was probably
mediated by hydroxyl radicals (16,17). 1,10-phenanthroline
is also a chelator of copper (and other metal ions) as well
as being an intercalator. It seems more than likely that
the chemical reactions outlined in Figure 1 occur at the
double helix itself and that this can account for the high
rate of DNA degradation at low concentrations of reagents.
Two important points were made by Marshall et al. (17).
First, the reaction is specific for double-stranded DNA

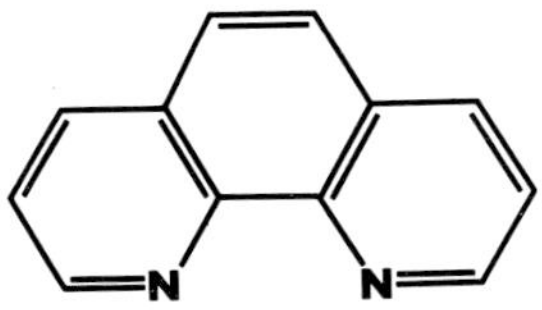

$$Cu^{++} + Phen \longrightarrow (Phen)_2 Cu^{++}$$

$$(Phen)_2 Cu^{++} + NADH \longrightarrow (Phen)_2 Cu^{+} + NAD^{+}$$

$$(Phen)_2 Cu^{+} + H_2O_2 \longrightarrow (Phen)_2 Cu^{++} + HO^{\bullet} + OH^{-}$$

$$OH^{\bullet} + DNA \longrightarrow degradation$$

FIGURE 1. Structure of 1,10-phenanthroline and probable chemistry of DNA cleavage reaction.

under the conditions used (although the strand cutting takes place by single-strand nicking), and second, digestion of a short piece of lac DNA indicated that there was no apparent base specificity in the cutting reaction.

We tested the cutting specificity of the 1,10-phenanthroline-cuprous complex on the plasmid 88B13, which contains a 12 kb Bam HI fragment from the locus 67B of Drosophila inserted into the Bam HI site of pBR322. Included in this fragment are three of the four small heat shock genes at this locus (coding for hsp 28, hsp 26 and hsp 23) as well as a developmentally regulated transcript (denoted R) that lies between hsp 26 and hsp 23. We used this plasmid so that there was no likelihood of contamination of the DNA with proteins of Drosophila. Figure 2 shows a comparison of digestion series of 88B13 by 1,10-phenanthroline-cuprous complex and by micrococcal nuclease. The digested plasmid was restricted to completion with Bam HI and a Southern blot of the electrophoresis products probed with a short fragment of DNA abutting the 5' Bam HI site. The similarity observed between the two patterns was particularly surprising in view of the reported random nature of DNA cutting by the 1,10-phenanthroline-cuprous complex (17). Clearly the intercalating complex could recognize some feature, presumably sequence-specified, of the DNA from this region of the Drosophila genome. In addition, the principal features of the preferential cutting pattern displayed the same unusual non-random nature with respect to functional units (i.e., transcribed regions) as had been previously shown for micrococcal nuclease cleavage (5). The patterns were reproducible whether or not the

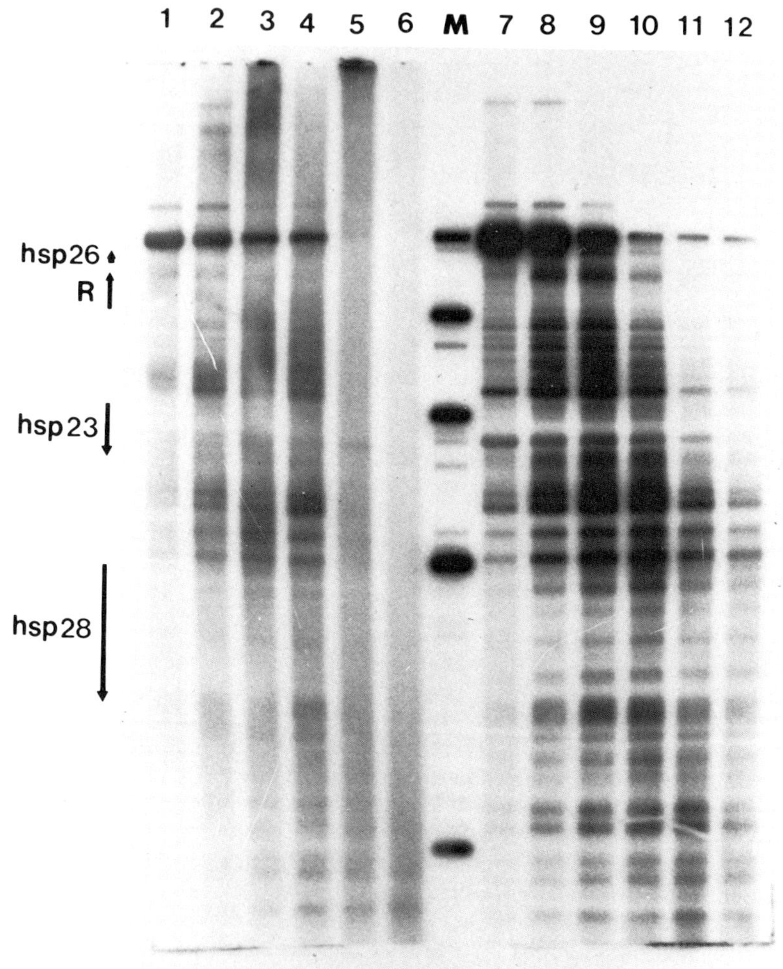

FIGURE 2. Digestion of 88B13 by 1,10-phenanthroline-cuprous complex (lanes 1-6) or micrococcal nuclease (lanes 7-12) for increasing periods of time. Samples were digested completely with <u>Bam</u> HI, run on a 0.9% agarose gel and probed with a short fragment abutting the 5'-Bam site of 88B13.

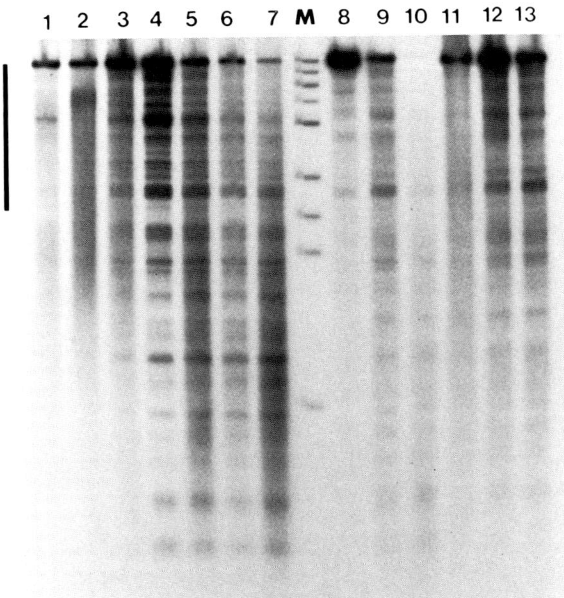

FIGURE 3. Digestion of pBR 322 with micrococcal nuclease (1-7) and 1,10-phenanthroline-cuprous complex (8-13). Total Bam digests were probed with the 275 bp Bam-Sal fragment of pBR 322. Plasmid initially supercoiled (1-7, 11-13) or Bam-linearized (8-10). Black line shows transcribed region.

88B13 plasmid was supercoiled initially. Mapping from the 3'-end of the insert to obtain more detailed information for hsp 26 and R confirmed the non-random nature of the cleavage patterns for both reagents (data not shown).

It was of interest to ask whether or not this broad correlation of susceptible regions of DNA with the spacer regions between clustered genes would be observed for prokaryotic DNA. Accordingly, a similar digestion experiment was performed using the 277 bp Bam HI-Sal I fragment of pBR 322 to probe Southern blots of a series of pBR 322 digestions done with either micrococcal nuclease or 1,10-phenanthroline-cuprous complex. By digesting the samples with Bam HI prior to electrophoresis, it is possible to map the preferential cleavage sites in the 3'-direction relative to the single Bam HI site of pBR 322. Both reagents produce abundant preferential cleavages in pBR 322 and for the most part the patterns are very similar (Figure 3). As previously noted for micrococcal nuclease (5), there seems to be no correlation in this instance between the location of preferential cleavages and known transcripts; in general the cleavage sites appear to be much more randomly distributed than for the Drosophila locus 67B DNA.

In Figure 4, the pattern of cleavages produced by the two reagents on nuclei are compared with those from purified genomic DNA from Drosophila, again using a short probe from the 5'-side of the region for mapping purposes. It is evident that the patterns for chromatin are remarkably similar in many ways to those from naked DNA, although there are some detectable chromatin-specific features particularly around the 5'-ends of the regions of transcription. These enhanced cleavages may correspond to the 5' DNase I hypersensitive sites previously mapped in this region (18). Specific features are again detected by both reagents. Hence it seems likely that the two cleavage reagents recognize some common structural features, not only at the chromatin level but also at the DNA level. As further discussed below, it seems very unlikely that sequence preferences per se for the two reagents are identical.

It was of interest to consider whether or not the patterns observed were related to patterns of DNA base modification. The state of nucleotide methylation in the Drosophila genome is relatively unknown, although it appears that there is little or no m^5Cyt, in contrast to the vertebrates. Possibly adenine methylation of the plasmid 88B13 grown in a dam$^+$ E. coli host could affect the cleavage by micrococcal nuclease giving rise to a non-random pattern. However, the pattern of cleavage from dam$^+$ and dam$^-$ hosts was observed to be identical (G. Fleischmann and S.C.R. Elgin, unpublished obervations).

The most important question raised by these results is whether the pattern of digestion in protein-free DNA across this region of 67B by both reagents is of structural significance, or whether it simply represents a preferential cleavage in regions of relatively high A-T content. Even if the latter were the case, it is possible that the reagents could have a predictive value for location of transcription units, given that spacers are often A-T rich when compared to the genes which they encompass.

Previously it was shown that micrococcal nuclease degrades regions rich in A-T residues with some preference. This has been interpreted as a consequence of the fact that the enzyme cuts single strands much more quickly than double strands; the much lower thermal stability of A-T regions in double helical DNA would predispose them toward early degradation (19). However, when analyzed at the sequence level in unique DNA, micrococcal nuclease shows a rather more restrictive sequence preference than originally

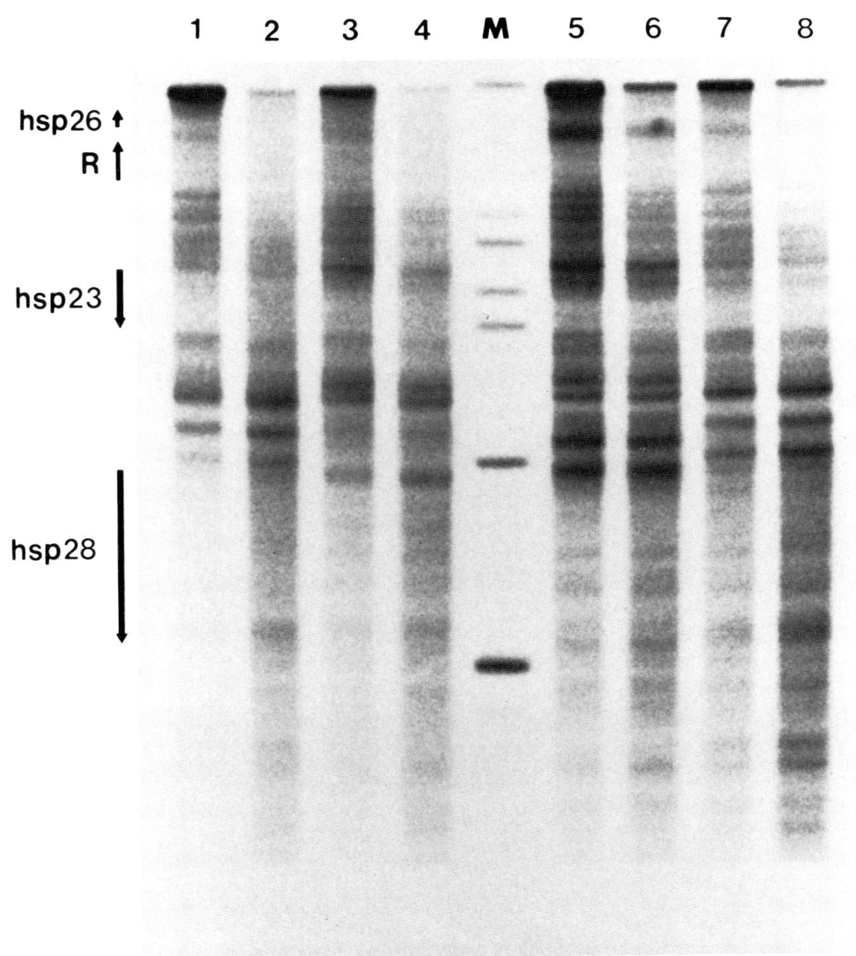

FIGURE 4. Partial digestions of 6-18 hour _Drosophila_ embryo nuclei (lanes 3-6) and genomic DNA (lanes 1,2,7,8) with 1,10-phenanthroline-cuprous complex (1-4) and micrococcal nuclease (5-8) at 25°. All samples were cut to completion with _Bam_ HI, run on a 0.9% agarose gel, and probed with a plasmid (88.3) that contains a short insert abutting the 5' -Bam site of the heat-shock region. Transcripts are denoted along the side.

envisioned. A-T richness is not necessarily of itself a
favored feature; preferential sites are often short
stretches (2 or 3 bases) of A and T, but surrounded by 5' C
and 3' G residues (8,9). However, the sum of the data
available [including our analysis of <u>Drosophila</u> 5S genes
(data not shown)] lead to the conclusion that there is as
yet no obvious way to predict whether a given sequence will
be cleaved or left untouched in the initial reaction with
micrococcal nuclease. This leads to the concept that other
features of a given DNA sequence, in particular the unique
secondary structure it will adopt (which is presumably a
function both of the immediate sequence and the overall base
composition) may well be involved in recognition events
between a specific region of DNA and a given macromolecule,
e.g., a nuclease or an intercalator.

     Considering the similarity of the cleavage patterns on
DNA, one might anticipate that the mode of cleavage and
sequence specificity of the 1,10-phenanthroline-cuprous
complex would be very similar to that of micrococcal
nuclease. This is, however, unlikely. The intercalating
complex does not cleave single-stranded DNA under these
conditions, and its mode of cleavage involves radical
abstraction, base release and subsequent β-elimination
leading to scission of the phosphodiester backbone (12).
This is very different to the presumed solvent displacement
reaction catalyzed by micrococcal nuclease (20). Little is
known about the sequence preference of phenanthroline in
intercalation, although in its crystal complexes with
dinucleoside monophosphates it resembles ethidium bromide
very closely (15). It has been observed that, in general,
intercalation seems to occur preferentially in DNA with a
higher G-C content (21,22). Ethidium bromide much prefers
to intercalate into a pyrimidine (3'→ 5') purine sequence
(CpG) rather than a purine (3'→ 5') pyrimidine sequence
(GpC) (23). Actinomycin D, which shows a marked preference
for binding to DNA of high G content, does not, however,
bind at every G residue. Depending on the neighboring
sequences that a given G-C base pair encounters, actinomycin
will either bind at this point or else completely ignore it
(24). It seems apparent that intercalators are very
sensitive to the conformational properties of a given base
sequence in DNA; they select their binding sites by
utilizing structural features of the double helix rather
than simply recognizing given sequences. When examined at
the sequence level directly, preferential cleavage sites for
micrococcal nuclease and 1,10-phenanthroline-cuprous complex
show some similarities, but also some major differences
(data not shown). No obvious concensus sequence for

1,10-phenanthroline-cuprous complex cleavage has been detected so far.

Recent results from many laboratories utilizing defined DNA sequences are showing that classical ideas about B-DNA conformations represented only an average of the specific conformations that unique DNA sequences adopt. The preference for pyrimidines to adopt different glycosyl torsion angles to purines leads naturally to a situation where sequence-dependent base-pair tilts allow more or less easy access for intercalators into a given groove and then between the bases (25). Again, the helical twist between adjacent base pairs appears to be very dependent upon the surrounding sequence (26). DNase I appears to recognize this by cleaving highly preferentially at the region with the highest helical twist in digestion of a dodecamer (27). Alternating pyrimidine-purine polymers have been found to exist in a conformation termed "alternating B", and this difference in structure along the sugar-phosphate backbone (28) can be clearly recognized by DNase I, a phenomenon that appears to be structurally based and independent of sequence preference considerations per se (27). Micrococcal nuclease has previously been seen to preferentially digest the sequence coding for the D-loop of tRNA in the gene, and it was conjectured that this was probably a structural recognition event (9). Our results lead us to believe that both micrococcal nuclease and the 1,10-phenanthroline-cuprous complex are capable of recognizing some common features of DNA in solution that are related to helix structure. The intriguing disposition of these structures relative to regions known to be used for transcription at some stage when present in a functioning cell prompt us to search further among other gene-spacer regions to ascertain if the correlation is more general, and whether in fact it may be an important informational structure for assembly and/or function of chromatin.

REFERENCES

1. Elgin, S.C.R. (1981). Cell 27, 413.
2. Bryan, P.N., Hofstetter, H., and Birnstiel, M.L. (1981). Cell 27, 459.
3. Pfeiffer, W. and Zachau, H.G. (1980). Nucleic Acids Res. 8, 4621.
4. Cartwright, I.L., Keene, M.A., Howard, G.C., Abmayr, S.M., Fleischmann, G., Lowenhaupt, K., and Elgin, S.C.R. (1982). CRC Crit. Rev. Biochem., in press.
5. Keene, M.A., and Elgin, S.C.R. (1981). Cell 27, 57.

6. Corces, V., Holmgren, R., Freund, R., Morimoto, R., and
    Meselson, M. (1980). Proc. Natl. Acad. Sci. USA 77,
    5390.
7. Sirotkin, K., and Davidson, N. (1982). Dev. Biol. 98,
    196.
8. Hörz, W., and Altenburger, W. (1981). Nucleic Acids
    Res. 9, 2643.
9. Dingwall, C., Lomonossoff, G.P., and Laskey, R.A.
    (1981). Nucleic Acids. Res. 9, 2659.
10. Huff, J.W., Sastry, K.S., Gordon, M.P., and Wacker,
    W.E.C. (1964). Biochemistry 3, 501.
11. Singer, B., and Fraenkel-Conrat, H. (1965).
    Biochemistry 4, 226.
12. Rhaese, H.-J., and Freese, E. (1968). Biochim.
    Biophys. Acta 155, 476.
13. Massie, H.R., Samis, H.V., and Baird, M.B. (1972).
    Biochim. Biophys. Acta 272, 539.
14. Lesko, S.A., Lorentzen, R.J., and Ts'o, P.O.P. (1980).
    Biochemistry 19, 3023.
15. Jain, S.C., Bhandary, K.K., and Sobell, H.M. (1979).
    J. Mol. Biol. 135, 813.
16. Que, B.G., Downey, K.M., and So, A.G. (1980).
    Biochemistry 19, 5987.
17. Marshall, L.E., Graham, D.R., Reich, K.A., and Sigman,
    D.S. (1981). Biochemistry 20, 244.
18. Keene, M.A., Corces, V., Lowenhaupt, K., and Elgin,
    S.C.R. (1980). Proc. Natl. Acad. Sci. USA 78, 143.
19. Tucker, P.W., Hazen, Jr., E.E., and Cotton, F.A.
    (1978). Mol. Cell Biochem. 22, 67.
20. Tucker, P.W., Hazen, Jr., E.E., and Cotton, F.A.
    (1979). Mol. Cell Biochem. 23, 67.
21. Müller, W., and Crothers, D. (1975). Eur. J. Biochem.
    54, 267.
22. Müller, W., Bünemann, H., and Dattagupta, N. (1975).
    Eur. J. Biochem. 54, 279.
23. Krugh, T.R., and Reinhardt, C.G. (1975). J. Mol. Biol.
    97, 133.
24. Wells, R.D. (1971). Prog. Mol. Subcell. Biol. 2, 21.
25. Dickerson, R.E., and Drew, H.R. (1981). Proc. Natl.
    Acad. Sci. USA 78, 7318.
26. Dickerson, R.E., and Drew, H.R. (1981). J. Mol. Biol.
    149, 761.
27. Lomonossoff, G.P., Butler, P.J.G., and Klug, A. (1981).
    J. Mol. Biol. 149, 745.
28. Klug, A., Viswamitra, M.A., Kennard, O., Shakked, Z.,
    and Steitz, T.A. (1979). J. Mol. Biol. 131, 669.

AN EXPOSED CHROMATIN STRUCTURE AT THE 5' END OF EUKARYOTIC GENES [1]

Carl Wu

Laboratory of Biochemistry, National Cancer Institute,
National Institutes of Health, Bethesda, MD 20205

ABSTRACT Several heat-inducible genes in _Drosophila_ exhibit
a DNase I-hypersensitive chromatin structure at the 5' terminus. This exposed structure is present before (and during)
induction, and is revealed by digesting gently chromatin
in isolated nuclei with DNase I and mapping the sites of
cleavage by a simple "indirect end-labelling" technique
that involves Southern blotting. The chromatin structure
of the rat preproinsulin II gene also shows a similar
DNase sensitivity at the 5' terminus in a rat insulinoma
(a beta-cell tumor) but not in rat kidney, liver, brain, or
spleen. The region of exposure for preproinsulin II gene
chromatin extends 250-300 bp upstream from the 5' end of
the mRNA. The several _Drosophila_ genes encoding the 70,000-
dalton heat-shock protein show DNase I hypersensitivity
extending to about 200 bp upstream from the 5' terminus.
Preferentially accessible sites in chromatin such as these
may function as entry sites for RNA polymerase and control
factors, and may be an essential component in the series of
regulatory events leading to the initiation of transcription.

INTRODUCTION

The orderly compaction of DNA in nucleosomes along the
chromatin fiber is punctuated by highly nuclease-sensitive
sites. Such hypersensitive sites in cellular chromatin can
be revealed by mild cleavage of chromatin in isolated nuclei
with a nucleolytic enzyme, DNase I, and Southern analysis of
the partially cleaved DNA using cloned hybridization probes.
In contrast to the smooth size distribution of cut fragments
seen in the cleavage pattern from the entire genome, we
unexpectedly found that the specific cleavage patterns from
each of five _Drosophila_ loci displayed a unique set of
preferred cleavages (1). This preference did not appear
on the naked DNA. Here, we summarize recent experiments

1
This work was supported by NIH Grant GM21514-06 to Walter
Gilbert, Harvard University.

147

which map hypersensitive sites to the 5'-terminal and
flanking regions of several Drosophila heat shock genes,
and to a similar position in rat preproinsulin II gene
chromatin from the expressing tissue. Such a surprising
correlation with a gene region reported by many others in
this volume to be important for in vitro and "in vivo" gene
expression suggests strongly a functional role for hypersensitive sites in chromatin.

RESULTS

We mapped the positions of DNase I-sensitive sites by
a novel indirect end-labelling technique that displays the
partial digestion products (2). After gentle treatment
of chromatin in nuclei with DNase I, the purified DNA is
cleaved completely with a rarely cutting restriction enzyme,
electrophoresed and blotted on to nitrocellulose in a typical
Southern transfer. The blot is then probed with a short
cloned ^{32}P-labelled DNA segment that abuts a chosen restriction cut; the lengths of the labelled subfragments greater
than the probe length define the distance between the restriction cut and the partial DNase I cuts. Nedospasov and
Georgiev (3) have independently used this same procedure to
map the location of micrococcal nuclease cuts on the SV40
minichromosome.

We applied this technique to the genes for two heat
shock proteins in D. melanogaster (Oregon R) embryos and in
cultured cells [Schneider line 2] derived from such embryos.
In Drosophila, an elevation in temperature from 25°C to near
37°C rapidly activates a small number of specific genes, and
sharply reduces pre-existing gene activity (see ref. 4 for
review).

The most abundant heat-induced product is a 70,000 molecular weight heat shock protein, hsp 70. There are five
copies of the uninterrupted 2.2-kilobase pair hsp 70 coding
sequence, two at the cytogenetic locus 87A and three at
87C. The five copies of the coding region and an adjoining
0.3-kilobase pair 5'-noncoding region are closely conserved;
however, the sequences near the 5' terminus diverge somewhat between the genes at the two loci, reflected in characteristic restriction site differences (5,6,7,9).

Figure 1 shows the partial DNase I cuts in noninduced
Schneider cell chromatin upstream from the internal Bam site
common to all five hsp 70 gene copies; the direction of electrophoresis is from left to right. The Bam digest alone,

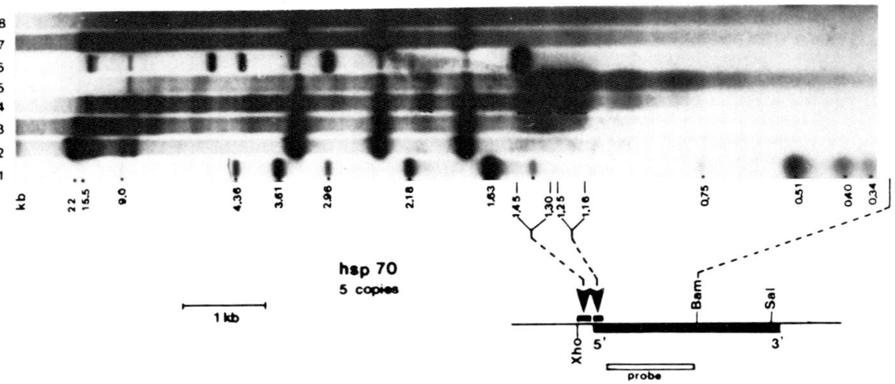

Figure 1.   Autoradiogram showing the partial DNase I
cuts in Schneider cell chromatin upstream from the internal
Bam site common to all hsp 70 gene copies.   The cloned
hybridization probe is indicated by the open bar and
represents approximately the 5' half of the hsp 70 coding
sequence (filled bar).   The lanes indicate: 1, marker DNA
fragments from pBR322.   On a gel of this agarose content
(1.3%), a linear relationship between log fragment length
and mobility is obtained in the region bounded by the 0.75-
and 3.61-kilobase pair markers; the resolution is about 30
base pairs for fragments not overexposed.   2, Bam digest
on Schneider cell DNA, purified directly from whole cells.
3-5, partial DNase I digests on chromatin at a final DNase
I concentration of 1, 2 and 4 units ml^{-1}, respectively,
followed by Bam cleavage.   Only the major regions of
preferential DNase I cleavage are indicated by the arrows
on the hsp 70 restriction map, aligned beneath the autoradiogram.   The filled bars directly under the arrows
indicate the measured span of hypersensitivity.   6, Bam
plus Xho digest on Schneider cell DNA.   The reaction did
not go to completion.   7,8, Partial DNase I digests on
naked DNA.   DNA at 0.4 mg ml^{-1} was digested with DNase I
at 0.1 and 0.3 units ml^{-1}, respectively, for 1 min at 25°C
before secondary Bam cleavage.   Reprinted with permission,
from ref. 2.

---

on purified DNA (Figure 1, lane 2) locates the different
upstream Bam sites corresponding to the five hsp 70 gene
copies.   (The fragments obtained in low yield are due either
to cross-reaction with partially homologous sequences or to

polymorphism in the organization of the sequences adjacent
to the hsp 70 coding region). Increasing extents of partial
DNase I cleavage on chromatin (Figure 1, lanes 3-5) reveal
major preferred cuts at two adjacent sites which span about
150 and 90 base pairs, flanking and extending through the 5'
end of the hsp 70 coding region. These two preferentially
cleaved sites are better resolved on a 50 cm long, 1.4%
agarose gel not shown. We have precisely mapped them to
within 10 bp by coelectrophoresis on the same gel lane using
fragments of known length determined from sequence data. The
5' upstream flank of the gene is hypersensitive from -38 to
-215 bp from the 5' end; the 5' downstream flank is hypersensitive from -8 to +100 bp. Interestingly, the 30 bp region
in between, from -8 to -38 bp, is relatively insensitive.
The TATAAAT sequence at -27 bp lies within this insensitive
site. There is also a point of greatest hypersensitivity
at position -93. The control DNase I cleavage on naked DNA
reveals a uniform fragmentation pattern (Figure 1, lanes 7,
8). Hypersensitivity to DNase I, is therefore, primarily a
function of chromatin structure.

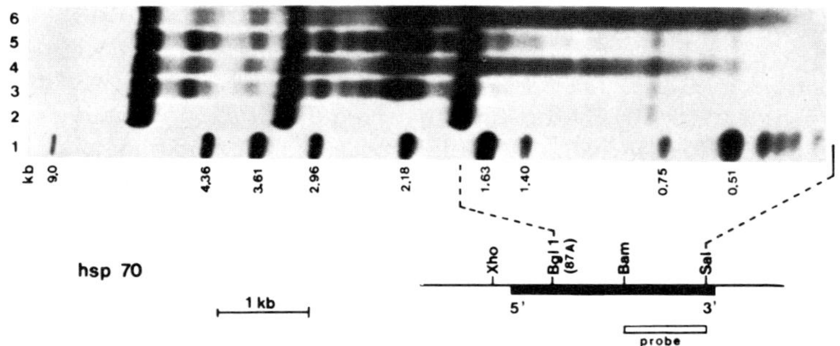

Figure 2. Autoradiogram showing the state of hsp 70
chromatin before and after heat shock induction. Sal and
Bgl I were used for secondary cleavage; the DNA fragments
were sized in a 0.8% agarose gel, and the labeled probe
(open bar) represents approximately the 3' half of the
hsp 70 coding sequence. The lanes indicate: 1, Marker DNA
fragments. 2, Sal plus Bgl I digest on Schneider cell DNA.
3,5, Partial DNase I digests on uninduced Schneider cell
chromatin at a final DNAse I concentration of 6 and 8
units ml^{-1}, respectively. 4,6, Partial DNase I digests on
chromatin in Schneider cells heat shocked at 35°C for 15
min. The two digestion points were chosen out of a series
for comparison against those in lanes 3 and 5 because of
their close equivalence to the latter in overall extends of
DNase I cleavage. Reprinted with permission, from ref. 2.

The 5'-terminal hypersensitive cuts shown here are averaged over the five hsp 70 gene copies, but further restriction analyses (ref. 2 and unpublished data) suggests strongly that the averaged hypersensitivity is exactly representative of each individual hsp 70 gene. We have also shown that neither the coding sequence nor the 3' end of the hsp 70 gene displays DNase I hypersensitivity (3).

The 5'-terminal hypersensitive chromatin structure is present in noninduced cells. Figure 2 shows the state of hsp 70 chromatin before and after 15 min of heat induction at 35°C. Here the DNase I cuts are mapped upstream from the 3'-terminal Sal site common to all five hsp 70 gene copies. (The two gene copies at locus 87A have been reduced to a 1.8 kb Sal-Bgl I fragment; any larger fragments are thus due to those copies at locus 87C). We find that essentially the whole of the hsp 70 transcription unit becomes sensitive to DNase I upon activation, as was shown originally in the globin gene by Weintraub and Groudine (8), but the level of sensitivity is less than the 5'-terminal hypersensitivity, which is still retained partly (compare Figure 2, lanes 3, 5 with lanes 4, 6). From Figure 2, lane 4, and from further mapping studies, it appears that the distal portion of the hypersensitive region remains essentially unchanged upon induction, whereas the proximal portion becomes less hypersensitive, possibly because of protection by bound polymerase or other factors (manuscript in preparation).

The sequences for the 83,000 dalton heat shock protein, hsp 83, reside at the cytogenetic locus 63BC (9,10). From a Sal restriction site within the single-copy coding sequence, we mapped the partial DNase I cleavages upstream and downstream in a region extending about 40 kilobase pairs (Figure 3a, b). A cluster of preferred cuts lie about 2.3-3.2 kilobase pairs upstream from the internal Sal site, and two clusters are observed starting from about 1.8 kilobase pairs downstream. No preferred cuts lie near the internal Sal site. The 5' terminus of a precursor to the hsp 83 mRNA lies about 160 base pairs downstream from the EcoRI site indicated in Figure 3. Thus, a DNase I-hypersensitive region flanks and extends into the sequences complementary to the 5' end of this RNA. Another gene, recently discovered by O'Connor and Lis (10) and by R. Morimoto and J. Jack (personal communication) lies just between the two clusters of hypersensitive sites on the downstream side of the hsp 83 gene. Both genes are transcribed at a low level in cultured cells under normal conditions (10, 11), thus accounting for the low sensitivity to DNase I shown by sequences homologous to RNA.

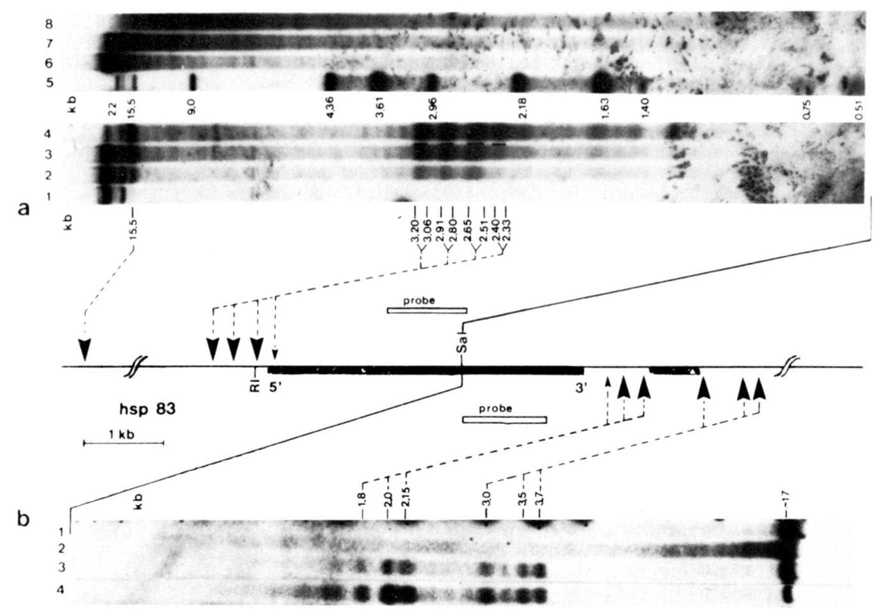

Figure 3.   a, Autoradiogram showing the partial DNase I cuts in Schneider cell chromatin upstream from the Sal site in the hsp 83 coding region. SalI was used for secondary cleavage and the DNA fragments (10 μg per track) were sized on a 1% agarose gel. The lanes indicate: 1, SalI digest on Schneider cell DNA. A minor polymorphism at the external Sal site is observed. 2-4, Partial DNase I digests on chromatin at a final DNase I concentration of 0.5, 1 and 2 units ml^{-1}, respectively. The more prominent cleavages are noted by the arrows on the restriction map. 5, Marker DNA fragments. 6-8, Partial DNase I digests on naked DNA. b, Autoradiogram showing the partial DNase I cuts in Schneider cell chromatin downstream from the Sal site in the hsp 83 coding region. The lanes indicate: 1, Sal digest of Schneider cell DNA. 2, Partial DNase I digest of naked DNA, 3,4, Partial DNase I digests on chromatin, at a final DNase I concentration of 6 and 8 units ml^{-1}, respectively. This experiment was separate from that in a, lanes 3,4, but the overall extents of cleavage are comparable. Reprinted with permission, from ref. 2.

We have extended these studies on chromatin structure to
the rat preproinsulin II gene (12). There are two equally
expressed, nonallelic genes for preproinsulin in the rat
(13). Whereas the two genes are very homologous, the surrounding sequences are not; furthermore, the preproinsulin
II gene possesses an additional 499-bp intron interrupting
its coding region. Within a 4.0-kb Bgl II chromosomal DNA
segment containing the preproinsulin II gene, the DNase I-
sensitive sites can be mapped by digesting rat chromatin
partially with DNase I, restricting purified DNA completely
with Bgl II, and probing a Southern blot of the electrophoresed DNA fragments with a ^{32}P-labeled segment abutting
the Bgl II site upstream from the 5' end of the gene (Figure
4).

When chromatin from five tissues: spleen, kidney,
liver, brain (not shown) and a pancreatic beta cell tumor
was digested with DNase I, only the insulinoma showed an
approximately 350 bp region of preferred cleavage that maps
at the 5' terminus of the gene (Figure 4, lanes 2, 3). The
exposed region extends upstream about 250-300 bp from the
putative cap site. The control digest of naked tumor DNA
did not reveal the 5'-terminal sensitivity (12). This rat
insulinoma is predominantly composed of well-differentiated
pancreatic beta cells, so it is likely that the chromatin
structure observed is representative of normal beta cells
in the islets of Langerhans. Furthermore, we expect that
this 5' feature will appear in the development of that tissue
before the preproinsulin gene is expressed. Intriguingly,
liver chromatin, but not insulinoma, spleen, or kidney
chromatin has a DNase I hypersensitive site just inside the
preproinsulin gene, near the 3' terminus (Figure 4, lanes 7,
8).

DISCUSSION

The experiments we have presented and those reported
in other laboratories (see ref. 14 for a recent review)
establish a new, probably non-nucleosomal chromatin structure, often located at the 5' terminus of eukaryotic genes.
This structure could function as entry sites for RNA polymerase and control macromolecules. Other hypersensitive
sites, such as the liver-specific site in preproinsulin II
gene chromatin, might signify the 5' ends of unknown or
unmapped genes, or they might be involved with different
chromosomal functions.

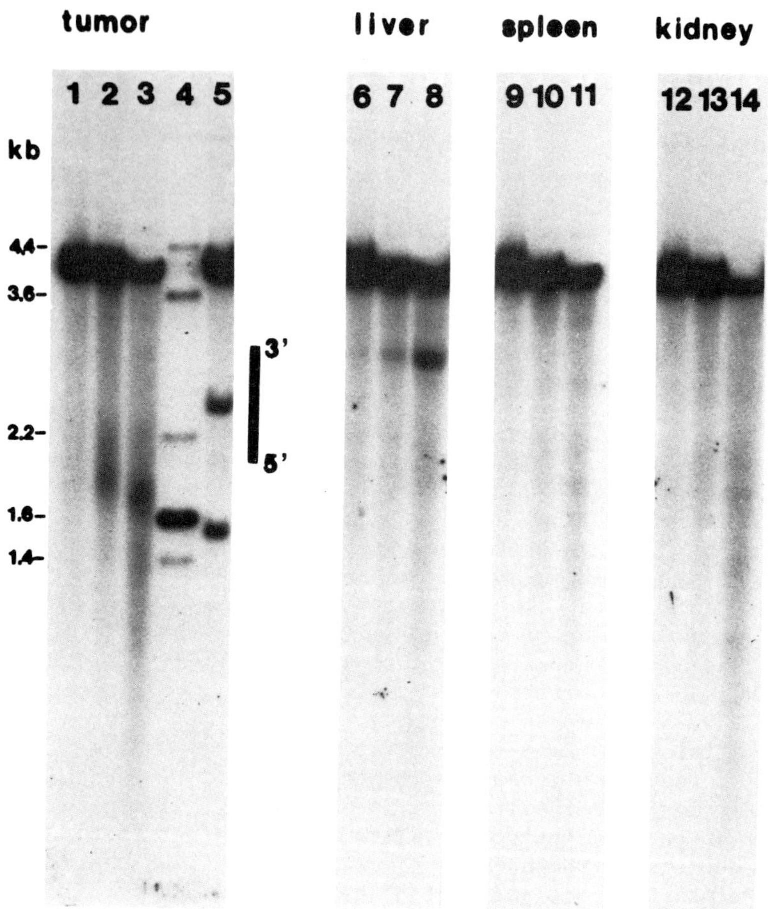

Figure 4.   Autoradiogram showing the DNase I fragmentation pattern downstream from the Bgl II cut on the 5' side
of the preproinsulin II gene.   The Southern blots show the
4-kb Bgl II segment containing the preproinsulin II gene and
the DNase I-generated subfragments from various tissues.   For
each tissue, DNase I cleavage increases from left to right.
Lanes: 1-3, tumor; 6-8, liver; 9-11, spleen; 12-14, kidney;
4, size markers from restriction fragments of pBR322; 5,
relative positions of the EcoRI sites downstream from the 5'
flanking Bgl II site, obtained from a partial EcoRI digest
after complete Bgl II cleavage of purified spleen DNA.   The
solid bar indicates the position of the preproinsulin transcription unit relative to the fragmentation pattern.
Reprinted with permission, from ref. 12

What could be the basis of this unique sensitivity?
A special DNA sequence at these regions could bind other
macromolecules such as specific proteins that cause the
sensitivity, or the sequence may possess a peculiar base
order that renders it sensitive when the DNA in the region
is coiled and supercoiled in nucleosomes and higher order
structures.

The tissue-specific hypersensitivity at the 5' end of
the rat preproinsulin II gene is especially interesting.
It has been widely recognized in developmental biology that
the state of the chromosome in differentiated cells must
possess some intrinsic determined differences that would, for
example, allow different cell types to respond to an inducing
signal by switching on different sets of genes. We speculate
that the tissue-dependent, hypersensitive chromatin structure
at 5' termini of developmentally regulated genes may serve
as the molecular basis of such determined differences. The
process by which such structures are created might then be
the molecular mechanism responsible for cell determination
in the ontogeny of multicellular organisms.

ACKNOWLEDGMENTS

I thank my colleagues in the laboratories of Walter
Gilbert and Matthew Meselson for generous gifts of plasmid
DNA and for helpful discussions. I am especially indebted
to W. Gilbert for providing much stimulus and for supporting
these experiments which were carried out in his laboratory
during my term in the Harvard Society of Fellows.

REFERENCES

1. Wu, C., Bingham, P.M., Livak, K.J., Holmgren, R. and
   Elgin, S.C.R. (1979). Cell 16, 797.
2. Wu, C. (1980). Nature (London) 286, 854.
3. Nedospasov, S.A. and Georgiev, G.P. (1980). Biochem.
   Biophys. Res. Commun. 92, 532.
4. Ashburner, M. and Bonner, J.J. (1979). Cell 17, 214.
5. Livak, K.J., Freund, R., Schweber, M., Wensink, P.C. and
   Meselson, M. (1978). Proc. Natl. Acad. Sci. USA 75, 5613.
6. Moran, L. et al. (1979). Cell 17, 1-8.
7. Ingolia, T.D., Craig, E.A. and McCarthy, B.J. (1980).
   Cell 21, 669.
8. Weintraub, H. and Groudine, M. (1976). Science 193, 848.

9. Holmgren, R., Livak, K., Morimoto, R., Freund, R. and Meselson, M. (1979). *Cell* 18, 1359.
10. O'Connor, D. and Lis, J.T. (1981). *Nucleic Acids Res.* 9, 5075.
11. Holmgren, R. (1981). Ph.D. Thesis, Harvard University.
12. Wu, C. and Gilbert, W. (1981). *Proc. Natl. Acad. Sci. USA* 78, 1577.
13. Lomedico, P., Rosenthal, N., Efstratiadis, A., Gilbert, W., Kolodner, R. and Tizard, R. (1979). *Cell* 18, 545.
14. Elgin, S.C.R. (1981). *Cell* 27, 413.

WORKSHOP SUMMARY: Chromatin Structure. Sarah C.R. Elgin,
Department of Biology, Washington University, St. Louis, MO.

The Workshop opened with a discussion of the role of DNA
structure in histone-DNA interactions. In particular, we
wished to consider how functional domains within the genome
might be recognized as such by virtue of the DNA organization
and chromatin structure, and what means might be available to
mark the boundaries of such domains. Regions of DNA that are
free of nucleosomes will thereby create a boundary for nucleosome organization. Several possibilities have recently come
to light. The DNase I hypersensitive site of chick chromatin
located 5' to one of the β globin genes has been shown to be
a nucleosome-free region (1), and one suspects that this is a
general characteristic of DNase I hypersensitive sites. Such
nuclease hypersensitive sites have been found at or near the
5' end of the region of transcription for a variety of genes
transcribed by RNA polymerase II, including the heat shock
genes of _Drosophila_, histone genes of _Drosophila_, the gene
encoding ribosomal protein 49 of _Drosophila_, the genes encoding alcohol dehydrogenase of yeast, the globin genes of
chick, and the preproinsulin gene in rat (see reference 2 for
a summary of published data). M. Chao reported that he and
his colleagues at Columbia have observed that the 180 bp upstream from the herpes virus thymidine kinase gene is sensitive to restriction enzymes. McKeon et al. (Abst. #0926)
have found that in chick embryo fibroblasts the region 5' to
the α 2(1) collagen gene is DNase I hypersensitive.
Beckendorf et al. (Abst. #0934) have initiated a genetic as
well as biochemical analysis of the Sgs-4 glue locus of
Drosophila; the results indicate that the cis 5' region within 1 kb of the gene, which has several DNase I hypersensitive
sites in salivary gland chromatin, is critical for gene expression. In the cases studied to date of genes expressed in
a tissue-specific fashion, such chromatin structures have
been observed only in differentiated cells capable of expressing that gene. One may advance the hypothesis that such
a 5' DNase I hypersensitive site is necessary, but not sufficient, for transcription _in vivo_. There is some confusion
and little data as to whether or not the DNase I hypersensitive site correlates with the "essential upstream sequences" or with the "enhancer" functions, differentiated by
Weissman (Workshop 2) because of the ability of the latter to
be effective in both orientations and in many positions relative to the transcribed region. However, it should be remem-
bered that the bulk of the data on enhancers comes from work
on circular viral DNA molecules, which will not show the
effects of "handedness" that one would expect to see in the
linear eukaryotic genome.

In addition to the 5' ends of regions of transcription, DNase I hypersensitive sites have been observed at origins of DNA replication (Palen et al., Abst. #0954) and in regions where DNA rearrangement events are thought to occur (3; A. Klar, personal communication). Such sites might in general represent either a necessity for the DNA therein to be more accessible, to be in an altered conformation, or both, for interaction with other macromolecules in the nucleus. That these sites may create order in the array of nucleosomes immediately upstream is suggested by the data of Wu (4), Worcel et al. (this volume), and Cartwright et al (this volume). The effect of discrete boundaries has been discussed in detail by Kornberg (5). Oddly enough, in the work of the above references on genes transcribed by polymerase II, while one observes considerable nucleosome order 5' to the gene, little is seen in the region of transcription, even when the gene is inactive.

DNA sequences with the potential for forming boundaries for a nucleosome array include Z-DNA and poly(dA)·poly(dT). Z-DNA can be generated from a sequence of alternating purine and pyrimidine residues. That it occurs in vivo is indicated by positive results in staining Drosophila polytene nuclei (primarily at the interbands) and other diploid nuclei using antibodies specific for Z-DNA, reported by A. Rich and his colleagues (MIT). During discussion of the controls and limitations of the cytological approach, D. Lohr (Arizona State University) noted that the acid treatment used in squashing polytene chromosomes will extract some basic proteins, and might therefore change the stress of winding, promoting formation of Z-DNA. Positive results with formaldehyde fixed material (6) indicate that the antibodies are detecting a pre-existing topological form. If indeed the anti-Z DNA reactivity were enhanced by additional topological torsion induced during squashing, the results would nevertheless indicate natural DNA segments with a high potential to form Z-DNA. Poly(dG-m^{5}dC)·poly(dG-m^{5}dC) will bind to histones when in the Z-form, but nucleosome structure cannot be detected. In contrast, the same polymer in the B form can interact with core histones to produce normal nucleosome cores. If these core particles are placed in solvents that would cause conversion of the protein-free polymer to the Z-form, no transition is observed, indicating a stabilization of the B form DNA (Nickol et al., Abst. #9029). These results suggest that Z-form DNA will disrupt normal chromatin structure, and could form a boundary for a nucleosome array. In reconstitution experiments it has been shown that histone cores will not bind to an 80 bp region of poly(dA)·poly(dT); shorter stretches of homopolymers may also show a significant effect, as discussed by Martinson (7). Thus we are now aware

of several cases where a particular DNA sequence and/or
conformation may dictate nucleosome organization.

Taking a larger perspective, we would like to know more
about DNA sequence organization and chromatin structure at
the level of transcribed region/spacer and at the level of
the larger DNase I sensitive domain of active genes. An interesting observation is that in the degradation of
Drosophila DNA at locus 67B by either micrococcal nuclease or
ortho-phenanthroline, Cartwright et al. (this volume) report
a striking pattern. Prominent DNA cleavage sites occur primarily in the spacer regions, allowing the regions encoding
transcripts to be identified as "gaps" in the pattern of
cleavage sites. That two such different reagents see such a
pattern suggests an interesting feature of DNA sequence
organization reflected in the structure of the double helix.
Much more data is needed, however. In the discussion, A.
Worcel (University of Rochester) pointed out that some cleavage is observed within the regions encoding H2A and H4 in
Drosophila DNA. Nonetheless, both groups have observed that
the chromatin structure of regions of transcription, even
when inactive, is less ordered than that of the adjacent non-
transcribed regions, suggesting a difference in the histone-
DNA interaction which could reflect an altered DNA structure.
One is reminded of the alternative chromatin fibers (beaded
and smooth) seen using electron microscopy to examine yeast
material (Hamkalo et al., Abst. #0920). B. Hamkalo (U.C.
Irvine) reported on recent progress in doing in situ hybridization to samples on an e.m. grid, an approach that allows
positive identification of the unit under study and therefore
should allow specific mapping of chromatin fiber structure.

The discussion then turned to questions of gene activation and transcription. A comparison of results from different systems suggests that in general the presence of a nuclease hypersensitive site at or near the 5' end of a gene and
an encompassing DNase I sensitive domain are both necessary
(although probably not sufficient) for transcription in vivo
(see reference 8 for a review of the evidence). U. Storb
(University of Washington) discussed the situation for im-
munoglobin genes, where it appears that activation is associated with the rearrangement event. In immune cells and certain precursor cells while unrearranged V_k genes are in an
inactive domain, rearranged V_k-C_k genes, as well as unrearranged C_k genes are in an active domain. The point in development at which the C_k open domain is established has not yet
been determined. The gene is activated when the V_k gene with
promoter is moved into the active domain of the C_k gene.
Interestingly, in cells in which C_μ is being expressed, the
switch region and C_μ gene lie in a DNase I sensitive domain
while the downstream C_γ gene is still insensitive. C. McKeon

(NIH) presented an analysis of the chick $\alpha 2(I)$ collagen gene
in a variety of tissues, where expression ranged from very
high to undetectable. In all cases the gene showed a similar
pattern of methylation, with no methylation in the 5' region
but significant amounts in the 3' region. The level of
expression is therefore independent of methylation as detected
by restriction enzymes (Abst. # 0926).

Finally, M. Wormington (Carnegie Institution of
Washington) reported on studies of transcription of the 5S
genes of <u>Xenopus</u>. Transcription <u>in vitro</u> of a chromatin template from somatic cell nuclei or from oocyte germinal
vesicles yielded the biologically appropriate 5S product,
regardless of the source of RNA polymerase III or S100 supernatant fraction. As 5S products can be obtained from somatic
cell chromatin with a depleted extract, the results can in
part be explained by the stability of the transcription
complex (see D. Brown et al., this volume). However, there
is also an implication that access to the oocyte 5S genes is
blocked in some way. These genes can be transcribed after
treatment of the chromatin with 0.6M NaCl provided that the
polymerase and all factors are supplied (9).

In summary, the recent work indicates that the structure
of chromatin in many ways may reflect not only the underlying sequence organization of DNA, but also its structure,
and that we have much to learn still about the tertiary
structure of DNA. We are beginning to gain some insight into
gene activation as a process involving multiple steps that
can be significantly separated in time. The interplay of DNA
sequence organization and chromatin structure provides
multiple possibilities for regulation of gene expression.

REFERENCES

1. McGhee, J.D., Wood, W.I., Dolan, M., Engel, J.D., &
   Felsenfeld, G. (1981) Cell 27, 45-55.
2. Elgin, S.C.R. (1981) Cell 27, 413-415.
3. Storb, U., Arp, B., & Wilson, R. (1981) Nature 294, 90-92.
4. Wu, C. (1980) Nature 286, 854-860.
5. Kornberg, R.D. (1981) Nature 292, 579-580.
6. Nordheim, A., Pardue, M.L., Lafer, E.M., Moller, A.,
   Stollar, B.D., & Rich, A. (1981) Nature 294, 417-422.
7. Kunkel, G.R., & Martinson, H.G. (1981) Nuc. Acids Res.
   9, 6869-6888.
8. Cartwright, I.L., Keene, M.A., Howard, G.C., Abmayr, S.M.,
   Fleischmann, G., Lowenhaupt, K., & Elgin, S.C.R. (1982)
   CRC Crit. Rev. Biochem., in press.
9. Bogenhagen, D.F., Wormington, W.M. & Brown, D.D. (1982)
   Cell 28, 413-421.

THE ROLE OF POLY(A) IN MAMMALIAN GENE EXPRESSION[1]

J. E. Darnell, Jr., J. R. Nevins,
and M. Zeevi[2]

The Rockefeller University
New York, N. Y. 10021

ABSTRACT   The role of poly(A) in eukaryotic mRNA metabolism may be largely (if not entirely) to increase the
half-life of cytoplasmic mRNA.   Those nuclear RNA transcripts that do not receive a poly(A) do not contribute
an appreciable quantity to cytoplasmic mRNA.

INTRODUCTION

Since 1970-71 it has been known that most mammalian
mRNAs contain a 3' terminal segment of polyadenylic acid (1-8).
This terminal structure is added in the cell nucleus (6,7,9)
to a region of the primary RNA transcript (an hnRNA molecule)
by elongation at a 3' OH derived most likely by endonucleolytic cleavage.   This conclusion follows from the demonstration that in a number of transcription units, RNA polymerase II,
the enzyme that synthesizes hnRNA, traverses the poly(A) site
and continues to synthesize RNA for a variable distance that
can be many thousand nucleotides long (10-14).   The conclusion that an otherwise unmodified primary transcript receives
the poly(A) is based on studies of adenovirus mRNA formation
showing that the receptor molecule was not yet spliced
(10,15).
All mRNA molecules do not contain poly(A).   Histone
mRNAs lack it (16), however, at present no other specific
cell mRNAs are known to arrive in the cytoplasm lacking
poly(A) (8,17,18).   Since poly(A) becomes shorter with age
(19) mRNA molecules that fail to bind to affinity chroma-
tography assays are found (17,18,20) but no mRNA other than
histones appears to be synthesized in the nucleus lacking
poly(A).

[1]This work was supported by grants from the National
Institutes of Health (CA 16006-09) and the American Cancer
Society (CD 123).
[2]Present address: Kiryat Weizmann, Rehovot, Israel

The metabolic role of poly(A) was originally explored in two types of experiments. First, a role in translation was sought for poly(A). Enzymatic elimination of the poly(A) (21, 22) or the presence of hybridized poly(U) (23) did not affect the <u>in vitro</u> translation of mRNA suggesting no obligatory role in protein synthesis. Later translation studies also allowed a test of the functional lifetime of mRNA lacking poly(A) (22,24). Globin mRNA from which poly(A) was enzymatically removed could still stimulate frog oocytes to make protein but the frog cells soon ceased globin synthesis. Restoration of the poly(A) to a size of 20 adenylate residues or longer or use of the normal mRNA led to continued globin synthesis for many days. This type of experiment was repeated in a mammalian cell cytoplasm. HeLa cells were injected with rabbit globin mRNA with and without poly(A). Globin synthesis for several days resulted from injection of mRNA containing poly(A) but the mRNA that lacked poly(A) failed to stimulate detectable globin synthesis for longer than a few hours. These experiments support the conclusion that the poly(A) has an important cytoplasmic role in stabilizing mRNA.

An earlier line of experiments on mRNA molecules lacking poly(A) was carried out by treating intact cells with the drug 3' deoxyadenosine (cordycepin) which inhibits poly(A) synthesis but still allows labeling of RNA synthesized by RNA polymerase II (6,7,25). The accumulation of newly labeled mRNA, either total HeLa cell mRNA or adenovirus mRNA, was reduced by a factor of 5 to 10 during a 30-minute exposure to ^{3}H-uridine. It was not clear from these early experiments whether the absence of poly(A) led to a failure of processing and transport of the mRNA or to an instability once the mRNA molecule reached the cytoplasm (26,27). With the detailed picture of adenovirus mRNA formation that was available by 1980 (28), this experiment was repeated examining the formation and transport of specific adenovirus mRNAs (29,30) after short periods of labeling. Splicing of adenovirus mRNA could be demonstrated in molecules labeled after 3'-deoxyadenosine addition. Moreover, the molecules were about 200 nucleotides shorter than their normal counterparts. Thus the proposed endonucleolytic cut at the poly(A) site apparently occurred as well as the splicing steps. Additional experiments showed that the shorter molecules contained sequences from both the 5' part as well as the 3' portion of the primary transcript indicating splicing did indeed occur. Finally, the appearance in the cell cytoplasm of spliced mRNA about 200 nucleotides shorter than normal was also detected. Moreover, this mRNA could enter polyribosomes. However, within the 40 minute time course of labeling in the normal cell, labeled mRNA rapidly accumulated whereas in the 3'-deoxyadenosine-treated cells it did not. This finding strongly indicates that the turnover rate of the mRNA lacking poly(A) is high and explains the failure to accumulate labeled mRNA in 3' deoxyadenosine-treated cells. Although poly(A) is synthesized in the nucleus, its role appears to be cytoplasmic, helping to achieve stability of the mRNA.

At least part of the nucleotide signal for the addition of poly(A) is AAUAAA as suggested from _in vitro_ mutagenesis (31) and sequence studies (32). Deletion of this sequence, which is found almost universally 10 to 20 bases upstream from the poly(A) addition site (32), results in the lack of accumulation of mRNA in the cytoplasm, although no studies have been done to determine the point in mRNA biosynthesis and accumulation that is affected by this deletion.

Do All Nuclear Transcripts Acquire Poly(A)? A separate line of experiments designed to determine whether every primary transcript receives poly(A) and the nature of poly(A) containing and poly(A) lacking (A^+ and A^-) molecules has also recently been completed. Whereas nuclear primary transcripts all appear to have a 5' cap structure ($m^7GpppN^1mpN^2p$) (33,34) only about one-quarter have poly(A) added (35). This conclusion, based on labeling studies of caps and poly(A), is not due to a group of hnRNAs that are delayed in receiving poly(A) but quickly exit to the cytoplasm or even acquire the poly(A) in the cytoplasm (36). All new cytoplasmic molecules appear both with newly made caps and poly(A) but virtually all new poly(A) is in the nucleus for the first two minutes of label (9,36) and the nucleus remains the predominant locus of labeled poly(A) up to 15 minutes after introduction of label. Thus the instantaneous ratio is 3 to 4 caps formed on long hnRNA for every unit of poly(A) added to the hnRNA.

Such a situation might arise if all active transcription units produced 3 to 4 primary transcripts for every transcript that is successfully processed (37,38). On the other hand, there could be an excess of different species of primary transcripts with only some types of molecules being processed. To distinguish these possibilities, poly(A)$^+$ nuclear RNA was separated from poly(A)$^-$ nuclear RNA. In confirmation of the rate studies, one-quarter of the caps were found in the poly(A)$^+$ molecules. RNA:DNA hybridization of labeled nuclear RNA to eight separate cloned DNA sequences prepared as recombinant DNA from CHO mRNA (37) gave a different distribution than the total caps in the nucleus. Instead of 75% of the specific sequences being in the poly(A)$^-$ group, 70% were found in the poly(A)$^+$ fraction. Since the average size of the labeled RNA in the poly(A)$^+$ and poly(A)$^-$ fractions was similar, about 5 KB in each, it seems that only a minority of the poly(A)$^-$ primary transcripts came from the same genes as the transcripts in the poly(A)$^+$ fraction (38). It is also obvious that the poly(A)$^-$ sequences do not contribute stable mRNAs to the cytoplasm, since in the cytoplasm almost all capped molecules are poly(A)$^+$ (35). It seems that well over half the total polymerase II output consists of molecules that are not polyadenylated and do not become stable mRNAs.

Without knowing the nature of these poly(A)$^-$ transcripts we cannot understand the meaning of this conclusion. If the transcripts are from genes that are correctly processed at some time in some other cell then a significant amount of gene control could lie at the level of processing, mainly in choosing to add poly(A). However, these nuclear RNAs could represent primary transcripts that cannot be used or at least are never used. Then they do not signal the possibility of control due to processing but might represent, for example, a certain laxity in transcriptional initiation in eukaryotic cells. It is important to try to determine more exactly the nature of these transcripts to settle these two general possibilities.

CONCLUSION

The results briefly reviewed in this note indicate the central role of poly(A) addition in mammalian gene expression. Without the addition of poly(A), a long lifetime in the cell cytoplasm seems unlikely for an mRNA molecule. And therefore nuclear transcripts that do not acquire poly(A) do not furnish stable mRNA. If the non-polyadenylated nuclear transcripts represent the products of "accidental" or "leaky" transcription that have incorrect starts and produce unusable sequences then we might suggest that when a transcript is started "on purpose", a part of the initiation complex at the correct origin of synthesis is the poly(A) adding capacity. If this suggestion is further strengthened by additional studies on the poly(A)$^-$ RNA, then the biochemistry of poly(A) addition should become a subject of much more serious attention.

REFERENCES

1. Kates,J.(1970).Cold Spring Harbor Symp. Quant. Biol. 35, 743.
2. Lim,L.,and Canellakis,L.S.(1970).Nature 227,710.
3. Edmonds,M.P.,Vaughan,M.H.,and Nakazato,H.(1971).Proc. Nat'l.Acad.Sci.USA 68,1336.

4. Darnell,J.E.,Jr.,Wall,R.,and Tushinski,R.J.(1971).Proc.
   Nat'l.Acad.Sci.USA 68,132.
5. Lee,Y.,Mendecki,J.,and Brawerman,G.(1971).Proc.Nat'l.Acad.
   Sci.USA 68,1331.
6. Darnell,J.E.,Jr.,Philipson,L.,Wall,R.,and Adesnik,M.
   (1971).Science 174,504.
7. Philipson,L.,Wall,R.,Glickman,R.and, Darnell,J.E.,Jr.
   (1971).Proc.Nat'l.Acad.Sci.USA 68,2806.
8. Brawerman,G.(1981).CRC Critical Reviews in Biochemistry
   10,1.
9. Jelinek,W.,Adesnik,M.,Salditt,M.,Sheiness,D.,Wall,R.,
   Molloy,G.,Philipson,L.and Darnell,J.E.,Jr.(1973).J.Mol.
   Biol. 75,515.
10. Nevins,J.R.and Darnell,J.E.,Jr.(1978).Cell 15,1477.
11. Fraser,N.W.,Nevins,J.R.,Ziff,E.,and Darnell,J.E.,Jr.
   (1979).J.Mol.Biol. 129,643.
12. Nevins,J.R.,Blanchard,J.-M.,and Darnell,J.E.,Jr.(1980).
   J.Mol.Biol. 144,377.
13. Ford,J.P.,and Hsu,M.-T.(1978).J.Virol. 28,795.
14. Hofer,E.,and Darnell,J.E.,Jr.(1981).Cell 23,585.
15. Weber,J.,Blanchard,J.-M.,Ginsberg,H.,and Darnell,J.E.,Jr.
   (1980).J.Virol. 33,286.
16. Adesnik,M.,and Darnell,J.E.,Jr.(1972).J.Mol.Biol. 67,397.
17. Minty,A.J.,and Gros,F.(1980).J.Mol.Biol. 139,61.
18. Harpold,M.,Wilson,M.,and Darnell,J.E.,Jr.(1981).Mol.and
   Cell.Biol. 1,188.
19. Sheiness,D.,and Darnell,J.E.,Jr.(1973).Nature New Biology
   241,265.
20. Kaufman,G.,Milcarek,C.,Berissi,H.,and Penman,S.(1977).
   Proc.Nat'l.Acad.Sci.USA 74,4801.
21. Bard,E.,Efrond,D.,Marcus,D.,and Perry,R.P.(1974).Cell
   1,101.
22. Nudel,U.et al.(8 authors)(1976).Eur.J.Biochem. 64,115.
23. Munoz,R.F.and Darnell,J.E.,Jr.(1974).Cell 2,247.
24. Huez,C.,Bruck,C.,and Clueter,Y.(1981).Proc.Nat'l.Acad.Sci.
   USA 78,908.
25. Penman,S.,Rosbash,M.,and Penman,M.(1970).Proc.Nat'l.Acad.
   Sci.USA 67,1978.
26. Adesnik,M.,Salditt,M.,Thomas,W.,and Darnell,J.E.,Jr.
   (1972).J.Mol.Biol. 71,21.
27. Darnell,J.E.,Jr.,Jelinek,W.,and Molloy,G.(1973).Science
   181,1215.
28. Darnell,J.E.,Jr.(1982).Nature,in press.
29. Zeevi,M.,Nevins,J.R.,and Darnell,J.E.,Jr.(1981).Cell
   26,39.
30. Zeevi,M.,Nevins,J.R.,and Darnell,J.E.,Jr.(1982).Mol.and
   Cell.Biol. 3,in press.
31. Fitzgerald,M.,and Shenk,T.(1981).Cell 24,251.

32. Proudfoot,N.J.,and Brownlee,G.G.(1976).Nature 263,211.
33. Salditt-Georgieff,M.,Harpold,M.,Chen-Kiang,S.,and
    Darnell,J.E.,Jr.(1980).Cell 19,69.
34. Babich,A.,Nevins,J.R.,and Darnell,J.E.,Jr.(1980).Nature
    287,246.
35. Salditt-Georgieff,M.,Harpold,M.,Wilson,M.,and Darnell,
    J.E.,Jr.(1981).Mol.and Cell.Biol. 1,179.
36. Salditt-Georgieff,M.,Harpold,M.,Sawicki,S.,Nevins,J.R.,
    and Darnell,J.E.,Jr.(1980).J.Cell Biol. 86,844.
37. Harpold,M.M.,Evans,R.M.,Salditt-Georgieff,M.,and Darnell,
    J.E.,Jr. (1979).Cell 17,1025.
38. Salditt-Georgieff,M.,and Darnell,J.E.,Jr.(1982).Mol.and
    Cell.Biol. 3,in press(June).

SMALL NUCLEAR RNPs AND RNA PROCESSING[1]

H. Busch, R. Reddy, D. Henning, D. Spector, P. Epstein,
N. Domae, M. Liu, S. Chirala, W. Schrier, and L. Rothblum

Department of Pharmacology, Baylor College of Medicine,
Houston, Texas 77030

ABSTRACT Small RNAs discovered about 16 years ago are
metabolically stable, conserved through evolution, and
localized to specific subcellular compartments. The sequencing of small RNAs has resulted in the discovery of
cap structures, and the nucleotide sequences of all the
six capped snRNAs are defined, in some cases for several
species. The discovery that patients with autoimmune
diseases produce antibodies directed against small RNPs,
and the hypothesis that U1 RNA may be involved in splicing hnRNAs, has brought small RNAs to the attention of
many investigators. All the capped snRNAs appear to be
synthesized by polymerase II, and other small RNAs by
polymerase III. Genes having identical sequences to U1,
U2, and U3 RNA have been isolated and, interestingly,
the human genome appears to contain more pseudogenes for
small RNAs than real genes.
   The role of snRNAs in the processing of hnRNA is
only a hypothesis, with which, however, some experimental evidence accumulated to date is consistent. There
are reports that suggest that the model is incomplete,
and alternative splicing mechanisms may exist that do
not utilize snRNA. An alternative theory is presented
which suggests snRNAs are binding elements for hnRNP and
the nuclear RNP network ("matrix") for transport to the
nuclear envelope where the hnRNP undergo further processing for release to the cytoplasm as mRNP.

INTRODUCTION

<u>Small Nuclear RNAs</u>. In early studies (1,2) on the nucle-
olar U-snRNA species, an RNA fraction was found in the 4-8S

[1]This work was supported by grants from Cancer Center
Program Grant 10893, awarded by DHEW; the Bristol-Myers Fund;
the Michael E. DeBakey Medical Foundation; the Pauline Sterne
Wolff Memorial Foundation; the Taub Foundation; and the
William S. Farish Fund.

region of a sucrose gradient that had different base composition from that of the rRNA or tRNA. Due to its high content
of uridylic acid, it was designated U-RNA (3-7). There was
initial concern that such small RNA species could be degradation products of higher molecular weight RNA as had been the
case for "chromosomal RNA" and other small RNA products (8,9).
To control nuclear RNases is difficult partly because a number of them have specific processing functions; they differ
from cytoplasmic contaminants in that they are released during isolation of nuclear subfractions. Some of these prob-
lems have been overcome using improved methods (10,11) and
RNase inhibitors.

By 1966, snRNA had been observed (12-14), but specific
RNA species had not been characterized although Reich et al
(15) had identified a small viral RNA. Subsequently, Pene et
al (16) isolated 5.8S RNA, and other snRNAs were identified
(1,2).

After studies were made on RNase contents of nuclear preparations, we chose the Novikoff hepatoma as an optimal tissue for study (10). When "RNase free" citric acid nuclei
were employed (11), single species of highly purified U1 and
U2 snRNAs could be isolated (3). The purification, terminal
nucleoside analysis, high uridylic acid content and evidence
that U-snRNA's are specifically localized were reported (3-7,
17). U3 RNA was found only in the nucleolus and not in the
nucleoplasm (3-7,17).

Nine lines of evidence showed that the RNAs were not
breakdown products of other RNA species (5). One line of
evidence (3,17) was that the turnover rates were very low by
comparison to other species, particularly rRNA (3,17). Despite the reproducibility of the isolation procedures, the
only satisfactory way to show that U-snRNAs were unique species, and not specific cleavage products or artifacts of degradation has been to determine their sequences (1,2).

<u>Evolutionary Conservation of SnRNA's</u>. Small RNAs are
present in viruses and in prokaryotic and eukaryotic cells,
but their gel electrophoretic patterns are different. After
snRNAs were found in human cells, rat tissues, Chinese hamster, and mouse, systematic studies concluded that snRNAs are
present in all vertebrates (1,2). snRNAs are independent of
cell type and malignancy (17,18). U1, U2, and U3 RNAs appear
to be highly conserved (19). Sequence analysis of U1 RNA
structures of chicken, rat, and human tissue show it is over
95% conserved. There are only two nucleotide differences
between rat and human U1 RNAs (19). Notable similarities
have been found for the sequence of U1 RNA of <u>Drosophila</u> (20)
with other U1 RNA sequences.

Detailed studies on nonvertebrates, e.g. cockroach, meal worm, blowfly, sea urchins, Tetrahymena, and Chironymus tentans revealed that they all contain small RNAs, but that their mobility on gels is different from vertebrate snRNAs (1,2). U1, U2, and U3 RNAs of Dictyostelium contain a cap structure similar to that found in rat U-snRNAs (21).

The U3 RNA sequence of rat (22) is homologous to that of U3 RNA of Dictyostelium (23), which indicates the similarity in structure of U-snRNAs of vertebrates and invertebrates. The characteristics of small RNAs are shown in Table 1. Table 2 lists the major small RNAs, their 5'-termini, chain lengths, and alternate nomenclature.

Most recently, we have found that dinoflagellates, which are considered to be eukaryotes that evolved 3 billion years ago, also contain U1-U6 snRNAs or very similar capped molecular (R. Reddy, D. Spector, D. Henning, and H. Busch, unpublished results). Diener (24) has suggested that base pairing of U1 RNA to intron "ancestors" may be part of a mechanism for viroid formation.

TABLE 1

GENERAL CHARACTERISTICS OF U SnRNAs

Size range 90-400 nucleotides

Copies per cell 1 x 10^6 molecules for the most abundant RNAs (e.g. U1 RNA); 1 x 10^4 to 5 x 10^5 for other RNAs

Half-lives U snRNAs are stable: half-lives of up to one cell cycle

Specific localization localized to specific subcellular compartments (e.g. U3 RNA in nucleolus; U1, U2, U4, U5, U6, and La 4.5 RNAs in nucleoplasm; and Y RNAs in cytoplasm)

Polymerases capped snRNAs U1 to U6 synthesized by pol II, and other small RNAs synthesized by pol III

Cap structure U1 to U6 snRNAs are capped, U1 to U5 RNAs contain a trimethylguanosine cap, and U6 RNA has a different cap

RNP particles all U snRNAs exist in RNP particles

Associated with precursor RNAs The U sn RNAs are associated with precursor hnRNAs and pre-rRNA

                                    H. BUSCH *et al.*

TABLE 2
TERMINI AND CHAIN LENGTHS OF snRNAs

RNA	5' terminus	Chain length	Alternative nomenclature
U snRNA			
U1	m_3GpppAmUmA	165	D
U2	m_3GpppAmUmC	188-189	C
U3	m_3GpppAmAG	210-214	A, D2
U4	m_3GpppAmGmC	142-146	F
U5	m_3GpppAmUmA	116-118	G'.5S III
U6	XpppGUG	107-108	H1.4.5 III
Other snRNA			
tRNA	pN	~80	
5S	pppN	121	G
5.8S	pN	158	E
La 4.5	pppG	90-94	
La 4.5 I	pppG	98-99	H2
7S	pppG	294-295	L
VAI, II	pppG	157-160	
7-1	pppN(?)	~260	M
7-2	pppN	~290	M
7-3	pppN	~300	K
8S	pppN(?)	~400	
Y1-Y3	pppN	~100	

RESULTS

<u>Subcellular Localization</u>. Many studies have been done
(1,2) to find the subcellular localization of different small
RNAs (Figure 1). The methods used to fractionate the nuclei
from cytoplasm include aqueous and nonaqueous methods and the
citric acid method (1,2). The results obtained by different
investigators for the location of small RNAs vary slightly;
however, the data point to the localization of small RNAs as
shown in Figure 1. The U-RNPs (U1, U2, U4, U5, and U6) are
nucleoplasmic, as shown by indirect immunofluorescence using
specific antibodies directed against these RNPs (25). Similar studies show La 4.5 and 4.5I RNPs to be located in the
nucleus and Y RNPs to be located in the cytoplasm (26).

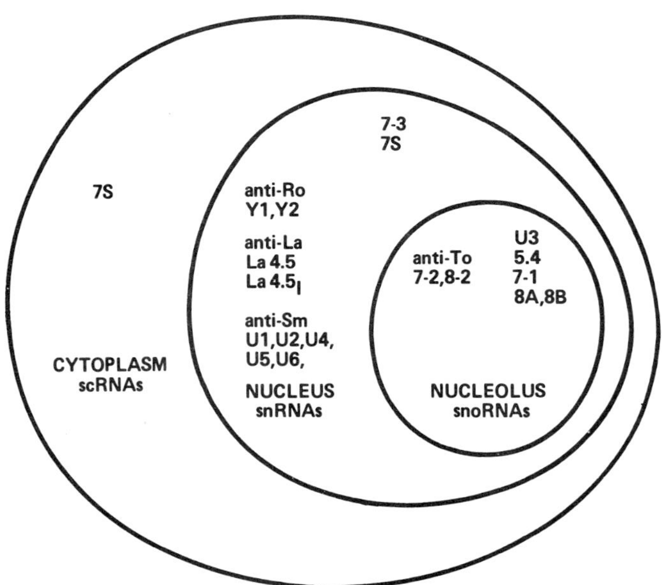

FIGURE 1.   Diagrammatic representation of subcellular
localization of small RNAs of rat. The three categories of
small RNAs, based on subcellular localization, are small cytoplasmic RNAs (scRNAs), small nuclear RNAs (snRNAs), and small
nucleolar RNAs (snoRNAs).

<u>Sequences of U-snRNAs: U1 RNA</u>. The initial sequence
data of U1 RNA (see Table 1 for alternative nomenclature),
which is in the highest concentration in the nucleus (Figure
2) revealed that its 5' cap contained the trimethyl G residue
(27). This was the first report of a cap structure on any RNA
species, and was followed by studies on cap of mRNA and viral
RNA.
The U1 RNA sequence contains a large number of uridine
residues, many of which are clustered (27). Gel analysis of
4-8S RNA separated U1 RNA into two species and the faster
moving U1A RNA was sequenced (27). Using rapid sequencing
methods (19,28), a number of modifications were made of the
original sequence of U1 RNA (Figure 2). The U1A RNA from the
chicken, rat, and HeLa cells is highly conserved (19). Of the
species examined chicken (19), rat (8,19) human, and insects
(29) contains two U1 RNAs, which differ in the center of the
molecule (30). The two forms of U1 RNA found in Novikoff
hepatoma appear to be isomers. In addition to establishing
its unique structure, U1 RNA sequence analysis led to a

```
pm3^2,2,7 G
p       10            20           30           40           50
pAUACΨΨACC   UGGCAGGGGA   GAUACCAUGA   UCACGAAGGU   GGUUUUCCCA
                              C                    G C

        60           70           80           90          100
GGGCGAGGCU   UAUCCAUUGC   ACUCCGGAUG   UGCUGACCCC   UGCGAUUUCC
             C        CC                    G

       110          120          130          140          150
CCAAAUGCGG   GAAACUCGAC   UGCAUAAUUU   GUGGUAGUGG   GGGACUGCGU
         U

       160          165
UCGCGCUCUC   CCCUG OH
         U
```

FIGURE 2. Nucleotide sequence of U1 RNA of Novikoff
hepatoma (27,28) or rat brain (19). Human (19) and chicken
(19) nucleotide variants are shown below the main sequence.

concept of hydrogen bonding to specific pre-mRNA consensus
IVS sequences and to analysis of the secondary structures of
U1 RNA, and later to analysis of homologies with other U-
snRNAs. Multiple secondary structures are possible with similar stability numbers. With the techniques developed for specific immunoprecipitation of the U1 snRNP particles, it is
possible to conduct direct studies on the conformational state
of the U1 RNA in snRNP particles (31). When T1 RNase digestion was followed by immunoprecipitation of the U1 snRNP particles, only one major site specific cleavage was observed at
position 107. The additional fragments produced by RNase A
cleavage permit more detailed analysis of the RNA structure
within the particles. The structure derived is in good agreement with one hypothetical structure determined by analysis
of stability numbers (31). For these analyses, the computer
program of Korn et al.(32) was invaluable.

     U2 RNA. The most remarkable feature of the U2 snRNA (33;
Figure 3) is the large number of modifications in its 5' end.
Detailed analysis of the 5' cap (35) produced both enzymatic
and chemical data that unequivocally established the presence
of the 5' trimethyl G and pyrophosphate linkage to the remainder of the molecule. The 5' structure of the U2 RNA (35) was
the first for which detailed chemical evidence was provided.
Among the unanswered questions raised in this study was
whether the turnover rate of trimethyl G is the same or different from the remainder of the molecule.
     The presence of 2'-O-methylated nucleotides and the re-
markable number of ψ residues on the 5' end of U2 RNA clearly

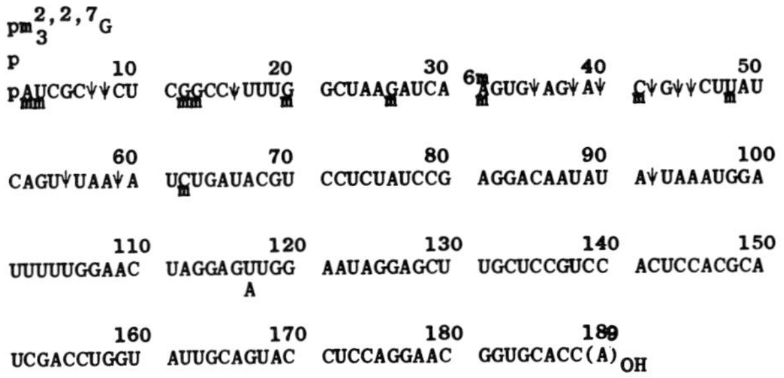

```
pm₃^2,2,7 G
p          10              20              30      6        40             50
pAUCGCψψCU   CGGCCψUUUG   GCUAAGAUCA   m GUGψAGψAψ   CψGψψCUUAU
   m  m        m m    m        m     m

            60              70              80              90             100
CAGUψUAAψA   UCUGAUACGU   CCUCUAUCCG   AGGACAAUAU   AψUAAAUGGA
               m

            110             120             130             140            150
UUUUUGGAAC   UAGGAGUUGG   AAUAGGAGCU   UGCUCCGUCC   ACUCCACGCA
                A

            160             170             180            189
UCGACCUGGU   AUUGCAGUAC   CUCCAGGAAC   GGUGCACC(A)OH
```

FIGURE 3. Nucleotide sequence of U2 RNA of Novikoff
hepatoma (2,33,34).

indicates its chemical distinction from U1 RNA, and further
that there is an extraordinary number of post-transcriptional
modifications. No special functions of U2 RNA are known.
Like U1 RNA, U2 RNA is in an Sm protein containing snRNP,
which is in the extranucleolar chromatin.

   U3 RNA. U3 RNA (36,37) is of interest because of its
specific localization in the nucleolus and its hydrogen bonding to nucleolar 35S and 28S RNA (38). U3 RNA is associated
with protein, and is uniquely present in nucleoli. No significant amount of U1 and U2 RNA is present in nucleoli.
   The hydrogen bonding of U3 RNA to the 28S and 35S RNA
was not stoichiometric and accordingly only suggestion was
that U3 RNA is involved in the processing of 28S RNA (38).
   Two of the three U3 RNA species found in Novikoff hepatoma cells have been sequenced (39), and the minor differences in their sequences are shown in Figure 4. The third
species of U3 RNA has not been completely sequenced, but it
appears to be a minor variant of U3B RNA. The U3 RNA species
are capped with trimethylguanosine (40,41). Despite the two
insertion/deletions in each RNA, U3A and U3B RNA have identical lengths of 214 nucleotides (39). When these two RNAs are
aligned for maximum homology, there are 17 base substitutions,
including purine → purine, pyrimidine → pyrimidine, and
purine → pyrimidine substitutions. The U3 RNA of HeLa cells
consists mainly of one species that is similar but not identical to U3 RNA of Novikoff hepatoma cells (18).

   U4 and U5 RNA. U4 RNA is another nucleoplasmic RNA that
has important homologies to U1 RNA (42,43). Its sequence
(42,43) contains the same m₃^2,2,7 G cap of other U snRNAs and

$pm_3^{2,2,7}G$

p

	10	20	30	40	50	60
pAAGACUAψA	CUψUCAGGGA	UCAUUUCUAU	AGUUCGUUAC	UAGAGAAGUU	UCUCUGACUG
m

	70	80	90	100	110	120
UGUAGAGCAC	CCGAAACCAC	GAGGACGAGA	CAUAGCGUCC	CCUCCUGAGC	GUGAAGCCGG
		G	GG U	U	

	130	140	150	160	170	180
CUCUAGGUGC	UGCUUCUGCC	UCUUGCCAUU	GGCAGCUGAU	GAUCGUCUUC	UCUCCUUCGG
	U	A U	A		GG CUU

	190	200	210	214
GGGGGUAAGA	GGGAGGGAAC	GCAGUCUGAG	UGGA
A	G	A	U_OH

FIGURE 4. Nucleotide sequence of U3 RNA of Novikoff
hepatoma (22,39). For <u>Dictyostelium</u> U3 RNA, see (23). Top
line, U3B RNA sequence. Nucleotide variants shown are in
U3A RNA.

it contains a ψ near the 5' cap (Figure 5). In addition,
position 63 is 2'-O-methylated and position 99 is m^6A. At
present, it appears that U4 RNA and possibly U5 RNA (Figure
6) are closely related to U1 RNA (43).

When U5 RNA was first purified, it was referred to as 5S
RNA III (1,2). It has at least two or more subspecies. Of
the U1 to U6 RNAs, U5 RNA is most enriched in uridine (35%);
it is capped with trimethylguanosine. Its complete sequence
has been defined (44,45; Figure 6).

$pm_3^{2,2,7}G$

p

	* 10	20	30	40	50
pAGCψUUGCG	CAGUGGCAGU	AUCGUAGCCA	AUGAGGUUUA	UCCGAGGCGC
m m			A	

	60	70	80	90	100
GAUUAUUGCU	AAUUGAAAAC	UUψUCCCAAψ	ACCCCGCCAU	GACGACUUGA
	m		G	C

	110	120	130	140	145
AAUAUAGUCG	GCAUUGGCAA	UUUUUGACAG	UCUCUACGGA	GACUG(G)_OH

(position 109 marked m^6)

FIGURE 5. Nucleotide sequence of U4 RNA of Novikoff
hepatoma (44) or rat (45). Mouse, human, and chicken nucleotide variants are shown below the main sequence.

$pm_3^{2,2,7}G$
p
```
p           10            20            30            40            50
pAUACUCUGG   UUUCUCUUCA   GAUCGUAUAA   AUCUUUCGCC   UUU↓AC↓AAA
 mm                                             m       m   m

            60            70            80            90            100
GAU↓UCCGUG  GAGAGGAACA   ACUCUGAGUC   UUAAACCAAU   UUUUUGAGGC
                                 U                             U

           110           118
CUUGUCUUGG  CAAGGCU(A)
UC    CUCCAA          OH
```

FIGURE 6.   Nucleotide sequence of U5 RNA of Novikoff
hepatoma or rat (44).   Mouse (45), human (44), and chicken
(44) nucleotide variants are shown below the main sequence.

U6 RNA.   U6 RNA is associated with purified perichromatin
granules (PCG , 46).   These dense granules are specifically
juxtaposed to chromatin and appear in electron micrographs as
dense cores surrounded by a white halo.   Because of their
association with newly synthesized mRNA (47), they could be
"carriers" or processing elements in or near the functional
chromatin.   Accordingly, U6 RNA and U6 RNP may be the first
U-snRNPs to combine with newly synthesized hnRNP.
     The structure of U6 RNA (48; Figure 7) differs from that
of the other U snRNAs in two major respects.   First, the cap
does not contain trimethyl G.   It is not known what structure
is linked to the nucleotide chain, but it is not a normal
nucleotide.
     Second, U6 RNA has several clusters of modified nucleo-
tides in the center of the molecule (Figure 7); in other U
snRNAs most modifications are in the 5' third of the structure.   The U6 RNA contains several 2'-O-methylated nucleotides
in addition to m^6A and m^2G.   This molecule appears to be

```
                                            6
            10            20            30            40   m    50
XpppGUGCUCGCU  UCGGCAGCAC  AUAUACUAAA  A↓UGGAACGA  ↓ACAGAGAAG
                                                           m

            60            70         2   80            90            100
AUUAGCAUGG   CCCCUGCGCA   AGGAUGACAC  GCAAAU↓CGU   GAAGCGUUCC
   mm        m  mm        m      m

           108
AUAUUUU(U)
         OH
```

FIGURE 7.   Nucleotide sequence of U6 RNA of Novikoff
hepatoma (48).

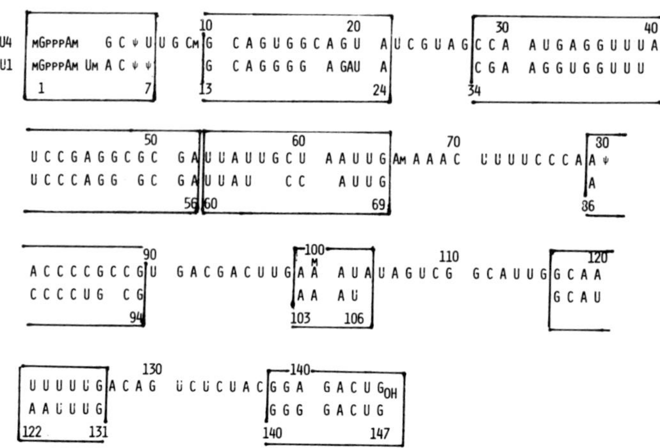

FIGURE 8. Homologies between Novikoff hepatoma U4 and U1 RNAs (44,45).

highly hydrogen bonded (48). Interestingly, the U1 to U5 RNAs have Am as their first nucleotide in the RNA sequence but U6 RNA has an unmodified Gp. When nucleotide sequences of the U snRNAs are analyzed, significant homologies are found (49), the most striking of which are between U1 and U4 RNA (Figure 8). Since U1, U2, and U4 to U6 RNPs contain one or more common proteins, these homologous regions may serve as binding sites for these protein(s). Although earlier studies concluded that U snRNAs of different tissues are similar, there may be minor differences in snRNAs of different tissues (1,2).

<u>Functions of snRNPs</u>. Sedimentation and electron microscopic studies suggest that there is a dynamic equilibrium between the free, small particles in the nucleus and the snRNP-hnRNP complexed particles. In cells producing large amounts of hnRNP, such as growing and dividing cells, there are millions of snRNPs interacting with the hnRNP particles, many of which have multiple cleavage sites. The lack of tight binding of the U-snRNP-hnRNP complexes has been noted by several groups (1,2). Many of the snRNPs are unattached, as are many hnRNP particles. At 37°, the hnRNP-snRNP complexes are unstable, which suggests that relatively weak binding forces exist between these two (50). When snRNP-hnRNP complexes are examined by electron microscopy, usually only one snRNP particle is found per hnRNP particle.

Do snRNPs Function in "Splicing" of hnRNA? The U-snRNPs
containing U1, U2, U4, U5, and U6 RNA are located in the
nucleoplasm. Their function could be in processes relating
to maturation of hnRNA in hnRNPs (1,2). The first clue to
the role of the snRNPs came from reports that U1, U2, and
other U snRNPs cosedimented with hnRNP particles (51). The
second came from studies showing that some snRNAs are found
hydrogen bonded to hnRNAs (52,53).

Lerner et al (29) analyzed all the available IVS sequence
near the splice consensus; sequences near the splice junctions
have been derived by others (reviewed in 54).

Mechanisms of splicing must account for Breathnach and
Chambon's "rule" that splice points have common sequences (GU
...AG) (54). This "consensus structure" has been a crucial
idea in the development of hypotheses relating to cleavage of
hnRNA. Two questions emerged from studies with hnRNA: How
does the cell make sense of the final product and why are mis-
takes relatively uncommon. When the sequences of known small
RNAs are analyzed, the 5' end of U1 RNA had good complementarity to consensus sequences (Figure 9) (29).

The evidence for the involvement of U1 RNP in splicing
of hnRNA is as follows: (a) the complementarity between conserved consensus sequences and U1 RNA; (b) U1 RNPs lacking
the 5' terminus sequence are not associated with hnRNPs (29);
(c) splicing of adenovirus hnRNA in HeLa cell nuclei is inhibited when they are preincubated with anti-Sm or anti-RNP
antibodies (56). When the nuclei are incubated with anti-Ro
or anti-La antibodies, splicing is not inhibited. Since other
cellular functions, e.g.,polyadenylation, are not inhibited,
and since anti-RNA or anti-Sm antibodies react with and presumably inactivate the U snRNPs, it is likely that U snRNPs
are required for splicing.

Do snRNPs Function in mRNP Transport? There is considerable uncertainty at present about the validity of the "splicing concept" of the function of snRNPs. Accordingly, it is
not surprising that alternative concepts have been proposed.
The idea that they are involved in "transport" is one we suggested some time ago. When the nuclear RNP network was first
described by Smetana et al (57), it was noted that "the con-
tinuity of the ribonucleoprotein network from nucleoli to the
nuclear membrane suggests that a defined pathway may exist for
transfer of ribonucleoprotein particles from the nucleolus to
cytoplasm". Later, Prestayko et al (37) noted "these small
molecules may serve a role in processing of nucleolar 28S RNA
and hence are associated only for a limited time. It is possible that these low molecular weight RNA's serve as essential
components for the movement of nucleolar products to the
nuclear ribonucleoprotein network".

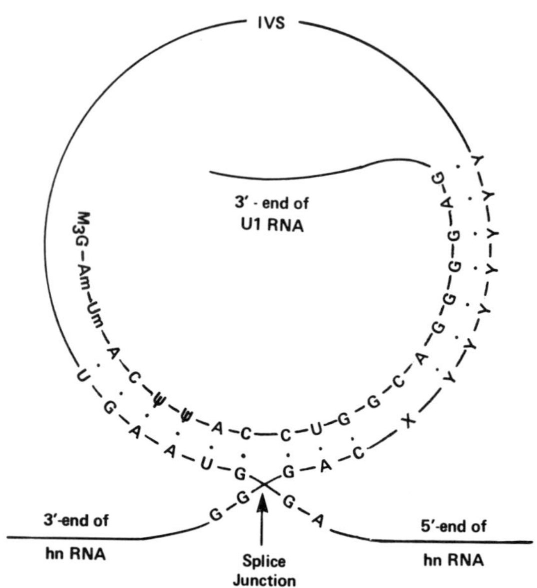

FIGURE 9.   A proposed model for splicing of hnRNA,
involving U1 snRNA.   The model is derived from others (29).
A similar model involving viral precursor RNA and viral snRNA
has been proposed (54).   The consensus sequences for the
splice junctions are known (29,54) and the sequence of U1 RNA
(2,19,27) were reported earlier.

Studies on nuclei immunostained with Sm-antibody showed
that dense particles the size of "interchromatin" granules
were distributed between the nucleolus and the chromatin lin-
ing the inner layer of the nuclear envelope (Fig. 10).   These

FIGURE 10.   A - Immunoelectron microscopic localization
of snRNP particles in the nucleus of Novikoff hepatoma ascites
cells utilizing Sm autoantibodies and the peroxidase-linked
second antibody method.   Immunostaining is localized to a re-
ticular network (arrowheads) extending between the perinucleolar chromatin around the nucleolus (No) and the juxta-nuclear
envelope chromatin (J).   The body of the nucleolus (No), cytoplasm (C) and the juxta-nuclear chromatin (J) were not immunostained.   X 25,000.   B - Novikoff hepatoma ascites cells
treated with normal human serum lack immunoreactive products
compared to cells treated with Sm antibody (Fig. A). X 33,000.

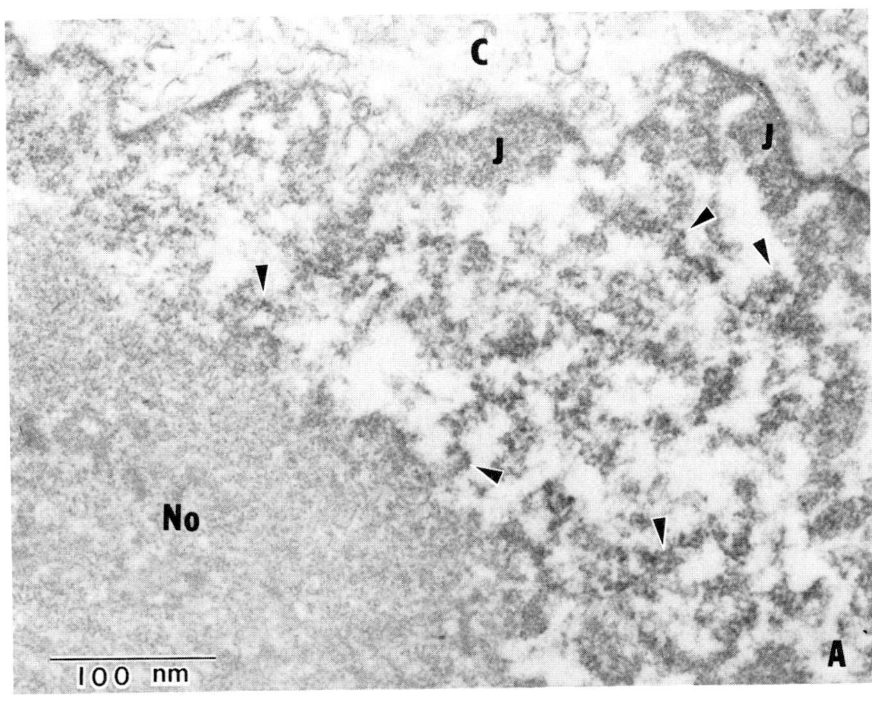
C
J
J
No
100 nm
A

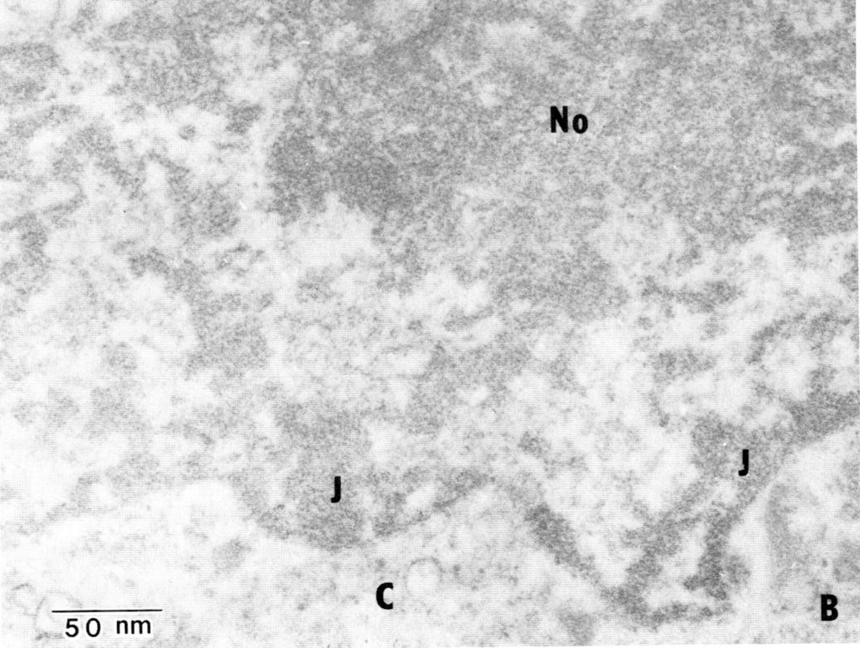
No
J
J
C
50 nm
B

dense particles have the size and density of the hnRNP-snRNP complexes and the "interchromatin granules" (Table 3) observed by electron microscopy (Fig. 11). The distribution of these particles throughout the nucleus was investigated (Table 4). No immunostained dense particles were observed either within the nucleolus or in the juxta-envelope chromatin (JEC). Moreover, in the areas containing large numbers of the particles, there were no differences in the particle concentration between the central, middle and peripheral portions of the nucleus (Table 4).

This equivalent density of the immunostained particles in this specific region of the nucleus along with the absence of such particles in the nucleolus and JEC is similar to previous observations in our studies on the nuclear RNP network or "nuclear matrix" (37). This evidence supports the idea that the nuclear envelope is the terminal binding site for the nuclear RNP network which is a dynamic terminus for the stream moving between the nucleolus and the vicinity of the nuclear pore.

Webb et al (58) expressed the idea that the nuclear RNP network is an "assembly line" or "conveyer belt" from the genes to nuclear pores or nuclear surface. Support for this idea derives from the recent report from our laboratory on the presence of a specific nuclear actin (59) and another protein now referred to as "nuclin" or Cp' which may have properties

TABLE 3
ARE THE snRNPs "TRANSPORT ELEMENTS"?

1. "Nuclear matrix" contains both hnRNP and
    snRNP in dense particles (=interchromatin
    dense granules).
2. RNases that decrease U1 snRNP and form
    U1*snRNP also decrease snRNP:hnRNP complexes.
3. Sm Ag is uniform in concentration in the
    central (core), middle and outer 1/3 nucleo-
    plasmic layers.
4. Sm Ag is absent from nucleolus and juxta-
    nuclear envelope chromatin.
5. Replacement of mRNA binding proteins occurs
    at nuclear-envelope:cytoplasmic junction.
6. Nucleus contains contractile elements -
    "nuclear actin" and "nuclin".

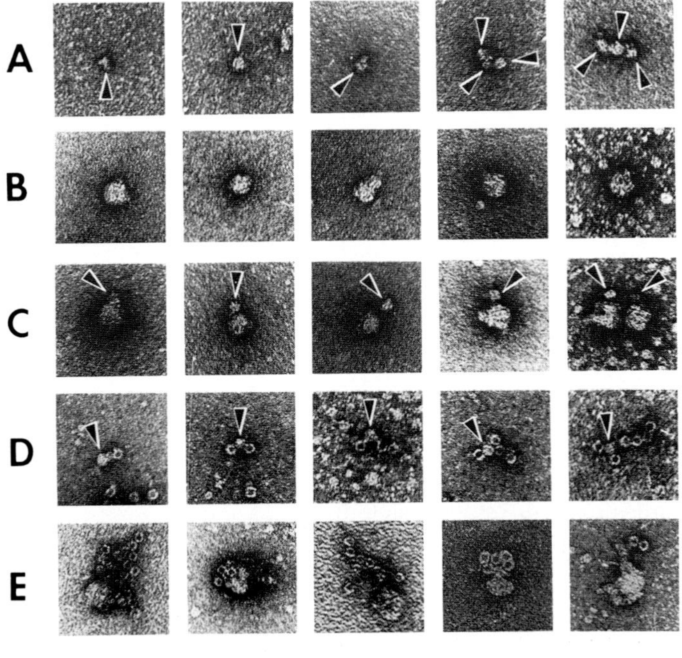

FIGURE 11.   Line A: Examples of electron micrographs of small particles (snRNPs) which were found in the 10-12S particle size fractions.   These particles are round or elliptical and are 115 Å in diameter; Line B: Examples of micrographs of large particles (hnRNPs) which were detected in the 30-40S particle size fractions.   These particles are 250 Å in diameter; Line C: Small and large complex particles which were also found in the 30-40S particle size fractions.   (Arrowheads indicate small particles.); Line D: Electron micrographs of small particles incubated with anti-Sm/anti-RNP antibodies and subsequently treated with ferritin-conjugated goat anti-human IgG.   Small particles with ferritin particles bridges result from antibody binding.   (Arrowheads show small particles.); Line E: Electron micrographs of clustered ferritin particles around small and large particle complexes.   Lines D and E indicate that small particles contain the determinants for anti-Sm/anti-RNP antibodies.

TABLE 4
NUMBER OF PARTICLES IMMUNOSTAINED BY
Sm-Ab IN VARIOUS NUCLEAR REGIONS

Region	Particles/cm^2
Nucleolus	0
Chromatin	
1/3 proximal to nucleolus	13.2±3.0
1/3 middle	14.4±2.4
1/3 peripheral	15.4±2
Juxta-envelope chromatin	0

similar to gelsolin. These two proteins may constitute critical elements of the elementary "conveyer belt" linking the
nucleolus to the nuclear envelope.

Sekeris and Guialis (52) suggested that snRNP particles
link the hnRNP particle(s) to the nuclear matrix; the hnRNP
complexes may consist of 1-5 hRNP particles (30-50S) which in
turn are linked to 1 snRNP. Presumably, these linkages occur
in or near the perinucleolar chromatin and after migration
through the nucleus on the RNP network, the complexes are dissolved by elements of the JEC where proteins of the hnRNP particles are removed and mRNP proteins are added.

The question then is, what are the mechanics of snRNP-
hnRNP linkage? At one point, the concept proposed by Lerner
et al (29) suggested that 20 nucleotides on the 5' end of U1
snRNA hydrogen-bonded to exon-intron junctions. It seems surprising and may be important that only U1 RNA and no other RNA
exhibits a similar "fit". Neither U2, U4, U5 nor U6 provide
corresponding complementarity to the exon-intron junctions.
In the studies of Epstein et al (31), it was concluded that
only the 5' 12 nucleotides of U1 RNA are available for complementarity, and of these, the 3 on the 5' terminal including 2
in the 5' cap did not exhibit complementarity. Accordingly,
only 9 nucleotides were involved in the hydrogen bonding
(which may be enough).

Studies in our laboratory have shown differing amounts of
U1*snRNP result from RNase cleavage, usually of the 6 5'-ter-
minal nucleotides of the U1 RNA (29). When U1*snRNP is found
in higher concentration, there is a decrease or absence of
electron microscopically observable snRNP-hnRNP complexes.
This result suggests that the major binding site of hnRNP-
snRNP complexes is in the terminal six nucleotides which

include the 3 methylated in the 5' cap. Accordingly, it may
be suggested that the 5' cap is the binding site rather than
12 of 21 5' terminal nucleotides. Since the cap structure is
common to all the U snRNPs, it would seem likely that they
represent the common junctional points which either bind to
protein, RNA or RNP of the hnRNP particles. It may well be
that the three methyl groups of m3G and other modified elements of the 5'-cap are the affinity points for hnRNP proteins, possibly the 40K "informatin" proteins.

DISCUSSION

General Concept of Assembly and Transport in the Nucleus
(Fig. 12). Figure 12 presents a scheme of assembly and transport of hnRNP-snRNP complexes. A likely site of the assembly
of hnRNP-snRNP complexes and their entry into the NRN-matrix
is the perinucleolar chromatin, a structure described in early
studies on the nucleolus (37) which is known to be composed
of highly repetitive DNA. At this site, snRNP which are
either free floating in the nucleoplasm or returned toward the
nucleolus by as yet undefined nuclear elements are juxtaposed
and bound to the hnRNP; the 5'-cap of snRNP would be the binding site for hnRNP proteins. Either through binding of the
snRNP, hnRNP or both this complex is shown linked to a "conveyer belt" or "moving stream" of nuclear contractile elements
such as nuclear actin which is polymerized in or adjacent to
the perinucleolar chromatin, either by unidirectional flow or
"pulsatile" effects.
     At the inner layer of the nuclear envelope in the JEC
(juxtaenvelope chromatin), several events are postulated to
occur. One is depolymerization of the moving or "conveyer
belt" system. The second is separation of the hnRNP and snRNP
which is returned to the nucleoplasm. The third is further
processing of the hnRNP to mRNP by displacement of the "informatin" proteins and addition of the specific mRNP proteins.
The mRNP then moves through the nuclear pore complex or the
nuclear envelope either alone or in complexes with ribosomes.
     This concept provides a basis for new experiments on the
polymerization and assembly reactions in the perinucleolar
chromatin, the mechanisms of transport of the hnRNP-snRNP complexes and the reactions involved in dissociation and maturation of the mRNP particles.

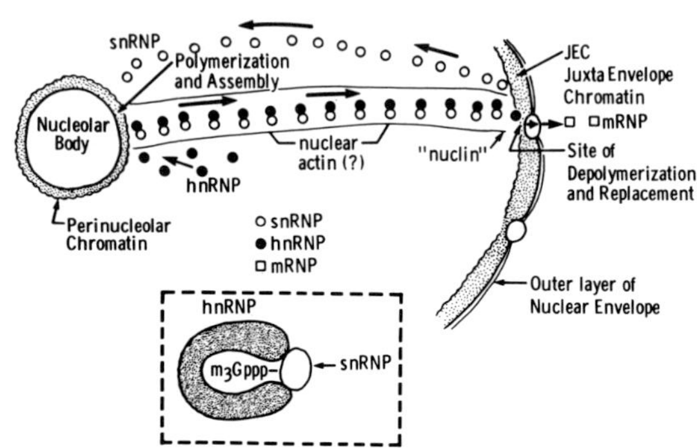

FIGURE 12. "Ski-lift" model of the role of snRNPs in transport of hnRNPs. Polymerization of matrix elements and entry of snRNPs into the matrix are considered to occur either in or adjacent to the repetitive DNA of the perinucleolar chromatin. There is unidirectional movement of matrix elements from the perinucleolar chromatin to the juxta-envelope chromatin. Entry of hnRNPs into this moving stream may occur as polymerization and assembly occurs or along the moving stream toward the nuclear envelope. Depolymerization, replacement and maturation of the hnRNP to mRNP are considered to occur either in the juxta-envelope chromatin or in or around the nuclear pore. It is also considered that the linkage between the snRNP (insert) occurs through the 5' cap and adjacent nucleotides of snRNPs and one or more protein or RNA regions of the hnRNPs.

REFERENCES

1. Busch, H., Reddy, R., Rothblum, L., and Choi, Y. C. (1982). Ann. Rev. Biochem. 51, 617-653.
2. Reddy, R., and Busch, H. (1981). In "The Cell Nucleus" (H. Busch, ed.), Vol. VII, pp. 261-305. Academic Press, New York.
3. Hodnett, J. L., and Busch, H. (1968). J. Biol. Chem. 243, 6334-6342.
4. Nakamura, T., Prestayko, A. W., and Busch, H. (1968). J. Biol. Chem. 243, 1368-1375.
5. Moriyama, Y., Hodnett, J. L., Prestayko, A. W., and Busch, H. (1969). J. Mol. Biol. 39, 335-349.
6. Prestayko, A. W., and Busch, H. (1968). Biochim. Biophys. Acta 169, 332-337.

7. Prestayko, A. W., Tonato, M., Lewis, C., and Busch, H. (1971). J. Biol. Chem. 246, 182-187.
8. Artman, M., and Roth, J. S. (1971). J. Mol. Biol. 60, 291-301.
9. Heyden, H. W., and Zachau, H. G. (1971). Biochim. Biophys. Acta 232, 651-660.
10. Chakravorty, A., and Busch, H. (1967). Cancer Res. 27, 789-792.
11. Higashi, K., Adams, H. R., and Busch, H. (1966). Cancer Res. 26, 2196-2201.
12. Sporn, M. B., and Dingman, W. (1963). Biochim. Biophys. Acta 68, 389-400.
13. Rosset, R., and Monier, R. (1963). Biochim. Biophys. Acta 68, 653-656.
14. Hadjiolov, A. A., Venkov, P. V., and Tsanev, R. (1966). Anal. Biochem. 17, 263-267.
15. Reich, P. R., Forget, B. G., Weissman, S. M., and Rose, J. A. (1966). J. Mol. Biol. 17, 428-439.
16. Pene, J. J., Knight, E., and Darnell, J. E. (1968). J. Mol. Biol. 33, 609-623.
17. Weinberg, R. A., and Penman, S. (1968). J. Mol. Biol. 38, 289-304.
18. Nohga, K., Reddy, R., and Busch, H. (1981). Cancer Res. 41, 2215-2220.
19. Branlant, C., Krol, A., Ebel, J. P., Lazar, E., Gallinaro, H., Jacob, M., Scri-Widada, J., and Jeanteur, P. (1980). Nucleic Acids Res. 8, 4143-4154.
20. Mount, S., and Steitz, J. A. (1982). Nucleic Acids Res. 9, 6351-6368.
21. Wise, J. A., and Weiner, A. M. (1981). J. Biol. Chem. 256, 956-963.
22. Reddy, R., Henning, D., and Busch, H. (1979). J. Biol. Chem. 254, 11097-11105.
23. Wise, J. A., and Weiner, A. M. (1980). Cell 22, 109-118.
24. Diener, T. O. (1981). Proc. Natl. Acad. Sci. USA 78, 5014-5015.
25. Lerner, E. A., Lerner, M. R., Janeway, C. A., Jr., and Steitz, J. A. (1981). Proc. Natl. Acad. Sci. USA 78, 2737-2741.
26. Hendrick, J. P., Wolin, S. L., Rinke, J., Lerner, M. R., and Steitz, J. A. (1982). Mol. Cell. Biol. 1, 1138-1149.
27. Reddy, R., Ro-Choi, T. S., Henning, D., and Busch, H. (1974). J. Biol. Chem. 249, 6486-6494.
28. Reddy, R., Henning, D., and Busch, H. (1981). Biochim. Biophys. Res. Commun. 98, 1076-1083.
29. Lerner, M. R., Boyle, J. A., Mount, S. M., Wolin, S. L., and Steitz, J. A. (1980). Nature 283, 220-224.

30. Lerner, M. R., and Steitz, (1979). Proc. Natl. Acad. Sci. USA 76, 5495-5499.
31. Epstein, P., Reddy, R., and Busch, H. (1981). Proc. Natl. Acad. Sci. USA 78, 1562-1566.
32. Korn, L. J., Queen, C. L., and Wegman, N. M. (1977). Proc. Natl. Acad. Sci. USA 74, 4401-4405.
33. Shibata, H., Ro-Choi, T. S., Reddy, R., Choi, Y. C., Henning, D., and Busch, H. (1975). J. Biol. Chem. 250, 3909-3920.
34. Reddy, R., Henning, D., Epstein, P., and Busch, H. (1981). Nucleic Acids Res. 9, 5645-5657.
35. Ro-Choi, T. S., Choi, Y. C., Henning, D., McCloskey, J., and Busch, H. (1975). J. Biol. Chem. 250, 3921-3928.
36. Busch, H., Ro-Choi, T. S., Prestayko, A. W., Shibata, H., Crooke, S. T., and El-Khatib, S. M. (1971). Perspect. Biol. Med. 15, 117-139.
37. Busch, H., and Smetana, K. (1970). In "The Nucleolus". pp. 285-314. Academic Press, New York.
38. Prestayko, A. W., Tonato, M., and Busch, H. (1971). J. Mol. Biol. 47, 505-515.
39. Reddy, R., Henning, D., and Busch, H. (1980). J. Biol. Chem. 255, 7029-7033.
40. Ro-Choi, T. S., Reddy, R., Choi, Y. C., Raj, N. B., and Henning, D. (1974). Fed. Proc. 33, 1548.
41. Reddy, R., Ro-Choi, Y. C., and Busch, H. (1972). J. Biol. Chem. 247, 7245-7250.
42. Reddy, R., Henning, D., and Busch, H. (1981). J. Biol. Chem. 256, 3532-3538.
43. Krol, A., Branlant, C., Lazar, E., Gallinaro, H., and Jacob, M. (1981). Nucleic Acids Res. 9, 2699-2716.
44. Krol, A., Gallinaro, H., Lazar, E., Jacob, M., and Branlant, C. (1981). Nucleic Acids Res. 9, 769-788.
45. Kato, N., and Harada, F. (1981). Biochem. Biophys. Res. Commun. 99, 1468-1476.
46. Daskal, Y., Komaromy, L., and Busch, H. (1980). Exp. Cell Res. 126, 39-46.
47. Puvion, E., and Moyne, G. (1981). In "The Cell Nucleus" (Busch, H., ed.), Vol. VIII, pp. 59-115. Academic Press, New York.
48. Epstein, P., Reddy, R., Henning, D., and Busch, H. (1980). J. Biol. Chem. 255, 8901-8906.
49. Busch, H., Reddy, R., Henning, D., and Epstein, P. (1980). In "International Cell Biology" (H. Schweiger, ed.), pp. 47-52. Berlin, Springer.
50. Zieve, G., and Penman, S. (1981). J. Mol. Biol. 145, 501-523.
51. Sekeris, C. E. and Niessing, J. (1975). Biochem. Biophys. Res. Commun. 62, 642-650.

52. Sekeris, C. E. and Guialis, A. (1981). In "The Cell
     Nucleus" (H. Busch, ed.), Vol. VIII, pp. 247-259.
     Academic Press, New York.
53. Northemann, W., Klump, H., and Heinrich, P. C. (1979).
     Eur. J. Biochem. 99, 447-456.
54. Breathnach, R., and Chambon, P. (1981). Ann. Rev.
     Biochem. 50, 349-383.
55. Yang, V. W., Lerner, M., Steitz, J. A., and Flint, S. J.
     (1981). Proc. Natl. Acad. Sci. USA 78, 1371-1375.
56. Murray, V., and Holliday, R. (1979). FEBS Lett. 106, 6-7.
57. Smetana, K., Steele, W. J., and Busch, H. (1963). Exp.
     Cell Res. 31, 198-202.
58. Webb, T. E., Schumm, D. E., and Palayoor, T. (1981).
     In "The Cell Nucleus" (H. Busch, ed.), Vol. IX, pp. 199-
     248. Academic Press, New York.
59. Bremer, J. W., Busch, H., and Yeoman, L. C. (1981).
     Biochemistry 20, 2013-2107.

ADENOVIRUS-2 MAJOR LATE PROMOTER : SEQUENCES UPSTREAM FROM THE TATA BOX ARE ESSENTIAL FOR EFFICIENT IN VIVO AND IN VITRO TRANSCRIPTION[1]

P. Sassone-Corsi, R. Hen, J. Corden[2], M.-P. Gaub, P. Chambon, and C. Kédinger

Laboratoire de Génétique Moléculaire des Eucaryotes du CNRS, Institut de Chimie Biologique, Faculté de Médecine, 11 Rue Humann, 67085 STRASBOURG Cédex, France

ABSTRACT We have analysed in vivo and in vitro the sequences required for specific transcription from the Adenovirus-2 major late promoter (Ad2MLP). Our results demonstrate that, in addition to the TATA box, sequences located further upstream between positions -97 and -34 (position +1 refering to the capsite), are necessary for efficient in vivo transcription. Recombinant plasmids containing wild-type or deleted Ad2MLP, linked to the SV40 T-antigen coding sequence, were introduced by microinjection or calcium-phosphate transfection into cells in culture. T-antigen expression was assayed by immunofluorescence and specific transcription was assayed by quantitative S1 nuclease mapping. Deletion of sequences located between positions -97 and -34 resulted in a strong reduction of T-antigen production and in a 30-fold decrease in specific RNA production. The effect of these upstream sequences was also investigated in vitro using a whole cell extract or an S100 transcription system and circular or linear templates. In the whole cell extract, the in vivo effect was faithfully reproduced when circular templates were used, but the effect was also seen with linear templates. With the S100 extract, some effect of the upstream sequences on specific in vitro transcription was seen with circular, but not with linear templates.

[1]This work was supported by grants from the CNRS (ATP 006520/50), from the INSERM (CRT 80.1029), from the "Association pour le Développement de la Recherche Française" and from the "Fondation Simone et Cino del Duca".
[2]Present address : Department of Microbiology, John Hopkins University School of Medicine, BALTIMORE, Maryland 21205.

INTRODUCTION

The comparison of DNA sequences in the 5' flanking region of a large number of genes coding for proteins has revealed two specific regions of homology. The highly conserved Goldberg-Hogness or TATA box sequence, is located about 30 nucleotides upstream from the cap site. Both in vitro and in vivo transcription experiments studies support the conclusion that this sequence is a promoter element required for efficient and selective transcription (see Ref. 1 and 2 for reviews). The second region of homology, sometimes called the CAAT box, is located at a variable position upstream from the TATA box and appears to be less conserved (3). This region is necessary for efficient in vivo transcription (4-9), but its effect in vitro is less clear : in some cases these upstream sequences had little or no effect (1, 10-14) while in others they seemed to be important for efficient transcription (15-18). In some viral genes a third promoter element has been found further upstream. This sequence, absolutely required in vivo, (19-23) appears to be dispensable in vitro (14).

In the case of the Adenovirus-2 major late promoter (Ad2MLP), previous in vitro studies using an S100 extract and a run-off assay (24), have shown that deletion of sequences located upstream from the TATA box had no effect on the specificity and the efficiency of transcription. We report here in vivo studies and in vitro experiments using a different transcription system, which demonstrate that such sequences are in fact required for efficient transcription from the Ad2MLP both in vivo and in vitro.

RESULTS

Microinjection in CV1 cell nuclei of recombinants containing the Ad2ML promoter reveals that sequences upstream from the TATA box are essential for efficient in vivo expression. Starting from a recombinant pMLA which contains the Ad2MLP region up to position -500, we constructed a series of deletions upstream from the TATA box of the Ad2MLP with endpoints located at -200, -150, -97, -62, -34, -29 and -21 (see Fig. 1 and Ref. 1). To study the function of the altered Ad2MLP sequences, we linked them to the coding sequence of the SV40 T antigen. The resulting recombinants constitute the pSVA series (see Fig. 1). After microinjection of the recombinant DNAs into CV1 cell nuclei, T antigen accumulation, as scored by indirect immunofluorescence (20) was used as a functional test for promoter activity. The results of such experiments are summarized in Table I. The same level of expression was obtained, irrespective of the amount of microinjected DNA, with the wild-type recombinant pSVA500 and recombinants pSVA200, 150 and 97 which were deleted in both directions from the unique XhoI site (position -260, see Fig. 1). In addition no decrease in T antigen expression was observed with a recombinant missing the sequences between -500 and -260 (not shown). Together, these results indicate that the sequences upstream from position -97 in the Ad2MLP are not required for efficient gene expression. However T antigen expression was significantly decreased at both DNA concentrations (especially at the lower DNA concentration which is likely to better reflect the relative promoter efficiency of the various deletion mutants) in pSVA34 and pSVA32, which have both an intact TATA box, but are deleted upstream from position -34 and -32,respectively (see Fig. 1). Deletion of the first two bases of the TATA box (pSVA29) or of the entire box (pSVA21) resulted in a further decrease. It is interesting to note that after the deletion of the TATA box the values obtained with pSVA29 and pSVA21, at the higher DNA concentration, are significantly greater than those obtained with pEMP which has no promoter sequence at all in front of the T antigen coding region. From these microinjection results it appears that at least two sets of sequences are important for the in vivo activity of the Ad2MLP. One corresponds to the TATA box region, whereas the other is located further upstream, between positions -34 and -97.

Insertion of the SV40 72 bp repeat in the pSVA recombinants allows detection of specific transcription from the Ad2ML promoter after transfection of HeLa cells. To analyse the effect of deletions in the Ad2MLP region at the transcriptional level, we decided to quantitate the amount of specific RNA synthesized in HeLa cells transfected with the various recombinants by S1 nuclease mapping. When pSVA500 (not shown) or pSVA34 (Fig. 2B, lane 1) were transfected in HeLa cells no specific band corresponding to RNA initiated from the Ad2MLP (position MLP+1) could be detected. This observation prompted us to investigate whether the SV40 72 bp repeat, which is known to enhance gene expression (20, 21), could be used to increase the amount of RNA transcribed from the Ad2MLP. The SV40 72 bp repeat was inserted in front of the deletion mutants pSVA34 and pSVA29, and the corresponding plasmids pSVBA34 and pSVBA29 (see Fig. 2A) were transfected in HeLa cells. It is clear from the results shown in Fig. 2B, lanes 2 and 8, that insertion of the 72 bp repeat results in a dramatic increase (perhaps several hundred fold) of the

amount of Ad2MLP specific RNA. In agreement with our previous results (20), insertion of the 72 bp repeat in the
opposite orientation [pSVBIA34 and 29, lanes 3 and 9 (Fig.2)]

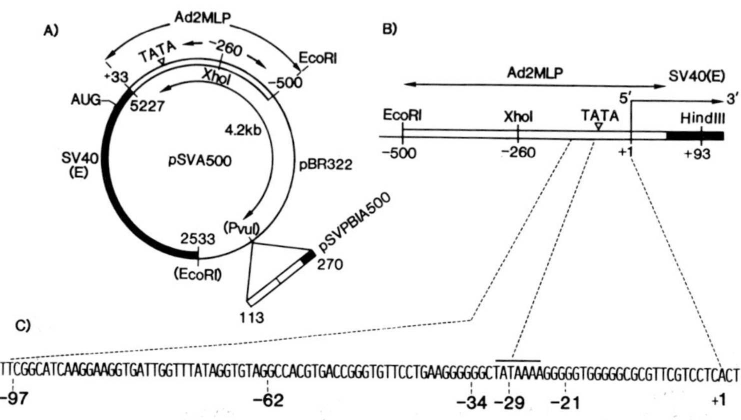

Figure 1 : Construction and structure of the pSVA recombinant series containing deletions in the Ad2ML promoter
region. Ⓐ pSVA plasmids were constructed by recombination
of parts of the previously described pEMP (27) and pMLA (1)
plasmids. pEMP contains the entire SV40 T antigen amino acid
coding sequence (without the early promoter region) between
coordinates 2533 and 5227 (20), inserted between the EcoRI
and BamHI sites of pBR322 (restriction sites in parenthesis
were lost during the constructions). pMLA contains the Ad2
BalI E fragment inserted into the EcoRI site of pBR322.
Deletions in the Ad2MLP region were created (arrows) from the
unique -260 XhoI site by limited exonuclease III digestion,
followed by S1 nuclease treatment and blunt-end ligation (1).
The wild type or deletion containing fragments were excised
from the corresponding pMLA recombinants with SalI (position
650 in pBR322) and PvuII (position +33 in the Ad2MLP) and
inserted between the repaired BamHI and the non-repaired SalI
sites of pEMP. pSVA500 contains the wild-type Ad2MLP region
from -500 to +33 (double line) inserted upstream from the
SV40 early (E) coding sequences (heavy line). Recombinants
pSVA200, 150, 97, 62, 34, 32, 29 and 21 contain 200, 150, 97,
62, 34, 32, 29 and 21 bp upstream from the initiation site
(+1), respectively, as determined by sequencing (pSVA97, 62,
34, 32, 29, 21) or restriction enzyme mapping (pSVA500, 200,
150). pSVPBIA500, 150, 97, 62, 34, 29 and 21 correspond to
the pSVA series in which a SV40 DNA fragment (coordinates 113
to 270, see Ref. 20), containing the 72 bp repeat, was inserted in the PvuI site of pBR322 in the orientation opposite

TABLE I

MICROINJECTION OF pSVA RECOMBINANTS IN CV1 CELL NUCLEI[a]

Recombinants	Immunofluorescent CV1 cells (%)			
	100 μg DNA/ml		33 μg DNA/ml	
pSVA500	100	(6)	100	(2)
pSVA200	89	(3)	140	(2)
pSVA150	95	(3)	127	(2)
pSVA97	110	(3)	138	(2)
pSVA34	32	(6)	6	(2)
pSVA32	31	(2)	5	(2)
pSVA29	7	(6)	0	(2)
pSVA21	9	(3)	0	(2)
pEMP	0	(6)	0	(2)

[a]About 2.10^{-11} ml of recombinant DNA at 33 μg/ml or 100 μg/ml was injected per cell nucleus as described previously (20). pEMP and the pSVA series are described in Fig.1. The results were expressed as the % of injected cells which were positive for T antigen, taking as 100 % the value obtained with pSVA500. Numbers in parentheses correspond to the number of independent assays carried out for each recombinant.

(Fig. 1 Continued)

to that of the SV40 (E) sequence. The direct repeat is re-presented by the open boxes and a flanking segment by the closed box. (B). Structure of the boundary between the Ad2MLP (double line) and the SV40 early region (heavy line) in pSVA500. The origin and direction of specific transcription from the Ad2MLP is shown by the arrow at +1. (C) Nucleotide sequence of the non-coding strand of the Ad2MLP region (28) from positions -99 to +3. The TATA box sequence is overlined and the 3'-boundary of 5 deletion mutants are indicated by vertical dashed lines.

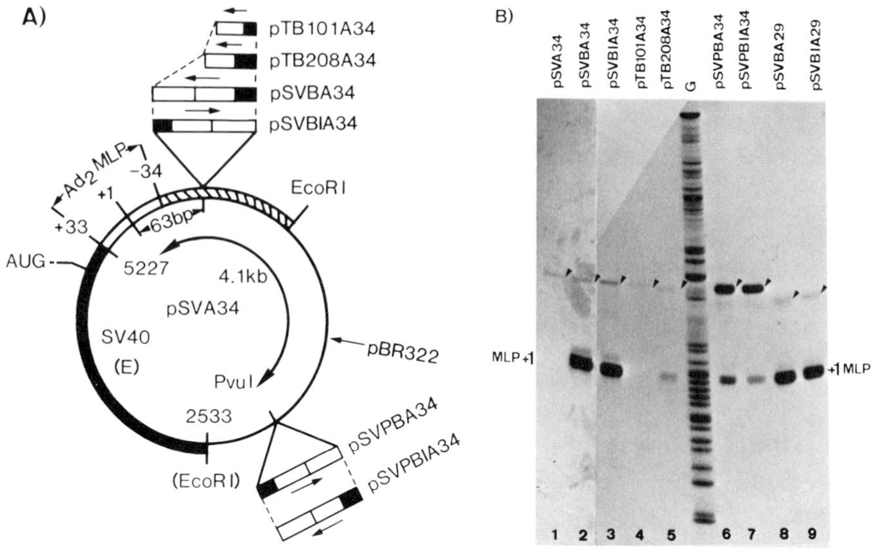

Figure 2 : The SV40 72 bp repeat enhances specific in vivo transcription from the Ad2ML promoter. Ⓐ Construction of derivatives of pSVA34 and pSVA29 containing the wild-type or mutated SV40 72 bp repeat. pSVA34 and pSVA29 were derived from pSVA500 as described in legend to Fig. 1. pSVA34 only is represented and the replacing adenovirus sequences upstream from -34 are hatched. The SV40 DNA segment containing the 72 bp repeat (see Fig. 1A) was then inserted in either orientation [the arrows indicate the natural orientation with respect to the SV40 (E) sequence] at 63 bp (SstI site, see Ref. 20) (pSVBA34 and pSVBIA34) or at 4.1 kb (pSVPBA34 and pSVPBIA34) from the Ad2ML capsite. Two other recombinants (pTB101A34 and pTB208A34) contain the fraction of the 72 bp repeat sequence present in the two deletion mutants, TB101 and TB208, previously described (20). pSVBA29 and pSVBIA29 were constructed as pSVBA34 and pSVBIA34, respectively. Ⓑ S1 nuclease mapping of the 5' end of RNA transcripts isolated from HeLa cells transfected with the pSVA34 and pSVA29 series. Calcium-phosphate transfection of HeLa cells, at 20-40 % confluence, was as previously described (20), using 20 µg of DNA per 10 cm Petri dish for each Ad2MLP recombinant. After 48 hr total cytoplasmic RNA was prepared from cells lysed with 0.3 % NP40, purified by phenol-chloforom extractions and ethanol-precipitated. The RNA (50 µg, equivalent to about one Petri dish) was dissolved in 10 µl of hybridization buffer containing 10 mM Pipes pH 6.5, 400 mM NaCl and an excess of the HindIII-XhoI single-stranded probe (5' end-labeled at the HindIII site, see Fig. 1B), heated 5 min at 85°C and hybridized at 68°C for 12 hr. The hybrids were di-

did not affect the magnitude of the 72 bp repeat effect. Deletions within the 72 bp repeat, known to drastically reduce its effect, resulted in a striking decrease of specific RNA [see Fig. 2 ; compare pSVBA34 (lane 2) with pTB101A34 (lane 4) and pTB208A34 (lane 5)]. It is interesting to note that no RNA could be detected with the deletion mutant pTB101A34, which suggests that the sequences deleted in this recombinant are essential for the activity of the SV40 72 bp repeat.

It is noteworthy that the pSVBA29 and pSVBIA29 mutants, which have lost the first two bases of the TATA box, were almost as efficient in promoting specific transcription from the Ad2ML capsite as the pSVBA34 and pSVBIA34 mutants, which possess an intact TATA box (compare in Fig. 2B, lanes 2 and 3 with lanes 8 and 9). This observation, which is in apparent contradiction with the results shown in Table I for the pSVA34 and pSVA29 recombinants, suggests that insertion of the SV40 72 bp repeat close to the Ad2ML capsite can mask the effect of deletion of promoter sequences. Since it was known from our previous studies (20) that the effect of the 72 bp repeat decreases with increasing length of intervening sequences, we inserted it in the PvuI site of pBR322, more than 4 kb upstream from the Ad2MLP region (pSVPBA34, pSVPBIA34 and pSVPBIA29, see Fig. 1 and 2A), hoping to minimize this masking effect. The following results were obtained : first, the amount of specific RNA present in cells transfected with the pSVPBA34 and pSVPBIA34 mutants was indeed much lower than that found in cells transfected with the corresponding pSVBA mutants (compare in Fig. 2B lanes 6 and

(Fig. 2 Continued)

gested with 2000 units of S1 nuclease at 25°C for 2 hr, in a final volume of 100 μl containing 30 mM sodium acetate pH 4.6, 0.3 M NaCl and 3 mM ZnCl$_2$. The S1 nuclease-resistant DNA probe was extracted with phenol-chloroform, precipitated twice with ethanol, and analyzed on a 8 % acrylamide-8.3 M urea gel as described (25). Lane G corresponds to a G sequence ladder (25) of the probe. The S1 nuclease-resistant probe fragments corresponding to specific transcription from the Ad2MLP are indicated by MLP + 1. Bands pointed out by the arrowheads correspond to nonspecific transcription initiated upstream from the deletion end-points (see Results).

7 with lanes 2 and 3). Second, as expected, a difference could now be seen between the recombinant possessing an intact TATA box (pSVPBIA34, lane 5, Fig. 3A) and a recombinant containing a partially deleted TATA box (pSVPBIA29, lane 6, Fig. 3A). We are confident that all of the differences described above are real since the same results were repeatedly observed in several independent transfection experiments performed with different preparations of recombinant DNAs.

It is striking that moving the 72 bp repeat 4 kb upstream from the Ad2MLP resulted in a dramatic increase in RNA initiated upstream from the deletion end-point (compare pSVPBA34 and pSVPBIA34 with pSVA34, pSVBA34 and pSVBIA34, arrowheads in lanes 6, 7 and 1-3, respectively). In fact S1 Nuclease mapping experiments have revealed that insertion of the SV40 72 bp repeat in the <u>Pvu</u>I site of pBR322 (pSVPBA recombinants) promoted RNA synthesis from several nearby sites (R. Hen, unpublished experiments).

S1 nuclease mapping of RNA synthesized in vivo, demonstrates that sequences located upstream from the TATA-box are essential for efficient transcription from the Ad2ML promoter. The pSVPBIA series of recombinants (Fig. 1) were transfected into HeLa cells and the RNA accumulated after 48 hrs was analyzed by quantitative S1 nuclease mapping (Fig. 3). The results shown in Fig. 3A (see also Table 2 for densitometric quantitation) indicate that deletions of sequences upstream from -97 have no effect on the amount of RNA initiated from the Ad2MLP (compare pSVPBIA500, 150 and 97 in lanes 1-3). Deletions extending further downstream resulted in a dramatic decrease of specific transcription : a 3-fold decrease when the deletion extended to -62, and a further 10-fold decrease when extended to -34 (compare lanes 3-5 in Fig. 3A and see Table 2) eventhough the TATA box was still intact. Deleting the first two bases of the TATA box generated an additional 2-fold reduction (compare pSVPBIA34 and 29, lanes 5 and 6), whereas deletion of the entire TATA box resulted in a further 2.5-fold decrease (compare pSVPBIA29 and 21, lanes 6 and 7). It is remarkable that even in the absence of the TATA box, the only band (excluding the deletion end-point band) which was seen on the original autoradiogram corresponds to RNA initiated at the Ad2ML capsite (pSVPBIA21, lane 7). This band was not an artefact of S1 nuclease mapping, since it was absent in the lane corresponding to pEMPB*, a recombinant which lacks all of the Ad2MLP region (see legend to Fig. 3A and lane 8).

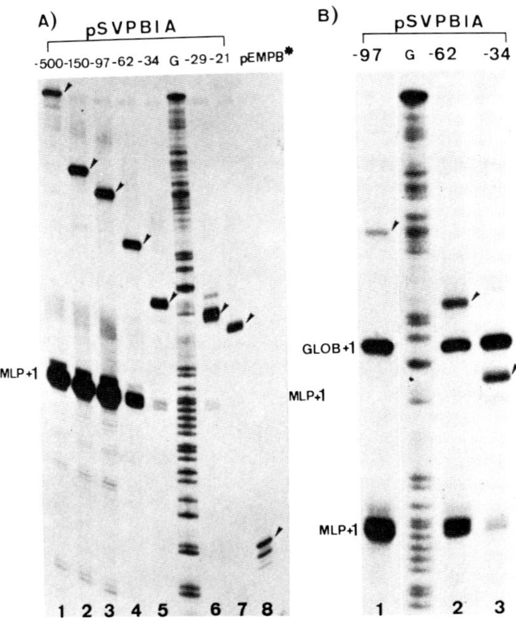

Figure 3 : Effect of deletions in the Ad2ML promoter region on specific in vivo transcription. (A) Quantitative S1 nuclease mapping of Ad2MLP transcripts produced in HeLa cells transfected with pSVPBIA500 and its deletion mutants pSVPBIA150, 97, 62, 34, 29 and 21 (Fig. 1A) as indicated. The specific transcripts were mapped with the HindIII-XhoI probe as in Fig. 2B. pEMPB* corresponds to pSVPBIA500 (see Fig. 1A) but lacks all the Ad2MLP sequence. (B) Same as in A, but the HeLa cells were cotransfected as indicated with the pSVPBIA mutants (20 µg DNA per dish) and a polyoma-β globin recombinant [pβ(244+)β in Ref. 22] (2 µg DNA per dish) used as an internal transcription control. The probe used to map the specific globin transcripts was a single-stranded BstNI restriction fragment (202 nucleotides), prepared from the globin recombinant. GLOB+1 refers to the position of the S1 nuclease-resistant band which corresponds to RNA initiated at the β-globin capsite. Experimental conditions and other symbols (G and arrowheads) are as in legend to Fig. 2.

Two sets of observations indicate that the differences found between the various pSVPBIA recombinants in Fig. 3A actually reflect the effect of the deletions and not variations in transfection efficiency. First, in all cases there was very little variation in the amount of RNA initiated upstream from the deletion end points, as shown by the intensity of the bands pointed out by arrowheads in Fig. 3A (for all recombinants, the same wild-type HindIII-XhoI probe was used, see legend to Fig. 3). Second, a rabbit β globin recombinant [pβ(244+)β, see Ref. 22] was cotransfected in HeLa cells as an internal control together with pSVPBIA97, 62 and 34. As shown in Fig. 3B, there was little variation in the intensity of the globin band (GLOB+1, in lanes 1, 2 and 3), whereas the differences between the various pSVPBIA recombinants were very similar to those seen in panel A [from the intensity of the deletion end-point bands (arrowheads, in lanes 1, 2 and 3) it is likely that the amount of pSVPBIA97 DNA which was actually transfected (lane 1) was lower than the other two pSVPBIA recombinants (lanes 2 and 3)].

The effect of the upstream sequences of the Ad2ML promoter can be faithfully reproduced in vitro. From the above results it was clear that sequences located between -97 and -34 were essential for efficient in vivo transcription from the Ad2MLP. Our previous in vitro studies using an S100 extract and a run-off assay did not demonstrate any effect of these sequences in specific in vitro transcription from the Ad2MLP (1). We therefore reinvestigated the promoter sequence requirements for in vitro transcription using the quantitative S1 nuclease assay and varying the method of preparation of the cellular extract and the form of the template.

In agreement with our previous run-off experiments (1), deletions down to position -34 did not affect specific initiation (MLP+1) from a linear template with the S100 extract (Fig. 4A, lanes 1-3), whereas deletion of the TATA box abolished it (pSVPBIA21, lane 4). In contrast, when the same experiments were carried out with a whole cell extract (WCE) the effect of the upstream sequences was clearly seen using the S1 nuclease or the run-off assays (Fig. 4B). The S1 nuclease mapping analysis shown in lanes 1-3 indicates that deletions of sequences upstream from -62 and -34 resulted,respectively,in a 3-and 20-fold decrease (see also Table 2) of specific transcription, relative to pSVPBIA97 which was as efficiently transcribed as the wild-type pSVPBIA500 (not shown). Similar results were obtained using the run-off assay (Fig. 4B, lanes 5-8). In this experiment transcription of an Ad5E4 linear template was used as an internal control. As with the S100 extract, no specific transcription was detected when the TATA box was deleted (lanes 4 and 8).

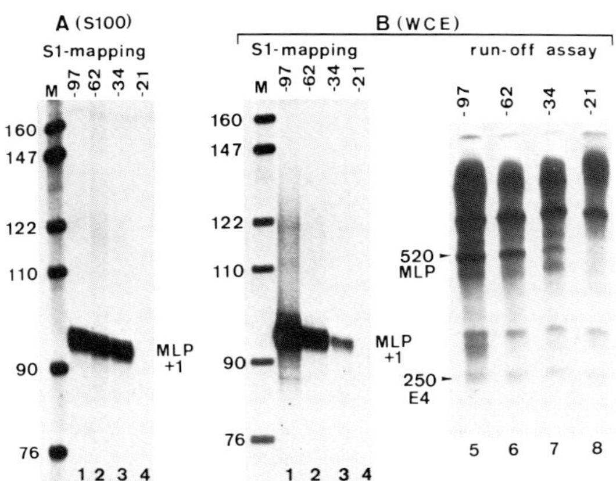

Figure 4 : Effect of deletions in the upstream region of
the Ad2ML promoter on specific in vitro transcription of
linear templates. (A) In vitro transcription with S100 extract. S100 extract was prepared from HeLa cells as described
(24).   Unlabelled RNA was synthesized in a three-fold standard reaction (29) in the presence of 10 µg/ml of the
pSVPBIA97, 62, 34 and 21 deletion mutants (lanes 1 to 4 respectively), cut with TaqI (which cuts the molecules into 8
fragments of which one, containing the Ad2MLP region, extends
from position 23 in pBR322 to position 4739 in SV40). After
purification the RNA was analyzed by quantitative S1 nuclease
mapping with the HindIII-XhoI probe as in Fig. 2B. (B) In
vitro transcription with a whole cell extract (WCE). The WCE
was prepared as described (26). RNA was synthesized in a
three-fold standard reaction (26) in the presence of 15 µg/ml
of the same TaqI-digested templates as in A. Lanes 1 to 4
show the S1 nuclease mapping analysis (as in A) of unlabelled
RNA synthesized on pSVPBIA97, 62, 34 and 21, respectively.
Lanes 5 to 8 show the electrophoretic analysis (5 % acryla-
mide-8.3 M urea) of labelled run-off RNA synthesized under
the same conditions but in the presence of $[\alpha-^{32}P]$ CTP.   A
constant amount (6 µg/ml) of a fragment containing the adenovirus type 5 (Ad5) E4 promoter region was added as an internal transcription control [the Ad5 SmaI restriction fragment
(98.4 to 100 map units, see Ref. 30) was cloned into pBR322;
after excision with SmaI and SphI, the resulting 0.7 kb DNA
fragment was purified and used as template]. The specific
transcripts from the Ad2ML and Ad5E4 promoters are indicated
by arrowheads (520 and 250 nucleotides in length, respectively). DNA size markers shown in lanes M are $[^{32}P]$-end-label-
led pBR322-MspI restriction fragments.

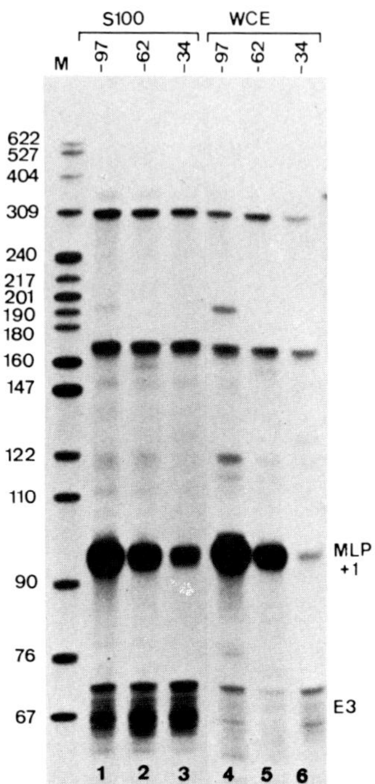

Figure 5 : Effect of deletions in the upstream region of the Ad2ML promoter on specific in vitro transcription of circular templates with a S100 or a whole cell extracts. The indicated pSVPBIA deletion recombinants were transcribed as circular templates in the presence of S100 (lanes 1 to 3) or WCE (lanes 4 to 6) extracts as described in the legend to Fig. 4. About 70 % of the DNA molecules were superhelical. As an internal transcription control, 3 μg/ml (lanes 1-3) and 7 μg/ml (lanes 4-6) of a recombinant plasmid containing the adenovirus type 5 (Ad5) E3 promoter region [the Ad5 EcoRI C restriction fragment (31), inserted into the EcoRI site of pBR322] was added and co-transcribed in each reaction. The synthesized RNA was processed and analyzed by S1 nuclease mapping with the HindIII-XhoI probe as described in Fig. 2B. The probe used to map the specific E3 transcripts was a single-stranded Sau3A restriction fragment (302 nucleotides) spanning the E3 capsites and 5'end-labelled with ^{32}P. MLP + 1 and E3 refer to the S1 nuclease-resistant bands which correspond to RNA initiated at the Ad2ML and the Ad5E3 capsites, respectively. M is as in Fig. 4B.

TABLE 2

COMPARISON OF THE IN VIVO AND IN VITRO EFFICIENCY OF
ADENOVIRUS-2 MAJOR LATE PROMOTER MUTANTS[a]

Recombinants	In vivo	In vitro			
		S100 extract		whole cell extract	
		linear template	circular template	linear template	circular template
pSVPBIA97	100	100	100	100	100
pSVPBIA62	33	100	53	30	33
pSVPBIA34	3	80	20	5	3
pSVPBIA21	0.6	0	0	0	0

[a]In vivo and in vitro efficiency of the Ad2MLP mutants
was determined by quantitative S1 nuclease mapping as described in Figs. 3-5. The intensity of the specific band
corresponding to initiation from the Ad2MLP (MLP+1, in Fig.
3-5) was determined by densitometry of autoradiograms exposed
for various periods of time. The values for the pSVPBIA21
recombinant used as circular templates were taken from unpublished experiments. In all cases the results were expressed
in % of the corresponding value obtained with pSVPBIA97.

It has been recently suggested (17) that free ends of
DNA could serve as entry sites for RNA polymerase B and thereby could mask the effect of promoter sequences with similar
function. Therefore circular recombinant DNAs (mainly superhelical, see legend to Fig. 5 and Discussion) were used as
templates with either the S100 or the whole cell extract
(WCE) (26) and specific transcription from the Ad2MLP was
measured by S1 nuclease mapping (Fig. 5). An Ad5E3 template
was included in the assay as an internal control (see legend
to Fig. 5). When the whole cell extract was used in the
presence of a circular template, deletion of the sequences
upstream from -34 appeared to have a more pronounced effect
than in the presence of a linear template (compare Fig. 5,
lane 4-6 with Fig. 4B, lanes 1-3). Using a circular template no specific transcripts were observed with the -21 mutant
(unpublished results). It is noteworthy that the effect of
the -62 and -34 deletions on in vivo transcription (Fig. 3
and Table 2) are faithfully reproduced in vitro with the
whole cell extract and circular templates. In addition it is
striking that with the S100 extract the use of circular
templates allowed us to see effects of the upstream sequences

which were totally masked when linear templates were used
(compare lanes 1 to 3 in Fig. 5 and 4A). However, the S100
extract appeared to be less efficient than the whole cell
extract at revealing this effect (compare lanes 1-3 with 4-6
in Fig. 5).

DISCUSSION

As previously shown for other RNA polymerase B promoters
(see introduction) our results demonstrate that sequences
located upstream from the TATA box are important for efficient in vivo promotion of RNA synthesis from the Ad2ML promoter.
Sequences upstream from -97 were not required for efficient expression, whereas sequences situated between -97 and
-34 played a major role in the efficiency of transcription.
Our results suggest either that two sets of sequences are
involved and that the effect of the -62 to -34 region is more
important than that of the -97 to -62 region, or that only
one sequence, centered on position -62, is implicated. This
second possibility is supported by a sequence homology
between the Ad2MLP sequence around -62 and the upstream sequence of the rabbit β-globin gene. This region of the globin
gene is homologous to the upstream consensus sequence previously pointed out (3) and was recently shown to be important
for efficient transcription in vivo :

<pre>
                                        -76
                                         |
Rabbit β globin   5' - C T G G T G T T G G C C A A T - 3'
                        ‾‾‾‾‾‾‾‾‾‾‾   ‾‾‾‾‾‾‾‾‾‾‾
Ad2MLP   5' - T A G G T G T A G G C C A C G - 3'
                                         |
                                        -62
</pre>

It has indeed been shown that mutations of both the two Gs
(at positions -77 and -76, see Ref. 32) or of either one of
the two Cs (at position -75 and -74, P. Dierks, A. Van Ooyen
and C. Weissman, personal communication) of the rabbit β
globin gene resulted in an about 10-fold reduction of in vivo
RNA synthesis. Additional mutations are currently being constructed to test the functional significance of the corresponding Ad2MLP sequence.
As expected from previous results (see Introduction)
the additional deletion of the TATA box induced a further
decrease in specific in vivo transcription. It is however
striking that even when the TATA box was completely deleted,
the residual transcription (0.6 % of the wild type) appeared
to be still mainly initiated from the Ad2ML capsite. This is

in contrast with the Ad2MLP in vitro situation [see Fig. 4
and 5 and our previous report (1)] where deletion of the TATA
box completely abolished the specific transcription and with
the SV40 in vivo situation where deletion of the TATA box
resulted in the appearance of multiple initiation sites.
However Dierks et al. (personal communication) have recently
shown, with the rabbit β-globin gene, that deletion of the
TATA box and of all further upstream sequences reduces the in
vivo transcription by 97 to 99 %, but that initiation is
still taking place predominantly from the normal capsite.
These observations suggest that some sequences located downstream from the TATA box are involved in the process which
directs the transcription machinery to initiate at the capsite.

The SV40 72 bp repeat element was required to obtain,
after HeLa cell transfection, an amount of RNA sufficient to
carry out analyses by S1 nuclease mapping. However it is
unlikely that its insertion at more than 4 kb from the adenovirus sequence differentially modulates the activity of the
elements of the Ad2ML promoter which appears to be stimulated
as a whole. Indeed, the effect of the various deletions was
roughly the same, whether it was measured indirectly, after
microinjection in CV1 cell nuclei of recombinants lacking the
72 bp repeat, or directly, at the RNA level, after transfection of HeLa cells with recombinants containing the 72 bp
element 4 kb from the Ad2MLP. However, it is clear from the
results shown in Fig. 2B that insertion of the SV40 72 bp
repeat in a position too close to Ad2MLP elements can mask
their function. These observations, and those which suggest
that the effect of the 72 bp repeat is the strongest on initiation from proximal sequences (see results), are in agreement with our previous hypothesis (20), namely that the SV40
72 bp repeat acts as an entry site for RNA polymerase B.
Obviously a similar element is not present in the first 500
bp located upstream from the Ad2ML capsite. Whether it is
present further upstream and plays a role in stimulating initiation of transcription from the Ad2MLP during viral infection remains to be established.

Our results demonstrate that the in vivo effect of the
upstream sequences (-97 to -34) of the Ad2MLP can be faithfully reproduced in vitro with a circular template and a
transcription system where the source of RNA polymerase B and
initiation factor(s) is a whole cell extract (see Figs. 3A
and 5). The circularity of the DNA is not absolutely required, since an effect of the upstream sequences was also seen,
with a linear template (Fig. 4B). A similar observation was
previously reported for in vitro transcription of a sea urchin H2A histone gene (17). It is striking that, in agreement

with our previous results, no effect of the upstream sequences on specific transcription of a linear template could be demonstrated when the whole cell extract was replaced by an S100 preparation. Assuming that the free ends of the linear template can functionally replace the upstream sequences [by providing entry sites as suggested by Grosschedl and Birnstiel (17)], one could interpret this observation by assuming that there are proteins present in the whole cell extract, but not in the S100 extract, which are capable of binding to the ends of DNA fragments. However such an explanation does not account for the observation that the whole cell extract system is more efficient at mimicking the in vivo situation than the S100 system, even on a circular template (Fig. 5). This difference cannot be ascribed to the routine addition of calf thymus RNA polymerase B to the S100 system, since the same results were obtained when transcription was performed by the RNA polymerase present in the S100 extract, in the absence of exogenous polymerase (results not shown). Further studies will show whether the whole cell extract contains some factor(s), present in limiting amount in the S100, and required to reveal the effect of the upstream sequences. It is unlikely that the role of such factor(s) would be to participate to the assembly of a chromatin structure possibly necessary to reveal the effect of the upstream sequences. In fact, we have analyzed the fate of the superhelical DNA template after various times of incubation and found that it was transformed within less than 5 min into covalently closed relaxed circles by the nicking-closing enzyme present in the extract (P. Sassone-Corsi, unpublished observation) whereas if nucleosomes would have been formed, at least some superhelicity would have been conserved (33). In any case the availability of an in vitro transcription system, which faithfully mimicks the in vivo requirement for promoter sequences located upstream from the TATA box, should be extremely useful for elucidating, at the molecular level, how the transcription machinery interacts with these sequences.

ACKNOWLEDGMENTS

We thank P. Moreau for gift of materials, W. Schaffner for his help with cell transfection and T. Leff for a critical reading of the manuscript. We gratefully acknowledge the technical assistance of K. Dott, C. Hauss, B. Boulay, E. Badzinski, C. Kutschis and C. Werlé. R. Hen is a fellow of the "Ligue Nationale Française contre le Cancer" and J. Corden of the "Université Louis Pasteur - Strasbourg".

REFERENCES

1. Corden, J., Wasylyk, B., Buchwalder, A., Sassone-Corsi, P., Kédinger, C. and Chambon, P. (1980) Science 209, 1406-1414.
2. Breathnach, R. and Chambon, P. (1981) in Ann. Rev. Biochem. (E.E. Snell, P.D. Boyer, A. Meister, C.C. Richardson, Eds.) Vol. 50, Annual Reviews Inc., pp. 349-383.
3. Benoist, C., O'Hare, K., Breathnach, R. and Chambon, P. (1980) Nucleic Acids Res. 8, 127-142.
4. Grosschedl, R. and Birnstiel, M.L. (1980) Proc. Natl. Acad. Sci. USA 77, 7102-7106.
5. Dierks, P., Van Ooyen, A., Mantei, N. and Weissmann, C. (1981) Proc. Natl. Acad. Sci. USA 78,1411-1415.
6. McKnight, S.L., Gavis, E.R., Kingsbury, R. and Axel, R. (1981) Cell 25, 385-398.
7. Mellon, P., Parker, V., Gluzman, Y. and Maniatis, T. (1981) Cell 27, 279-288.
8. Struhl, K. (1981) Proc. Natl. Acad. Sci. USA 78, 4461-4465.
9. Grosveld, G.C., de Boer, E., Shewmaker, C.K. and Flavell, R.A. (1982) Nature 295, 120-126.
10. Wasylyk, B., Kédinger, C., Corden, J., Brison, O. and Chambon, P. (1980) Nature 285, 367-373.
11. Grosveld, G.C., Shewmaker, C.K., Jat, P. and Flavell, R.A. (1981) Cell 25, 215-226.
12. Hu, S.-L. and Manley, J.L. (1981) Proc. Natl. Acad. Sci. USA 78, 820-824.
13. Mathis, D.J. and Chambon, P. (1981) Nature 290, 310-315.
14. Tsai, S.Y., Tsai, M.-J. and O'Malley, B.W. (1981) Proc. Natl. Acad. Sci. USA 78, 879-883.
15. Myers, R.M., Rio, D.C., Robbins, A.K. and Tjian, R. (1981) Cell 25, 373-384.
16. Tsuda, M. and Suzuki, Y. (1981) Cell 27, 175-182.
17. Grosschedl, R. and Birnstiel, M.L. (1982) Proc. Natl. Acad. Sci. USA 79, 297-301.
18. Lee, D.C., Roeder, R.G. and Wold, W.S.M. (1982) Proc. Natl. Acad. Sci. USA 79, 41-45.
19. Benoist, C. and Chambon, P. (1981) Nature 290, 304-310.
20. Moreau, P., Hen, R., Wasylyk, B., Everett, R., Gaub, M.P. and Chambon, P. (1981) Nucleic Acids Res. 9, 6047-6068.
21. Banerji, J., Rusconi, S. and Schaffner, W. (1981) Cell 27, 299-308.
22. De Villiers, J. and Schaffner, W. (1981) Nucleic Acids Res. 9, 6251-6264.

23. Levinson, B., Khoury, G., Van de Woude, G. and Gruss, P. (1982) Nature 295, 568-572.

24. Weil, P.A., Segall, J., Harris, B., Ng, S.-Y. and Roeder, R.G. (1979) J. Biol. Chem. 254, 6163-6173.

25. Maxam, A. and Gilbert, W. (1980) Meth. Enzym. 65, 499-580.

26. Manley, J.L., Fire, A., Cano, A., Sharp, P.A. and Gefter, M.L. (1980) Proc. Natl. Acad. Sci. USA 77, 3855-3859.

27. Benoist, C. and Chambon, P. (1980) Proc. Natl. Acad. Sci. USA 77, 3865-3869.

28. Akusjärvi, G. and Pettersson, U. (1979) J. Mol. Biol. 134, 143-158.

29. Sassone-Corsi, P., Corden, J., Kédinger, C. and Chambon, P. (1981) Nucleic Acids Res. 9, 3941-3958.

30. Hérissé, J., Rigolet, M., Dupont de Dinechin, S. and Galibert, F. (1981) Nucleic Acids Res. 9, 4023-4042.

31. Hérissé, J., Courtois, G. and Galibert, F. (1980) Nucleic Acids Res. 8, 2173-2192.

32. Grosveld, G.C., Rosenthal, A. and Flavell, R.A. (1982) Nucleic Acids Res., in press.

33. Germond, J.E., Hirt, B., Oudet, P., Gross-Bellard, M. and Chambon, P. (1975) Proc. Natl. Acad. Sci. USA 72, 1843-1847.

TRANSFER RNA PROCESSING AND GENE REGULATION[1]

Sidney Altman, Madeline Baer, Cecilia Guerrier-Takada, and Robin Reed

Department of Biology, Yale University
New Haven, CT 06520

ABSTRACT Transcripts of tRNA genes in prokaryotes include sequences from adjacent genes coding for rRNA, protein or other tRNAs. RNA processing events do not merely tailor the gene transcripts to the final size of mature tRNAs but also must play a role in the regulation of expression of cotranscribed genes as is especially evident in mammalian mitochondria. A role in gene regulation is not so apparent for nuclear tRNA genes of eukaryotes. An important enzyme in the processing of the 5' termini of tRNAs is RNase P, which is a ribonucleoprotein. The gene coding for the RNA subunit of this enzyme from E. coli has been cloned and the nucleotide sequence determined. The gene transcript has five contiguous nucleotides which are complementary to an invariant region in tRNA and which may be important in determining the substrate recognition properties of RNase P.

INTRODUCTION

It has long been recognized that tRNA genes are transcribed with sequences flanking both the 5' and 3' termini of the mature tRNA sequences (1,2,3). Since the extra nucleotides must be removed in order to make a mature tRNA molecule, enzymes must exist which carry out these processing functions. The study of such enzymes has been an absorbing and fruitful endeavor over the past several years (4,5,6). More recently, it has also become apparent that transcripts which include tRNA sequences may also include sequences coding for other gene products and that the

[1]This work was supported by USPHS grant GM19422 and NSF grant PCM79-04054 to S.A.

ISBN 0-12-525960-3

previously regarded "simple" functions of RNA processing
enzymes may actually have deeper significance for gene
regulation than was first suspected. In this paper we
review some aspects of the relation between tRNA processing
and gene regulation and we further describe details of the
structure of one RNA processing enzyme, RNase P, which is a
ribonucleoprotein (7,8). We do not deal with the excision
of intervening sequences from certain tRNA genes because
this is examined in another chapter in this volume (9).

RESULTS AND DISCUSSION

<u>Transfer RNA processing in prokaryotes</u>: Transfer RNA
genes occur in many different genetic contexts in <u>E. coli</u>.
They may be embedded in rRNA cistrons (6,10; between the
sequences coding for 16S and 23S rRNA or distal to 3'
terminus of the sequences for 5S RNA), or lie adjacent to
genes coding for protein (11,12,13) or other tRNAs (5,14)
(Fig. 1). In each of these cases, it was not possible to
determine that the tRNA sequences were transcribed with

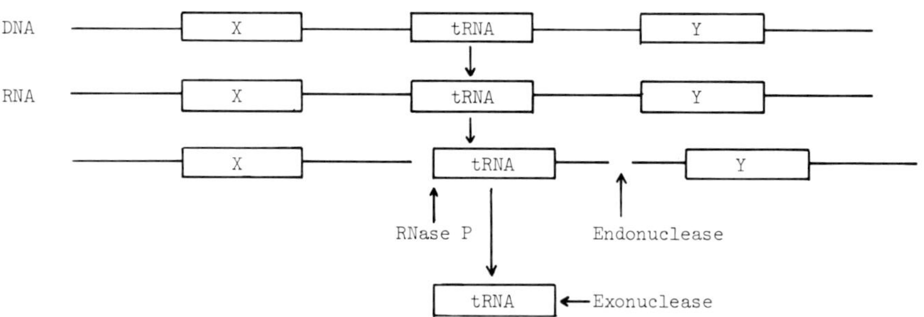

FIGURE 1. A generalized scheme for the processing of
tRNA gene transcripts by nucleases. X and Y denote
sequences coding for protein and tRNA, rRNA and tRNA, tRNA
alone, but not protein and rRNA. (Reprinted with permission
permission from MIT copyright (c) 1981.)

those coding for other genes or to understand the required
RNA processing events until certain mutants affecting RNA
processing were identified.

   With the isolation of mutants of _E. coli_
thermosensitive for the enzyme RNase P (15,16), which
processes the 5' termini of all tRNA precursors, it became
possible to identify multimeric tRNA precursor molecules
containing as many as seven tRNA sequences. Some of these
precursors had already been processed at their 5' termini.
In many other cases it was not clear whether or not the
transcripts had been processed at their 3' termini. The
identities of sequences adjacent to processed precursors and
the enzymes involved in processing prior to RNase P action
were unknown. The fact that processing was needed at all
for tRNA biosynthesis in _E. coli_ was surprising but could be
explained by the obvious notion that processing could
control the level of various tRNA species by cleaving
different sequences in tRNA precursors with different
efficiencies. Another possible rationale arose from the
need to remove defective sequences from the cell to prevent
jamming of the protein synthetic machinery with
non-functional molecules.

   With the advent of recombinant DNA methods and DNA
sequencing techniques, protein coding sequences were
identified at the 3' side of two tRNA gene clusters in
E. coli (11,12) and upstream from a T4 encoded tRNA gene
cluster (17,18; see below). Furthermore, Northern blot
analysis of _E. coli_ RNA showed that transcripts could be
detected which include both the tRNA and protein coding
sequences (12,13). Why _E. coli_ tRNA genes should be
transcribed with proteins, in one case with an elongation
factor in protein synthesis and in the other with a small
basic protein of unidentified function, is not known. Nor
is there as yet any information which indicates whether or
not processing events must occur between the two kinds of
gene sequences in the transcript to facilitate or to prevent
expression of the protein coding genes. This topic is
currently under investigation[2].

[2]Deletion mutants are available which remove the
intergenic processing site beween the _TyrT_ and _tpr_ loci
(19).

There is another case, that of the cotranscription of
T4-encoded tRNA genes with a small basic internal protein of
T4 (17,18), which does seem to indicate that cotranscription
of tRNA and protein is somehow related to cellular
physiology. In this latter case, the size of the
transcription product depends on the time after infection,
i.e. on the phage transcription program.

Only a few cases are now known in which tRNA genes are
cotranscribed with protein. However, the possibility that
this is a widespread phenomenon cannot be ruled out because
sequences flanking tRNA genes are still insufficiently
characterized.

<u>Cotranscription of tRNA and rRNA</u>: Certain tRNA genes
are located within the seven rRNA cistrons of <u>E. coli</u>
(10,20). These observations were made initially by
hybridizing tRNAs to cloned rRNA cistrons. Subsequently,
characterization of the sequences of 30S transcripts from
rRNA cistrons, which can be isolated in RNase III⁻ mutants,
confirmed the initial results. The number, location and
identity of the tRNA genes is variable and no underlying
pattern can be discerned in the choice of tRNA genes which
are cotranscribed with rRNA. These particular tRNA species
are consequently present <u>in vivo</u> in relatively large amounts
but why their production should be related to that of rRNA
is not known. There is no apparent obligatory role for tRNA
processing enzymes in the excision of rRNA sequences from
the large transcript of rRNA cistrons (6). A function for
these enzymes in the regulation of rRNA does not appear to
be implicit. In other systems, however, such a role for
processing enzymes is obvious and of paramount importance.

<u>tRNA processing in mammalian mitochondria</u>: The genomes
of human and mouse mitochondria and their transcription
patterns have been characterized (21-23). Genes coding for
rRNA or protein are abutted on either side by tRNA genes.
Transcription units span large segments of the mitochondrial
genomes. Through the use of tRNA processing enzymes only
(i.e. endonucleases which cut either at the 5' or 3'
termini of tRNA sequences) all coding sequences are reduced
to monocistronic species. Poly A addition to the 3' termini
of the released protein coding sequences generates UAA
nonsense codons (which are not encoded in the DNA) and
prepares these sequences for translation. In these
organisms it is apparent that requirements of genome economy
have led Nature to employ processing signals, rather than
multiple transcription start and stop signals, to yield
individual gene transcripts. Different gene products must

be produced in direct relation to the efficiency with which
they are processed. Such an efficient and clear-cut use of
tRNA processing as "punctuation" in programs for gene
expression is not achieved in bacteria, where space
limitations on the genome are not as stringent as in
mammalian mitochondria. A general role for tRNA processing
in gene regulation is not apparent at all if one considers
nuclear genes. In the latter case, a completely different
solution has been used for providing transcripts of genes
coding for different products.

     tRNA processing and nuclear genes of eukaryotes: Aside
from two possible exceptions in yeast (24,25) all tRNA genes
in the nuclei of eukaryotes appear to be transcribed as
monomeric species (26). Furthermore, they are transcribed
by an RNA polymerase (pol III) which is reserved
specifically for the transcription of some small RNA species
which do not code for proteins (27). Nature has used
different polymerases, with different promoter signals, to
discriminate between genes coding for protein, rRNA or tRNA
rather than rely on a single polymerase with a multitude of
interlocking or overlapping transcription signals. It is
not necessarily the case, however, that the use of tRNA
genes or other genes coding for small RNAs as signals for
transcription programs is completely dispensed with. For
example, it has yet to be proven that tRNA genes, or pseudo
genes, are never transcribed by pol II with other coding
sequences for the purposes of controlling
post-transcriptional events. It is apparent, however, that
the organization and transcription of tRNA genes in
eucaryotic nuclei is different from that in prokaryotes and
mitochondria. The compartmentalization of pol I and rRNA
biosynthesis makes teleological sense. However, why the
enzymology of mRNA synthesis should be so distinctly
separated from that of tRNA or 5S RNA biosynthesis is
puzzling. Perhaps these latter sequences must be under the
control of a separate, unique class of strong promoters
(recognized by pol III) to assure a workable level of mature
products in the cell. Most proteins are not needed in as
vast quantities as, for example, is tRNA.

     The role of RNase P in tRNA processing: RNase P, an
endoribonuclease, generates the 5' termini of mature tRNA
sequences from tRNA gene transcripts in many, if not all,
organisms (5,7,8,28,29). In addition, as discussed above,
RNase P-like activity is essential in regulating the
expression of genes in mammalian mitochondria. This enzyme
must have a unique mode of recognizing its substrates since

there is no sequence homology around any of its cleavage
sites. The discovery that RNase P was composed of an RNA
and protein subunit led to speculation that the RNA moiety
might be involved in substrate recognition through
hydrogen-bonded interaction with invariant regions of all
tRNA precursor substrates (30). New data concerning the
nature of the RNA subunit of RNase P shed further light on
this matter. The characteristics of RNA-protein
interactions in the RNase P scheme of action have also
become clearer recently.

RNA-protein and RNA-enzyme interactions in the RNase P
reaction: Three reactions involved in the action of RNase P
are shown in Fig. 2. The first is the association of the
RNase P RNA moiety (M1 RNA) with the protein (C5), the
second is the association of the enzyme with the substrate
and the third the inhibition of the enzymatic reaction by
tRNA, one of the products of the enzymatic reaction. The
dissociation constants governing these reactions are also
shown in the Fig. 2 and indicate a satisfying progression of
strengths of interaction. That is, the dissociation
constant governing the formation of the enzyme complex from
RNA and subunits is characteristic of a highly specific
interaction. Further, the inhibition of the enzymatic
reaction by one of its end products, tRNA, is at least two
orders of magnitude lower than the K_m for the real
substrate. The first dissociation constant was determined
by measuring the association of isolated M1 RNA and C5
protein in cesium salt density gradients and the latter two
by more conventional enzyme kinetic measurements (31). The

$$\text{RNA (MI) + Protein (C5)} \rightleftharpoons \text{RNase P} \qquad K_D \sim 10^{-11} M$$

$$\text{Enzyme (RNase P) + Substrate (tRNA Precursor)} \rightleftharpoons \text{ES} \longrightarrow \text{tRNA + 5' fragment} \qquad K_M \sim 10^{-8} M$$

$$\text{Enzyme (RNase P) + Inhibitor (tRNA)} \rightleftharpoons \text{EI} \qquad K_I \sim 10^{-6} M$$

Figure 2. Equations governing RNA protein interactions
in the action of RNase P on tRNA precursors.

ease with which M1 RNA can be isolated has facilitated the cloning of the DNA coding for this gene. Cloning could only be achieved by overcoming the problem of low level contaminating fragments of rRNA in M1 RNA preparations. The cloned DNA enabled us to determine the sequence of the corresponding M1 RNA (which could not be determined by direct RNA sequencing) and led to new ideas concerning the role of M1 RNA in the RNase P reaction.

<u>Nucleotide sequence and secondary structure of M1 RNA</u>: Using M1 RNA as a probe, a DNA fragment containing the region coding for M1 RNA was isolated from a λgt 7 library (32) of the E. coli genome[3], subcloned into a derivative of pBR325 and DNA sequence analysis of the subcloned fragment was carried out. The sites of initiation and termination of the mature M1 RNA sequence were located in the DNA sequence by comparing the terminal nucleotides of the M1 RNA and its RNaseT$_1$-generated oligonucleotide content with the DNA sequence. The corresponding sequence for M1 RNA can be drawn in a number of possible secondary structures. One such structure, which maximizes the number of hydrogen bonds, is shown in Fig. 3. For our purposes, the major point of interest is the loop region of stem III. Note that five contiguous nucleotides are complementary to the invariant region GTψCPu found in all procaryotic tRNAs (33). We know from separate experiments that loop III is essential for function of M1 RNA and is accessible for hydrogen bonding in the enzyme complex. It is possible that RNase P recognition of its substrates is a two step reaction in which the first step involves recognition of a roughly tRNA-like structure in tRNA precursors and the second step is invasion of the TψC loop by M1 RNA. RNase P is consequently rigidly positioned on its substrate.

Many experiments can now be planned to ascertain whether the possible hydrogen bonding scheme we envision does indeed occur during substrate recognition. Do ribonucleoprotein complexes with catalytic activity have a widespread occurrence in nature?

<u>Ribonucleoprotein complexes</u>: Ribonucleoproteins (RNPs) can be broadly divided into two classes as shown in Table 1: transient complexes and permanent complexes (we neglect RNA-activated enzymes for the present discussion). We

[3]In collaboration with H. Donis-Keller of Biogen, Inc., Cambridge, Mass.

TABLE 1

CLASSIFICATION OF RIBONUCLEOPROTEIN COMPLEXES

Class	Members
Transient (non catalytic)	RNA viruses hnRNP mRNP
Permanent (catalytic)	ribosomes RNase P amylose isomerase snRNP (?)

assert that transient complexes have no individual catalytic
activity. In this category we group RNA viruses and hnRNP
and mRNP particles. In both these cases the complex
performs either a temporary packaging or transport function.
The internal proteins of some viruses may remain associated
with the viral RNA <u>in vivo</u> but there is no evidence that
these proteins have any catalytic function (37). The second
class of RNP involves complexes in which, once formed, the
RNA and protein stay together and perform some catalytic
function which neither moiety has by itself. Ribosomes and
RNase P are members of this class. There is some evidence
that rabbit amylose isomerase (34; a branching enzyme which
moves fragments of amylose chains a fixed number of sugar
residues along another chain before making a branch) is also
in the same class.[4] While it would also be satisfying to
find catalytic function related to mRNA splicing associated
with snRNPs (35), this has not yet been shown to be the
case. (Certainly most snRNP preparations have non-specific
ribonuclease activity on tRNA precursor substrates. It is
conceivable that this activity might be manifest as a highly
specific endonuclease with the right substrate and under the
right conditions.) More RNPs will be characterized and
their function determined in the coming years. In the
meantime, RNase P remains a paradigm for study not only from

[4]There are some suggestive data that an <u>E. coli</u>
photoreactivating enzyme may also belong in this group (36).

the point of view of enzyme mechanisms but also with respect
to its role in the regulation of gene expression.

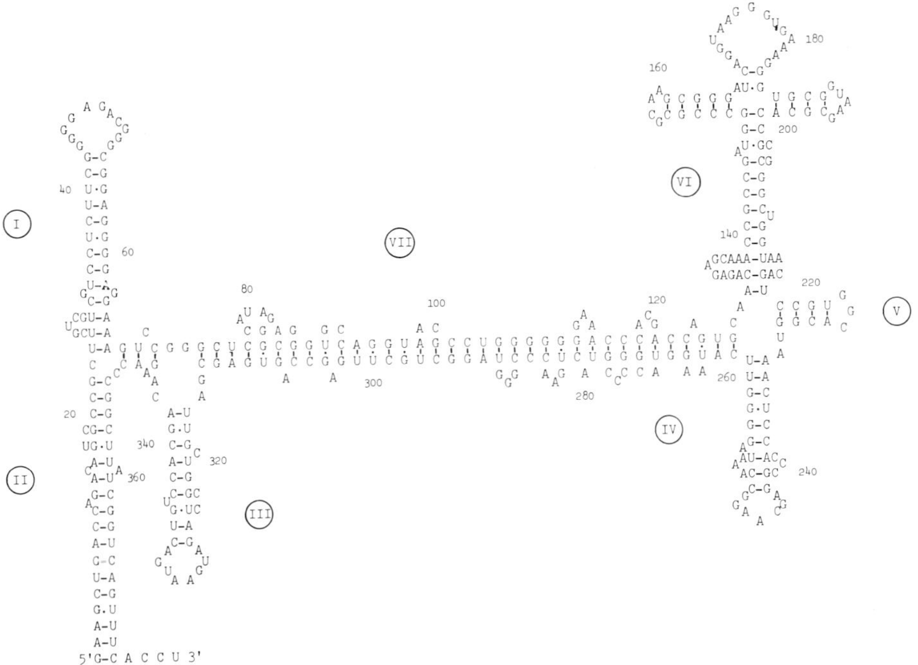

Figure 3.   A possible secondary structure for M1 RNA.

ACKNOWLEDGMENTS

We thank Dr. Henryk Cudny, Mr. Paul Furdon and Dr. W.H.
McClain for help of various kinds.

REFERENCES

1.   Burdon, R. H.   (1971).   Prog. Nucl. Acid Res. Mol.
     Biol. 11, 33.
2.   Altman, S. (1975).   Cell 4, 21.
3.   Perry, R. P. (1981). J. Cell Biol. 91, 28s.
4.   Abelson, J. N. (1979). Ann. Rev. Biochem. 481, 1053.
5.   Altman, S. (1981) Cell 23, 3.
6.   Gegenheimer, P. and Apiron, D.   (1981). Microbio.
     Revs. 45, 502.

7.   Kole, R., Baer, M. F., Stark, B. C. and Altman, S.
      (1980). Cell 19, 881.
8.   Kole, R. and Altman, S. (1981). Biochem. 20, 1901.
9.   Abelson, J. N. (1982). In "Gene Regulation" (B.
      O'Malley and C. F. Fox, eds.), in press, Academic
      Press, New York.
10.   Lund, E., Dahlberg, J. E. and Guthrie, C. (1980). In
      "Transfer RNA: Biological Aspects" (J. Abelson, P.
      Schimmel, and D. Soll, eds.), pp. 123-137, Cold Spring
      Harbor, New York.
11.   Altman, S., Model, P., Dixon, G. and Wosnick, M.A.
      (1981). Cell 26, 299.
12.   Lee, J. S., An, G., Friesen, J. D. and Fiil, N. P.
      (1981). Cell 25, 251.
13.   Rossi, J., Egan, J., Hudson, L. and Landy, A. (1981).
      Cell 26, 305.
14.   Nakajima, N., Ozeki, H. and Shimura, Y. (1981). Cell
      23, 239.
15.   Schedl, P. and Primakoff, P. (1973). Proc. Nat. Acad.
      Sci. 70, 2091.
16.   Sakano, H., Yamada, S., Ikemura, T., Shimura, Y. and
      Ozeki, H. (1974). Nucl. Acids. Res. 1, 355.
17.   Velten, J. and Abelson, J. (1980). J. Mol. Biol. 137,
      235.
18.   Fukada, K., Gossens, L. and Abelson, J. (1980). J.
      Mol. Biol. 137, 213.
19.   McCorkle, G. M. and Altman, S. (1982). J. Mol. Bol.
      155, 83.
20.   Nomura, M. and Post, L.E. (1980) In "Ribosomes" (G.
      Chambliss, G.R. Graven, J. Davies, K. Kavis, C. Kehen,
      M. Nomura, eds.), pp. 671. University Park, Baltimore.
21.   Anderson, S., Bankier, T., Barrell, B. G., de Bruyn, M.
      H. L., Coulson, A. R., Drouin, J., Eperon, I. C.,
      Nierlich, D. P., Roe, B. A., Sanger, F., Schrier, P.
      H., Smith, A. J. H., Standen, R. and Young, I. G.
      (1981). Nature 290, 457.
22.   Ojala, D., Merkel, C., Gelfand, R. and Attardi, G.
      (1980). Cell 22, 393.
23.   Bigg, M. J., Von Etten, R. A., Wright, C. T., Walberg,
      M.W. and Clayton, D. A. (1981). Cell 26, 167.
24.   Mao, J., Schmidt, O. and Soll, D. (1980). Cell 21,
      509.
25.   Schmidt, O., Mao, J., Ogden, R., Beckmann, J., Sakano,
      H. and Abelson, J (1981). Nature 287, 750.
26.   Feldmann, H. (1977). Nucl. Acids. Res. 4, 2831.

27. Roeder, R. G. (1976). In "RNA Polymerase" (R. Losick
     and M. Chamberlin, eds.), pp. 285-329. Cold Spring
     Harbor, New York.
28. Akaboshi, E., Guerrier-Takada, C. and Altman, S.
     (1980). Biochem. Biophys. Res. Comm. 96, 831.
29. Kline, L., Nishikawa, S. and Soll, D. (1981). J.
     Biol. Chem. 256, 5058.
30. Altman, S., Bowman, E. J., Garber, R. L., Kole, R.,
     Koski, R. A. and Stark, B. C. (1980). In "Transfer
     RNA: Biological Aspects" (J. Abelson, P. Schimmel and
     D. Soll, eds.), pp. 71-82. Cold Spring Harbor, New
     York.
31. Stark, B. C. (1977). Ph.D. Thesis, Yale University
     (New Haven).
32. Davis, R. W., Botstein, D. and Roth, J. R. (1980).
     "Advanced Bacterial Genetics", Cold Spring Harbor, New
     York.
33. Rich, A. and RajBhandary, U. L. (1976). Ann. Rev.
     Biochem. 45, 805.
34. Korneeva, G. A., Petrova, A. N., Venkstern, T. V. and
     Bayev, A. A. (1979). Eur. J. Biochem. 96, 339.
35. Lerner, M.. R., Boyle, J. A., Mount, S. M., Wolin, S.
     L. and Steitz, J. A. (1980). Nature 283, 220.
36. Snapka, R. M. and Sutherland, B. M. (1980). Biochem.
     19, 4201.
37. Wagner, R. R. (1975). In "Comprehensive Virology" (H.
     Fraenkel Conrat and R. Wagner, eds.), Vol. 4, pp. 1-94.
     Plenum, New York.

THE _IN VITRO_ TRANSCRIPTION OF _DROSOPHILA_ tRNA GENES

T. Dingermann, S. Sharp, J. Schaack, D. DeFranco,
D.L. Johnson, L. Cooley, and D. Söll

Department of Molecular Biophysics and Biochemistry,
Yale University, New Haven, Connecticut 06511

ABSTRACT   Extracts derived from _Drosophila_ Kc cells,
Hela cells and _Xenopus_ oocytes support accurate _in
vitro_ transcription of cloned _Drosophila_ tRNA genes.   By
analyzing the transcription activity of native and mutant tRNA genes we have identified the sequences required for transcription initiation and termination.
The primary transcripts have purine triphosphates at
their 5'-ends and leader sequences of 3-15 bases.   The
exact initiation site is dependent on the nature of the
5' flanking sequence. T_5 sequences are efficient
termination signals.   Less efficient termination occurs
in T-rich stretches.   Our results show that transcription of tRNA genes is controlled by two regions within
the mature tRNA coding sequence.   In the tRNA these two
regions code for the D-stem/D-loop (D-control region)
and the T-stem/T-loop (T-control region).   Promotion of
tRNA gene transcription is mediated by tRNA gene
specific transcription factors.   These factors bind cooperatively to the D- and the T-control regions; short-
ening or lengthening the spacer distance between the two
control regions, reduces the ability of factors to bind
cooperatively.   Although the binding of a putative factor to the D-control region can occur in the absence of
the T-control region, factor binding appears to be se-
quential.   Cooperativity is dependent upon factor binding to the T-control region.   Once factor binds to the T-
control region a transcription complex is formed which
is stable for many rounds of transcription initiation.

INTRODUCTION

In recent years there has been a rapid growth in the
understanding of eukaryotic gene expression.   Mostly this is
attributable to the availability of methods for gene isolation and subsequent manipulation.   _In vitro_ transcription
systems have been used to test the effects on transcription
of predetermined alterations to genes.   Our efforts have been
concentrated on studying the transcription properties of tRNA
genes of _Drosophila_, using RNA polymerase III activity pre-

sent in various <u>in vitro</u> systems. The availability of faith-
ful transcription systems and gene manipulation <u>in vitro</u> has
allowed identification of the functional domains of eukaryotic tRNA genes. Herein we describe some of these studies and
give an overview of current results. Although regions functional for promotion, initiation and termination of tRNA gene
transcription have been identified, information on the mechanism of tRNA gene transcription is still limited.

RESULTS AND DISCUSSION

<u>Transcription Promoter Sequences</u>. We have constructed
deletion mutations within a <u>Drosophila</u> tRNA^{Arg} gene by
using the processive nuclease BAL-31 (1). In the resulting
mutants, successively larger sequences of <u>Drosophila</u> DNA from
either the 5' or 3' side of the tRNA gene have been replaced
by pBR322 DNA (Fig. 1). By analyzing the ability of the 5'-
and 3'-deletion templates to direct the synthesis of RNA <u>in</u>

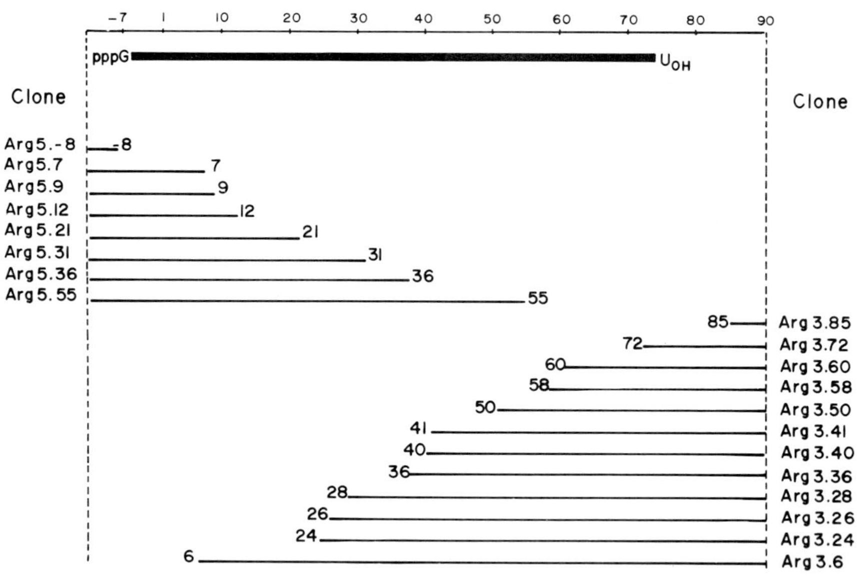

FIGURE 1. Schematic representation of deletions within
the tRNA^{Arg} gene. The nomenclature of 5'- and 3'-deletion
mutants is as follows: 5'-deletion clones are described as
pArg5. and 3'-deletion clones as pArg3.; the second number in
the designation (e.g. pArg5.7) indicates the extent of deletion, the number being the first (5' deletion) or the last
(3' deletion) nucleotide that remains from the wild-type <u>Dro-
sophila</u> pArg sequence.

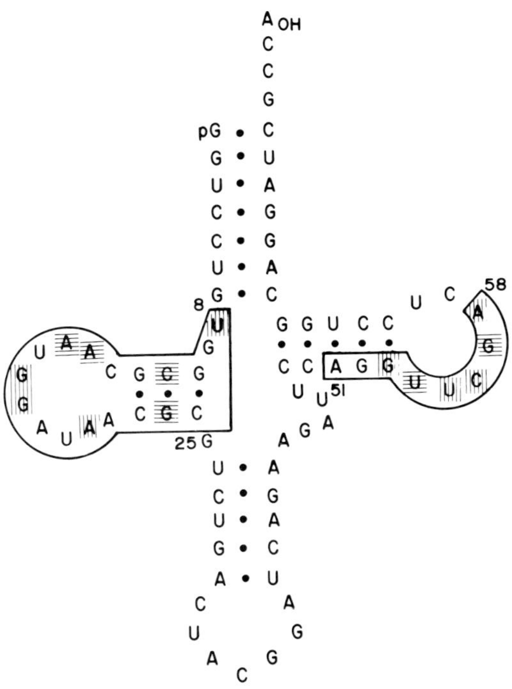

FIGURE 2. Internal transcription control regions (box-
ed) drawn into the cloverleaf structure of <u>Drosophila</u>
tRNA^Arg. Crosshatched nucleotides represent invariant or
semi-invariant positions in tRNA (2).

<u>vitro</u>, we have defined the space for borders of the two in-
tragenic transcriptional control regions required for tRNA
gene transcription (Fig. 2). One control region, bounded by
nucleotides 8 and 25, encodes the D-stem and D-loop and the
invariant uridine at position 8 (2). The second control region is bounded by nucleotides 50 and 58 and includes the sequence that codes for the semi-invariant sequence G-T-ψ-C in
loop IV of tRNA. The D-stem/D-loop control region of the
gene displayed "classical" gene promoter qualities in that
removal of the single nucleotide, thymidine-8, or nucleotide
25, a cytidine, resulted in complete loss of transcriptional
activity <u>in vitro</u>. Although the T-loop control region enhanced transcriptional activity, its presence was not an absolute requirement for tRNA gene transcription. Templates
devoid of the T-loop region were transcribed <u>in vitro</u>, whereas templates containing this region but lacking the D-stem/D-
loop region showed no transcriptional activity.

    The 5' and 3' deletion mutant tDNAs, which were used in
the analysis of the intragenic control region, retain one or
the other end of the tRNA gene and its adjacent flanking sequence. These additional tRNA gene sequences may supplement
the transcriptional function of the intragenic control regions. To determine whether the intragenic control regions
alone can direct transcription initiation, we have cloned
tRNA gene residues 7–26 and also 7–58 without any additional
tRNA gene sequences. These sub-clones are referred to as
tRNA minigenes and each was examined for its ability to support RNA synthesis <u>in vitro</u>. These regions were selected for
testing since the 5' deletion mutant pArg5.7, which has a deletion up to position 7 within the mature tRNA coding sequence, supports transcription. Similarly the 3' deletion
mutants pArg3.26 and pArg3.58, which have deletions up to positions 26 and 58,respectively, support RNA synthesis (3).
    The ability of DNA from pArg7/26 and pArg7/58 to support
RNA synthesis was examined in extracts derived from <u>Xenopus</u>
whole oocytes (see Fig. 3). Their transcription abilities
were compared to that of the DNA from the parental clones.
Since the normal termination site of pArg in the cloned
pArg3.26 and pArg3.58 has been removed, termination occurs at
approximately 40 bp from the tDNA/pBR322 fusion site. The
primary transcript of the pArg3.26 DNA is therefore 73 nucleotides long, while that of pArg3.58 DNA is even longer
(105 nucleotides) (Fig. 3). The minigene pArg7/26 is not
transcribed while the minigene pArg7/58 gives rise to a 105
nucleotide long RNA which initiates predominantly, with a G
14 bases before residue T_8 in the D-control region.   This
shows that pArg7/58 contains all the necessary signals for
initiation of tRNA gene transcription. As Figure 3 shows,
transcription of the minigenes results in the formation of a
single transcription product which is not processed. This
extends earlier observations that the primary transcripts
from DNA templates which contain deletions within the 5' or
3' sequences of the mature tRNA coding region are not substrates for processing nucleases (1). The region between
gene residues 7–58 contains the transcriptional promoter sequence for the <u>Drosophila</u> tRNA^{Arg} gene.   The promoter sequence consists of two sequence domains, the D-control region
and the T-control region, one located at each end of the 7–58
region.
    These constructions and transcription analyses were not
able to demonstrate the importance of the sequence found between the two control regions. To examine the function of
this region several deletion mutant tRNA genes were resected
in such a way that the sequence between the D- and T-control
regions of the native tRNA gene was shortened or lengthened.
The deletion mutant pArg3.26, which supports a low level of

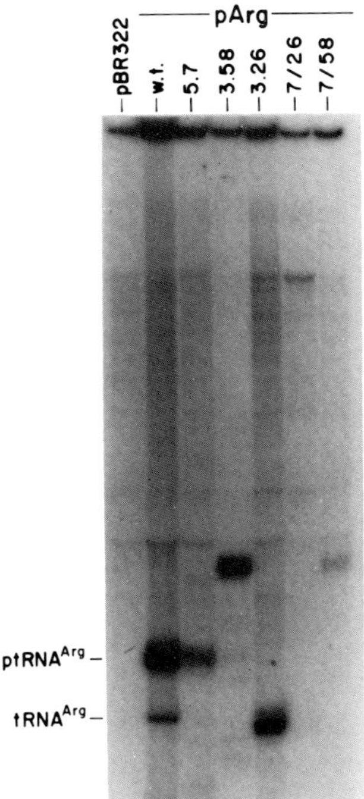

FIGURE 3. Transcription of tRNA Minigenes. Autoradiograph of a polyacrylamide gel electrophoretic separation of
^{32}P-labeled RNAs which have been synthesized in an $\underline{X}$.
$\underline{laevis}$ oocyte extract programmed with either a plasmid containing a wild type tRNA^{Arg} gene or a plasmid lacking portions of the wild type tRNA^{Arg} sequence. pArg tRNA corresponds to precursor tRNA^{Arg} which is 84-86 nucleotides in
length and tRNA^{Arg} is mature sized tRNA^{Arg} which is 73
nucleotides in long.

RNA synthesis, was used as the 5' portion of the gene (D-control region). The mutant genes pArg5.21, pArg5.36 and
pArg5.55, which do not support RNA synthesis $\underline{in}$ $\underline{vitro}$, were
used as the 3' portions of the mutant genes pArg26x21,
pArg26x36 and pArg26x55 respectively. The structure of these
DNAs is shown compared to the wild type tRNA^{Arg} gene (Fig.
4). Although these DNAs were able to support RNA synthesis
(Fig. 5), increasing or decreasing the length of the sequence
between the two control regions resulted in decreased levels
of transcription (Fig. 5). Preliminary studies of transcription of templates with longer DNA sequences inserted between
the two control regions indicate that, for cooperativity to
occur, there is an upper limit as to the distance separating
the two promoter elements. A mutant tRNA^{Arg} gene which has

200 nucleotides between the D- and the T-control regions is transcribed <u>in vitro</u> at a very reduced level. This level is similar to that of pArg3.26. These results suggest that the region located between the two intragenic control regions in the native tRNA gene does not contain sequences essential for promotion of transcription. This region appears to function in the gene as a spacer-region maintaining a certain distance between the two promoter elements such that these can direct efficient initiation of transcription.

```
              26                                      50
              ʼ                                       ʼ
pArg        --GCG _______ 24 nucleotides _______ CAGGTTCGA--

pArg26x21   --GCG _______ 51 nucleotides _______ CAGGTTCGA--

pArg26x36   --GCG _______ 36 nucleotides _______ CAGGTTCGA--

pArg26x55   --GCG _______ 17 nucleotides _______ TCCGGTCGA--
```

FIGURE 4. Schematic representation of mutated pArg templates with a different number of nucleotides located between the two internal control regions. These genes contain all parental tRNA^{Arg} gene sequences used for initiation and termination of transcription. The distance between the two control regions in wild-type pArg is 24 nucleotides. The coordinates 26 and 50 refer, respectively, to the 3' end of the D-control and to the 5' end of the T-control regions.

<u>Initiation of tRNA Gene Transcription</u>. As discussed above the minimum DNA sequences required to direct initiation of tRNA gene transcription are contained within the D- and T-control regions. These two regions direct transcription initiation to occur upstream within the sequence immediately adjacent to the 5' end of the mature tRNA coding sequence. For a number of <u>Drosophila</u> tRNA genes tRNA synthesis initiates with a purine nucleotide, pppA or pppG, which can be located between nucleotide positions -3 and -8 from the mature tRNA coding sequence. Transcription of tRNA genes which have altered DNA sequences within the immediate 5' flanking region however results in formation of RNA products initiating between nucleotide positions -2 and -15. The initiation sites of transcripts formed in a HeLa cell extract programmed with the pArg 5' deletion mutants, are shown in Fig. 6. Two precursor RNA products are formed from transcription of pArg in the HeLa cell extract. One product initiates with a G residue encoded at nucleotide position -7 relative to the 5' end of the mature tRNA. The other transcript initiates with an A at nucleotide position -4. Removal of <u>Drosophila</u> sequences

up to nucleotide positions -10 and -8, which are upstream of
the transcription initiation site for the wild-type gene, resulted in major changes in the selection of the initiating
nucleotide. Fig. 6 demonstrates the change in the sites of
initiation for pArg5.-10 and pArg5.-8 and also shows the
array of sites of initiation for the other pArg 5' deletion
mutants which efficiently support RNA synthesis. Initiation
of transcription for the 3' deletion mutant genes occurs at
the same sites as the transcript of the wild-type gene, i.e.,
at nucleotide positions -7 and -4. The minigene pArg7/58
supports formation of an RNA product which initiates at the
same nucleotide positions as pArg5.7.
   These studies show that, although transcription initiation is directed by the intragenic control sequences, the ac-

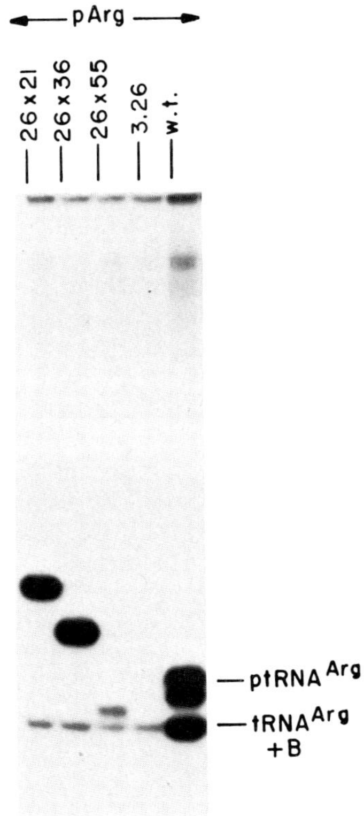

FIGURE 5. Autoradiograph of a gel electrophoretic separation of ^{32}P-labeled RNA transcribed from the wild-type
pArg clone and from pArg clones having alternations within
the sequence separating the D- and T-control regions (see
Fig. 4).

tual RNA initiation site is governed by the nature of sequences located in the 5' flanking region adjacent to the mature tRNA coding sequence. The mechanism involved in the selection of the initiating nucleotide does not appear to depend upon the presence of any particular DNA sequence in the 5' flanking region. Genes which have been fused at the 5' end of the mature tRNA coding seqeunce to various plasmid DNA sequences still support efficient RNA synthesis *in vitro*.

```
                       -20   -15   -10    -5      1       8
                        '     '     '     '      '       '

                                             •  •
pArg                   --TGTTACACTCGCACGTCAAGC GGTCCTGT--
                                               •  o
pArg5.-10              --AGAGGATCCGGCACGTCAAGC GGTCCTGT--
                                             ••  •  o
pArg5.-8               --GTAGAGGATCCGGCGTCAAGC GGTCCTGT--
                                               o  •  •
pArg5.-3              --CCGGCGTAGAGGATCCGGAGC GGTCCTGT--
                                             o••
pArg5.1               --CGTCCGGCGTAGAGGATCCGG GGTCCTGT--
                                               •   o•
pArg5.7               --ACGATGCGTCCGGCGTAGAGG ATCCGGGT--
                                               •  o
pArg5.8               --CACGATGCGTCCGGCGTAGAG GATCCGGT--
                                               •  •
pArg3.(72,60,58)  --TGTTACACTCGCACGTCAAGC GGTCCTGT--
                                               •  o•
pArg7/58              --ACGATGCGTCCGGCGTAGAGG ATCCGGGT--
```

FIGURE 6. Initiation Sites for tRNA^{Arg} 5' Deletion Mutants. The major (•) and minor (o) transcription initiation is indicated above the nucleotide position. Each DNA was transcribed using a HeLa cell extract using either a GTP, UTP, ATP or CTP radiolabel. Precursor tRNA was recovered and the 5' terminus (pppNp) was determined after RNase T_2 digestion. Terminus analyses were also performed by S1 mapping. Transcripts formed in the presence of (α-^{32}P)GTP were hybridized to 5'-^{32}P-labeled DNA fragments derived from pArg or mutant DNAs. The RNA/DNA hybrids were digested using 1500 units/ml of nuclease S1 according to the procedure of Weaver and Weissmann (10).

Modulation of tRNA Gene Transcription by 5' Flanking Sequences. Quantitation of the *in vitro* transcription efficiencies of various tRNA genes revealed that all tRNA genes are not transcribed with the same efficiencies. In particular, transcription of *Drosophila* tRNA^{Lys} genes demonstrated

that even though these genes share identical mature tRNA coding sequences, their template abilities are widely different
(4). The 5' flanking regions of two cloned <u>Drosophila</u>
tRNA^{Lys} genes, which have different transcription efficiencies, were exchanged, leaving the mature coding sequence intact (5' switching experiment). Transcriptional analysis of
the altered and unaltered tRNA^{Lys} gene revealed that the
region 5' flanking the mature tRNA coding sequence modulates
the template ability of a tRNA gene (4). A similar 5'
switching experiment has been performed using two <u>Drosophila</u>
tRNA^{Arg} genes.   In these experiments an oligonucleotide sequence has been identified such that removal of this sequence
leads to relief of transcription repression. This sequence
does not appear to be a site of recognition for a specific
transcription modifying factor. Instead repression of transcription seems to result from an inability of the DNA to direct transcription initiation by interact with RNA polymerase
III.

In general the 5' flanking sequence appears to function
in the selection of the nucleotide that initiates transcription. The examples of the tRNA^{Lys} and the tRNA^{Arg} genes
may reflect an inability of RNA polymerase III to select an
initiating nucleotide within a certain DNA milieu. Whether
or not the 5' flanking sequences affects some function within
the intragenic control regions is currently being investigated.

<u>Termination of tRNA Gene Transcription</u>. Initiation of
transcription occurs upstream to the region that codes for
the mature tRNA. This results in a precursor tRNA which has
a 5' leader sequence. Transcription termination occurs downstream of the mature tRNA coding sequence and results in the
precursor tRNA also having a 3' trailer sequence. Both the
5' leader and 3' trailer sequences are subsequently cleaved
by processing-nucleases leading to maturation of the tRNA.

Unlike 5' flanking sequences of tRNA genes there is a
distinct sameness of 3' flanking sequences. In general 3'
flanking sequences are AT-rich with multiple stretches of
oligothymidylate sequences. The 3' terminus of precursor
tRNAs is a heterogeneous array of uridylate residues which
results from the transcript terminating within an oligothymidylate stretch. Transcription usually terminates within
the oligothymidylate stretch nearest to the 3' end of the
mature tRNA coding sequence. From these observations it has
generally been concluded that the termination 'signal' for
transcription by RNA polymerase III is an oligothymidylate sequence, consisting of at least four T residues, within the
non-coding strand of the DNA. From studies using <u>Xenopus</u> 5S
DNA genes, oligothymidylate sequences have been shown to be

the termination signals for transcription by RNA polymerase III (5).

In the construction of the 3' deletion mutants of pArg the wild type terminator sequence, an oligothymidylate sequence of seven T residues (T-seven), was removed. In order to study transcription termination of these truncated tRNA genes, <u>Drosophila</u> DNA was fused through an <u>Eco</u>RI site to pBR322 DNA. This construction strategy resulted in several possible terminator sequences being introduced by the plasmid DNA. Potential termination signals include T-five sequences within pBR322 sequences at positions 4233-4237 and 4323-4327 (numbering according to Sutcliffe, 6).

Transcription of the tRNA gene 3' deletion mutants proceeded even after deletion of all tRNA gene sequences except the regions which code for the 5' half of the acceptor stem and the D-stem/D-loop (D-control region). Termination of these templates was efficient and occurred within a T-five sequence (pBR322, 4323-4327). The site of termination remained the same, independent of the regions of the tRNA gene which were deleted. Therefore it appears that the minimum sequence requirements for termination of tRNA gene transcription are the presence of the D-control region, to direct specific initiation, and that an oligothymidylate stretch is present downstream of the initiation site.

Changing the concentration of nucleoside triphosphates, especially UTP, in transcription reactions resulted in a varying ability of transcription to terminate within some sequences. The results of these experiments showed that while a minimum sequence of T-four was required for termination, the four T residues need not be contiguous. Indeed if the oligothymidylate stretches of 4 or 5 T residues were 'interrupted' by a single nucleotide insertion, termination still occurred at this site. The following are the sequences derived from the plasmid pBR322, (showing the flanking nucleotides) which serve as sites of transcription termination for mutant tRNA genes:

```
5' - ATTCTTG
5' - CTATTTTTATA
5' - GTTTCTTA
5' - CTTTTC
5' - CTATTTGTTTATTTTTCTA
5' - CTTATTC
```

<u>Cooperative</u> <u>Factor</u> <u>Binding</u>. To examine quantitatively the regions of a tRNA gene that contribute to the overall transcription efficiency of a template, we used the transcription-competition approach. In this method a transcription extract is programmed with two different genes. One

gene is a standard gene whose transcriptional properties have
been well characterized. The effect of the competitor gene
on transcription of the standard gene is monitored. An important feature of the transcription-competition assay is
that the test gene need not necessarily support RNA synthesis. This approach has been applied to the analysis of mutant tRNA genes from Bombyx (7), mutant VA RNA genes (8), and
more recently, to quantitate the contribution of the Xenopus
5S RNA gene intragenic control region to 5S RNA gene transcription (9).

The competition assay is thought to be based upon the
ability of genes to compete with each other for specific
transcription factors which, in unfractionated cell-free extracts, are present in concentrations that limit tRNA gene
transcription. Evidence suggests that the factor limiting
transcription is not RNA polymerase III but instead specificity factors for transcription of tRNA genes (7) or for
transcription of 5S RNA genes (9).

We have established a competition assay for Drosophila
tRNA gene transcription in cell-free extracts derived from
either Drosophila Kc cells or HeLa cells. Quantitation of
the transcription efficiency has been used to demonstrate
that both intragenic control regions of the tRNA gene are required for maximum competitive ability.

The results of the competition experiments demonstrate
that maximum competition is only observed when two regions
within the mature tRNA coding sequence are present (Fig. 7).
These regions, for the Drosophila tRNA[Arg] gene, include
nucleotide positions 8-12 and nucleotide positions 50-65. A
notable feature, however, is that as deletions from either
end of the gene extend into the tRNA coding sequence the competitive strength progressively decreases. In addition, removal of any Drosophila sequences between the coordinates
tested, -10 and +85, which are outside the mature tRNA coding
sequence, leads to a lowered competitive strength. Thus,
while the regions bounded by nucleotides 8-12 and 50-65 represent the essential sequences for competition, the sequences required for maximum competitive strength include the
regions that code for the 5' and 3' arms of the acceptor stem
of the tRNA as well as the adjacent 3' flanking sequence
(Fig. 7).

The regions of the Drosophila tRNA[Arg] gene that have
been shown to be responsible for the competitive ability of
this gene include the intragenic transcription control regions. These regions may be involved in recognition and
binding of tRNA gene transcription factors. Once the nucleotides between coordinates 7 and 12 have been deleted, the remaining D-control region contributes little to the competitive ability of the tDNA. If a tRNA gene specific transcrip-

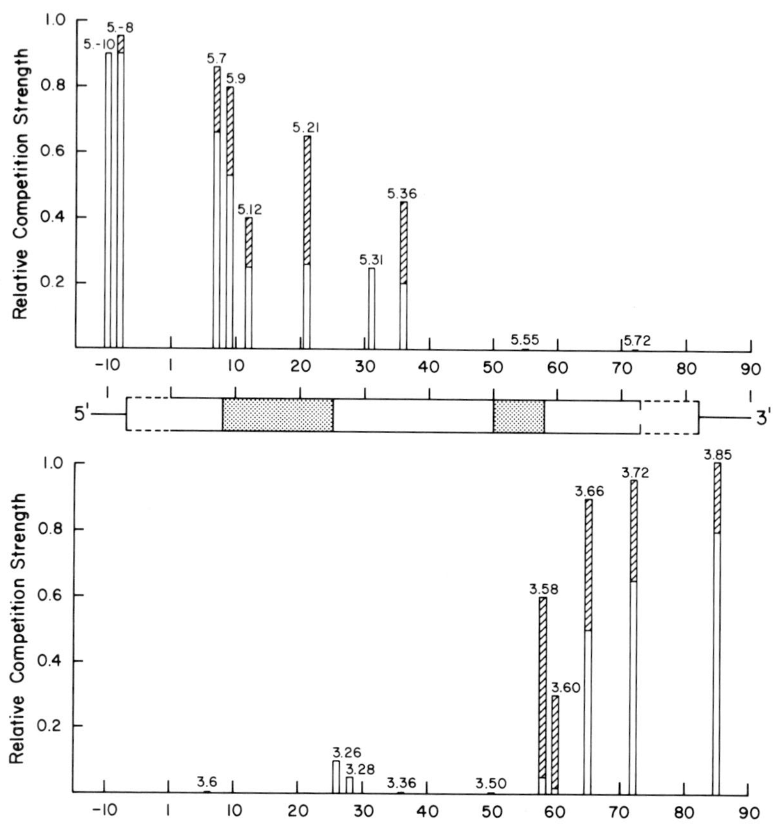

FIGURE 7.   Summary of Competitive Strengths of 5' and 3'
Deletion Mutants of the <u>Drosophila</u> tRNA^{Arg} Gene.   The results of the transcription-competition experiments are compiled in histogram form to demonstrate the competitive abilities of each region of the tRNA gene.   The competitive
strengths of each deletion mutant has been graphed using the
coordinates of the <u>Drosophila</u> tRNA^{Arg} gene as the axis.
The horizontal rectangle represents the tRNA^{Arg} gene and
the two intragenic control regions are indicated by the
stipled regions.   The dashed portions of the rectangle represent the 5' leader and 3' trailing sequences of the precursor tRNA (primary transcript).   The upper and lower his-
tograms show the competitive strengths of 5' and 3' deletion
mutant DNAs,respectively.   The competitive strengths in the
<u>Drosophila</u> extract as open vertical bars; the crosshatched
vertical bars represent competitive strengths in the HeLa
extract.   When the competition strength is the same in both
systems only the open vertical bar is drawn.

tion factor recognizes sequences within the D-control region
it is conceivable that the region 7-12 would be a major site
for recognition or binding of this factor. The 5' deletion
mutations with more than eight nucleotides of the mature tRNA
coding sequence deleted are not transcribed <u>in vitro</u>. However these DNAs maintain an ability to compete in the transcription-competition assay. A model to explain these observations could include two tRNA gene specific transcription
factors. One factor would interact within the D-control region. The other factor, which accounts for the level of competition of DNAs devoid of the D-control region, would interact within the T-control region. Owing to the presence of
the T-control region in the 5' deletion mutants that extend

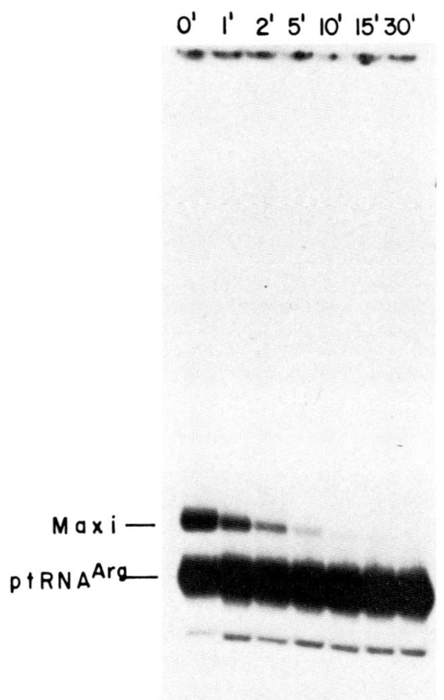

FIGURE 8. Time Dependence of Stable Complex Formation.
At time 0, 0.6 μg of pArg (competing template) plus 0.3 μg
pBR322 were added to each 40 μl transcription reaction (HeLa
cell extract). At the times indicated 0.3 μg of maxigene
(reference template) was added. The reactions were allowed
to proceed for 90 min after the addition of maxigene. The
^{32}P-labeled RNA products were analyzed by gel electrophoresis.

between nucleotide positions 9 and 36, these DNAs are able to
compete for this second factor. However, owing to the absence of the essential D-control region, these are poor templates for <u>in vitro</u> transcription reactions.

   <u>A Stable Complex is Formed During tRNA Gene Transcrip-
tion</u>. In the competition experiments discussed above the two
DNAs (competitor and reference templates) were added simultaneously. Under these conditions the transcription efficiencies of maxigene (pArg26x36) and pArg were equal. After
a lag period, transcription of these templates proceeded
linearly for several hours. When one template is added to a
transcription reaction at varying times after addition of a

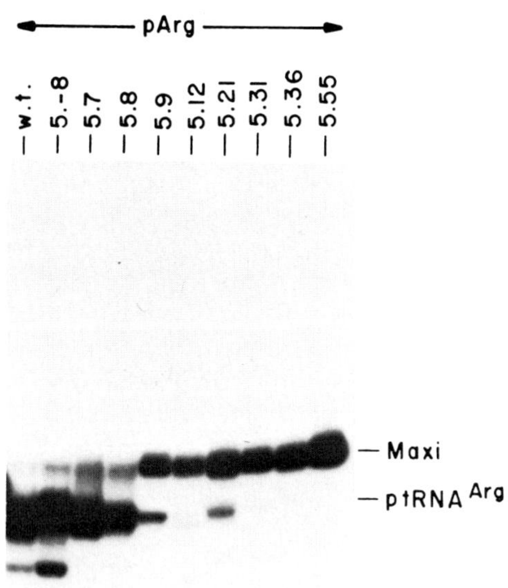

   FIGURE 9. Analysis of 5' Deletion Mutants in Stable
Complex Formation. At time 0, 0.6 µg of the competing 5'
deletion mutant as indicated plus 0.3 µg pBR322 were added to
each transcription reaction (HeLa cell extract). After 15
min incubation, 0.3 µg maxigene (reference template) was
added to each reaction and incubation continued until 105
min. The ^{32}P-labeled RNA products were analyzed by gel
electrophoresis.

first template, the transcription level of the second template is greatly affected. Fig. 8 shows the effects of adding maxigene to a transcription reaction which has been transcribing pArg for varying times. Pre-incubation of pArg in this manner, before addition of maxigene, leads to a reduction in transcription of the maxigene. Since the templates are kinetically similar and tRNA synthesis is linear for at least 2 hrs, we conclude that the first template sequesters an essential transcription factor in the extract. Sequestering of this factor occurs within 10 min of incubation resulting in factor not being available for the transcription of a template added to the reaction after this time. Further studies have shown that sequestering of this factor occurs on addition of the first template regardless of which template is added first. Moreover, this factor appears to be responsible for the formation of a stable transcription complex because it remains unavailable for transcription of a 'second' template for at least 6 rounds of re-initiation of transcription.

To determine the regions within a tRNA gene responsible for binding the factor which forms the stable transcription complex, each of the pArg 5' deletion mutants was pre-incubated in a transcription reaction. These reactions were challenged after 15 min by the addition of maxigene DNA. By measuring the level of maxigene transcription we have been able to determine the 5' deletion DNAs that no longer bind the factor. Each of the 5' deletion DNAs sequesters the transcription factor. However, as the deletion extends into the mature tRNA coding sequence the level of maxigene transcription increases (Fig. 9). This indicates a reduction in the ability of these DNAs to sequester the factor. Deletion up to gene residue 55 results in a DNA that has a very low ability to bind the factor as is evidenced by the increase in the level of maxigene transcription (Fig. 9). These results indicate that the presence of the T-control region is essential for formation of the stable transcription complex. In fact the results obtained by this approach mirror the results obtained using the transcription-competition assay.

ACKNOWLEDGMENTS

This work was supported by grants from the National Institutes of Health and the National Science Foundation. T.D. is supported by a Fellowship from the Deutsche Forschungsgemeinschaft. D.L.J. is supported by a Fellowship from the American Cancer Society.

REFERENCES

1. Sharp, S., DeFranco, D., Dingermann, T., Farrell, P. and Söll, D. (1981) Proc. Natl. Acad. Sci. USA 78, 6657-6661.
2. Sprinzl, M. and Gauss, D.H. (1982) Nucl. Acids Res. 10, r1-r55.
3. Sharp, S., Dingermann, T. and Söll, D. (1982) submitted for publication.
4. DeFranco, D., Schmidt, O. and Söll, D. (1980) Proc. Natl. Acad. Sci. USA 77, 3365-3368.
5. Bogenhagen, D. and Brown, D.D. (1981) Cell 24, 261-270.
6. Sutcliffe, J.G. (1978) Cold Spring Harbor Symp. Quant. Biol. 43, 77-90.
7. Sprague, K.U., Larson, D. and Morton, D. (1980) Cell 22, 171-178.
8. Fowlkes, D.M. and Shenk, T. (1980) Cell 22, 405-413.
9. Wormington, W.M., Bogenhagen, D.F., Jordan, E. and Brown, D.D. (1981) Cell 24, 809-817.
10. Weaver, R.F. and Weismann, C. (1979) Nucl. Acids Res. 7, 1175-1193.

THE GROWTH HORMONE GENE FAMILY: STRUCTURE, EVOLUTION, EXPRESSION, AND REGULATION[1]

Norman L. Eberhardt, Mark Selby†, Guy Cathala[2], Michael Karin[3], Arthur Gutierrez-Hartmann, Synthia H. Mellon, Nancy C. Lan, David Gardner and John D. Baxter

The Howard Hughes Medical Institute Laboratories, the Metabolic Research Unit, and the Departments of Medicine and Biochemistry and Biophysics, and †Microbiology and Immunology, University of California, San Francisco, California 94143

ABSTRACT: The genes for growth hormone (GH), prolactin (Prl) and chorionic somatomammotropin (CS) are a family of related genes. Multiple hGH and hCS genes are linked on about 100 kb of DNA on chromosome 17. The hGH and hCS genes exhibit >90% homology in coding, intron and flanking regions, suggesting that they are evolving by a concerted mechanism. Marked 5'-flanking homology between hCS and hGH genes also suggests that minor nucleotide differences in the 5'-flanking DNA or structures located in the distal flanking DNA may regulate differential expression of these genes. Portions of the proximal 5'-flanking DNA including possible regulatory structures and Exon I may have been inserted into the rGH, rPrl and hCS genes in separate evolutionary events. The hGH and hCS genes are flanked by Alu family sequences and the rPrl gene contains an Alu-related sequence within Intron D. The rGH gene contains a repetitive sequence within intron B related to tRNA genes; this structure can be expressed and accounts for abberrant rGH transcripts. Alternate pre-mRNA processing results in variable proteins for hGH and rPrl and polyadenylation of rGH mRNA is regulated differentially with possible functional consequences for mRNA translation. The regulation of the rGH gene by thyroid and glucocorticoid hormones occurs at the level of transcription; however, glucocorticoids may also affect mRNA stability.

[1]This work was supported by NIH Grants AM19997 and AM18878 and NSF Grant 8004735.
[2]Present Address: U.S.T.L. Laboratoire de Biologie Moléculaire, Place Eugene Bataillon, 34060 Montpellier cedex, France.
[3]Present Address: Department of Microbiology, University of Southern California School of Medicine, Los Angeles, California 90033

INTRODUCTION

Growth hormone (GH), prolactin (Prl) and chorionic somatomammotropin (CS, placental lactogen) are a family of related proteins that share amino acid homology (1), common immunoreactivities and partially overlapping biological functions (2). On the basis of these properties Niall et al. (1,2) proposed that the genes for these proteins evolved from a common evolutionary precursor gene. Consequently, the study of these genes is of interest with respect to the molecular evolution of a complex gene family. Since GH and Prl are synthesized in the anterior pituitary by somato-trophs and lactotrophs, respectively and CS is synthesized by the syncytiotrophoblast of the placenta, a comparison of these genes may provide insights into tissue-specific gene expression. In addition, the genes are regulated by a varie-ty of hormones and other effector molecules, epidermal growth factor for example (3) and provide an excellent system for the study of gene regulation. In the current studies selected aspects of the structure, evolution, expression and hormonal regulation of this family of related genes is discussed.

STRUCTURES OF THE GROWTH HORMONE-RELATED GENES

<u>Analysis of Complementary DNA Structures</u>. The struc-tures of the cloned cDNAs for human(h), rat(r) and bovine(b) mRNA for GH (4-6) and Prl (7-9) and for hCS (10-12) have been determined. The hCS and hGH genes maintain 93% nucleic acid sequence homology, whereas the hPrl gene has 42 and 41% nucleotide sequence homology with the hGH and hCS genes, respectively. These data support the concept derived from amino acid sequence comparisons (1) that these hormones are evolutionarily related. Moreover, the data suggest that the common ancestral gene was duplicated, yielding a gene that became the precursor to hPrl and another gene that subse-quently duplicated to form precursors for hGH and hCS. The coding sequences for hGH and hCS are also more related to each other than those for hGH and rGH. Although this result suggests that duplications leading to GH and CS genes occurred after the divergence of humans and rats (convergent evolution), it is also possible that intragenic exchange between hCS and hGH genes (concerted evolution) accounts for their high degree of homology. Moreoever, if the hGH and hCS genes are evolving by a concerted mechanism, the actual divergence of the two genes may have occurred before the time predicted based on sequence homology (8,9).

Further evidence that the GH-related genes are evolving by different mechanisms is suggested from a comparison of the Prl and GH genes from human, bovine and rat species (9). Thus, the human and bovine Prl genes are more related to each other than to the rPrl gene, whereas the bGH and rGH genes are more similar to each other than the hGH gene. In addition, the evolutionary divergence analysis (based on the unit evolutionary period or neutral theory) of both genes predicts markedly different divergence patterns for the three species, both of which disagree with the fossil record (9). One explanation of these findings may be that the individual genes within different species are evolving by different mechanisms.

Analysis of Chromosomal Genes. Chromosomal DNA analysis indicates that the rGH (13,14) and rPrl (15) genes are single copy sequences. In humans there are at least 2 nonallelic hGH genes and 5 nonallelic hCS genes (Table I) (16). All of

Table I. Characterization of hGH and hCS Genes.

Gene	Remarks	EcoRI	BamHI	BglII	PvuII	Sequence
hGH-1	normal GH	2.6kb	1	1	2	(12,19)
hGH-2	variant GH	2.6kb	2	1	2	(12)
hCS-1	normal CS	2.9kb	1	0	2	(16)
hCS-2	normal CS	2.9kb	1	1	2	(12)
hCS-3[a]	———	2.9kb	1	0	3	N.D.
hCS-4[a]	———	2.9kb	1	0	N.D.	N.D.
hCS-5	———	9.8kb	2	3	N.D.	N.D.

[a]Flanking DNA of the hCS-3 and hCS-4 genes differs from the other CS genes.

the hGH and hCS genes are located on the long arm of chromosome 17 (17) and as discussed below are closely linked, a finding which lends further support to the concept that these genes may be evolving by intragenic exchange mechanisms (concerted evolution). By contrast, there appears to be a single hPrl gene (N. Cooke, unpublished observations) which is located on chromosome 6 (18).
    Examination of a number of overlapping bacteriophage λ clones containing hGH and hCS genes reveals that several of the genes are linked (16). One linkage group contains four

genes within 57 kb of DNA in the order: 5'-(hCS-2)-(hCS-4)-
(hGH-2)-(hCS-3)-3'. Another linkage group contains the hCS-1
and hCS-5 genes within 12.7 kb of DNA. The hGH-1 gene has
also been linked to the hCS-5 gene (D. Moore, personal
communication). However, except for localization on the same
chromosome, the linkage between these two arrays has not been
determined. The established linkage indicates that the hGH
and hCS genes are interspersed and not segregated in individual groups. In addition, at least two of the genes are
oriented in opposite directions. The data suggest that all
of the hGH and hCS genes are linked in an array spanning at
least 100 kb of DNA on chromosome 17.

The primary structures of the chromosomal genes for hGH
(12,19), hCS (12,16), rGH (13,14) and rPrl (15) have been
determined. All of the GH-related genes which have been
sequenced contain five exons separated by four introns. Additionally, the intron/exon boundaries occur at identical
locations when the coding sequences are aligned for maximum
homology, providing further support for the common origin of
these genes. However, except for the hGH and hCS genes,
equivalent introns share little nucleotide sequence homology
and differ significantly in size.

One of the hGH genes (designated hGH-1) contains coding
and untranslated sequences which agree with the hGH cDNA and
conforms to the hGH amino acid sequence (12,19). The other
hGH gene, a variant (designated hGH-2), codes for a protein
with 13 amino acid substitutions relative to hGH (12); however, it is not known whether this gene is expressed in vivo.
Nevertheless, the hGH-2 gene can be expressed after ligation
to an SV40 vector and subsequent infection of African Green
monkey cells and the variant GH that is produced binds
efficiently to the GH receptor (20). Thus, the hGH-2 gene
maintains the potential for expression. However, analysis of
patients with primary hGH deficiency indicates a complete
deletion of the hGH-1 gene with retention of the hGH-2 gene
(21). Thus, if the hGH-2 gene is expressed in vivo, this
gene product is not capable of replacing the essential
functions of hGH.

Currently, two hCS genes have been sequenced. One of
these genes, designated hCS-1, contains coding and untranslated sequences which conform to the hCS cDNA (16). The
other gene, designated hCS-2, contains a single amino acid
substitution within the signal peptide and differs from the
cDNA in the 3'-untranslated regions (12). Thus, the hCS-1
gene may be responsible for most of the hCS synthesis in
placenta. Both genes appear to be completely functional in
that they contain appropriate promoter structures and have
the correct splice junctions at the intron/exon boundaries.
Recently, a patient which lacked immunoreactive hCS during an

otherwise normal pregnancy was found to have a deletion of
all of the hCS genes except hCS-5; the hGH-2 gene was also
deleted in this individual (22). This observations raises
the possibility that hCS may not have an essential function.
Alternatively, the remaining hCS-5 gene may code for a protein (possibly different from hCS) which provides essential
hCS functions.

   Homology of the hCS and hGH Genes. Comparison of the
hCS-1 and hGH-1 genes revealed striking nucleotide sequence
homology (>90%) in the 5'-flanking region extending 464
nucleotides 5' from the transcription initiation site and
3'-flanking regions extending 167 nucleotides 3' from the
translation termination signal. Marked nucleotide sequence
homology (>90%) was also found in each of the four introns
(16). Similar homologies were obtained in comparisons of the
hCS-1, hCS-2, hGH-1 and hGH-2 genes. The marked homology in
the flanking regions and introns among these four genes provides strong additional support for the concept that the hGH
and hCS genes may be evolving by a concerted mechanism. To
our knowledge the sequence conservation in the 5'-flanking
regions of these two differentially expressed genes exceeds
that found in any genes studied to date. Such extensive
sequence conservation in the 5'-flanking regions implies that
structures important for differential expression of these
genes could be accounted for by subtle nucleotide differences
within the flanking DNA. Alternatively, such structures
could be located more distally from the regions which have
been sequenced.

   Possible Origin of the 5'-End of the Genes. An examination of the structure of the rGH, rPrl, hGH and hCS genes revealed that a portion of the 5'-end of these genes, including
the TATAAA sequence and Exon I was flanked by direct repeats
(15,16). For rGH and rPrl the repeats consisted of deca- and
undeca-nucleotides, respectively (15) and for hGH and hCS the
repeats were three-part structures consisting of octa-, undeca-
and hexa-nucleotides separated by small insertions (16). In
all cases, the direct repeats are located a few nucleotides
5' to the TATAAA sequence and within Intron A. Based on an
analogy with transposition, the presence of the direct repeats
suggests that the proximal 5'-end of the genes, including
Exon I were inserted into the genes in separate events; the
direct repeats would have been generated during the insertion
event. In the case of the rGH and rPrl genes, there is little
homology in the putatively inserted region, but there is a
region of homology 5' to the TATAAA sequence, supporting the
possible separate insertion of Exon I in these two genes.

Since the region of this putative insertion includes the
TATAAA sequence and portions of the 5'-flanking and 5'-
untranslated regions as well as the codons specifying the
first three amino acids of the signal sequence, it is possible
that this structure may be important for some aspect of the
expression of the rGH, rPrl, hGH and hCS genes. Moreover,
these proteins are regulated by different hormones and are
expressed in different tissues; thus it is possible that this
proximal 5'-end of the gene contains structures important for
these differentially regulated properties. In addition, as
shown in Figure 1, there is significant conservation of
nucleotide sequence homology between the rGH and hGH genes
in the proximal 5'-flanking region which includes the region
proposed to have been inserted. Such sequence conservation
suggests that this region may be of importance for some
aspect of the expression of these two genes. However, it is
likely that structures 5' to the TATAAA sequence are also
important for the differential regulation of these genes.

```
rGH:    ATGTGTGGGAGGAGCTTCTAAATTATCCATCAGCACAAGC:TGTCAGTGGCT
        *********************************** ********* *********
hGH:    ATGTGTGGGAGGAGCTTCTAAATTATCCATTAGCACAAGCCCGTCAGTGGCC

rGH:    CCAGCCATGAATAAATGTATAGGGAAAGGCAGGAGCCTTGGGGTCGAGGAAA
        *   ***** ********* *   ****      * *          *  ** *
hGH:    C:::CCATGCATAAATGTACACAGAAACAGGTGGGGGCAACAGTGGGAGAGA

rGH:    :ACAGGTATGGGTATAAAAAGGGCATGCAAGGGACCAAGTCCAGCACCCTCGAG
        *  **    ***************    **** ****     ** ** * **     *
hGH:    :AGGGGCCAGGGTATAAAAAGGGCCCACAAGAGACCGGCTCAAGGATCCCAAGG
                                                              Met
rGH:    CCCAGATT:CCAAACTGCTCAGG:TCCTGTGGACAGATCACTGAGTGGCG[ATG]
        ****    *   ** ***   ****** **********    ****  **  **  Met
hGH:    CCCAACTCCCCGAACCACTCAGGGTCCTGTGGACGC:TCACCTAGCTGCA[ATG]
```

FIGURE 1. Nucleotide sequence comparison of the proximal 5'-ends of the rGH (13) and hGH (19) genes. Homology
was maximized by the introduction of suitable insertions (:).
Homologies are indicated by asterisks.

<u>Repetitive DNA Associated with the GH-Related Genes.</u>
Several different classes of dispersed-middle repetitive
DNA are associated with the GH gene family. Sequences related to the Alu family are found within Intron D of the rPrl
gene (15) and in the DNA which flanks all of the hGH and hCS
genes (16). Another class of repetitive DNA is found in the
3'-end of the rPrl genes and remains to be characterized (16).
Finally, sequences which appear to be related to tRNA genes
are present in the rat genome at a level of approximately
100,000 copies and are located within Intron B and the 3'-
flanking DNA of the rGH gene (13,14).

The class of repetitive DNA represented within Intron B of the rGH gene appears to be expressed, since it hybridizes to RNA of multiple size classes (23). Nevertheless, it is not known if the specific sequence within Intron B of the rGH gene is expressed _in vivo_. This repetitive DNA is designated tRNA-like DNA, since the secondary structure of the putative transcript is similar to tRNAs. In addition, the conserved sequences present in various rat tRNA genes are also present in the tRNA-like DNA (23). The primary structure of the tRNA-like DNA found in Intron B of the rGH gene is shown in Figure 2. The structure contains two 195 bp tandemly duplicated sequences followed by a 73 bp sequence which is homologous to the 5' end of the tandemly duplicated sequences. The entire repeated structure is flanked by 18 bp direct repeats. Each of the duplicated sequences contains an open reading frame which could code for related proteins of 40 amino acids; the significance of this is unknown. However, as discussed later, abberrant rGH-related transcripts which initiate within Intron B have been observed. Moreover, S1 mapping analysis suggests that there is a splice site which maps to the proximal 5'-end of the first tandem repeat structure within the region containing the open reading frame (Fig. 2). The data further suggest the presence of multiple splice sites within Exon III of the rGH gene. Thus, it is possible that some of the abberrant transcripts which are observed could code for polypeptides which are related to rGH.

The tRNA-like DNA contains three RNA polymerase III promoter structures that are active in an _in vitro_ transcription system (Gutierrez-Hartmann, A., unpublished results). In addition, there is a putative RNA polymerase II promoter which is located immediately 5' to the 18 bp direct repeats. As discussed later, this promoter appears to be preferentially utilized by mouse L cells that have been transformed with the rGH gene and may account for truncated rGH transcripts in this system.

The dispersed middle-repetitive DNA (probably alu sequences) associated with the hGH and hCS genes flanks the 5'- (within 500 bp) and 3'-ends (within 300-700 bp) of all of these genes which have been examined. These repetitive sequences may extend the overall homology observed between these genes (discussed earlier). The function of these sequences is not known; however, possible functional roles which may be considered include: (i) the tissue-specific and hormone-regulated expression of these genes; (ii) the duplications giving rise to this gene family; or (iii) the recombinatorial events that have been proposed to occur between these genes.

```
▼   5'-Splice
gtaagttcctcggtgttgggtgcctgactgtggaagcaggaaggggcagatcccaccctcgccccgagtcc

ctgcccctaggaagtcataggaggaaactatgccgttagatgagcagaaacagaatgggtcgtccatAACA
                                                                    ‾‾‾‾
────────────────────►*                  Met Ala Gln Trp Leu Arg Ala Pro Asp Cys Ser
GTAATGACAGAGAGGGCTGGAGAG                ATG GCT CAG TGG TTA AGA GCA CCC GAC TGC TCT
        *--First Repeat----------------------------------------------------------
Ser Lys Gly Pro Glu Phe Asn Ser Gln Gln Pro His Gly Gly Ser Gln Pro Ser
TCC AAA GGT CCT GAG TTC AAT TCC CAG CAA CCA CAT GGT GGC TCA CAA CCA TCT
----------------------------------------------------------------------------

Val Lys Arg Ser Asp Ala Leu Phe Trp Cys Val OP
GTA AAG AGA TCC GAT GCC CTC TTC TGG TGT GTC TGA  AGACAGCTACAGTGTACTTATA
----------------------------------------------------------------------------
                                               *                  Met Ala Gln Arg Leu
TAATAAACAAATAAATCTTTAAAAAAAAAAAAACAAAAACGGGGCTGGAGAG  ATG GCT CAG CGG TTA
----------------------------------------------*--Second Repeat---------------
Arg Ala Pro Asp Cys Ser Ser Arg Gly His Glu Phe Asn Ser Gln Gln Pro His
AGA GCG CCC GAC TGC TCT TCC AGA GGT CAT GAG TTC AAT TCC CAG CAA CCA CAT
----------------------------------------------------------------------------

Gly Gly Ser Gln Pro Ser Val Lys Arg Ser Asp Ala Leu Phe Trp Cys Ile OP
GGT GGC TCA CAA CCA TCT GTA AAG AGA TCT GAT GCC CTC TTC TGG TGT ATC TGA
----------------------------------------------------------------------------
                                                                  *
   AGACAGCTACAGTGTACTTATATATATAATAAATAAATAAATCTTTAAAAAAAAAACAAAACAGGGGCTGGGG
-----------------------------------------------------------------------*--Third

ATTTAGCTCAGTGGTAGAGCGCTTACCTAGGAAGCGCAAGGCCCTGGGTTCGGTCCCCAGCTCCGAAAAAA
Repeat----------------------------------------------------------------------

                                          ────────────────────────►
AGAACCAAAAAAAAAAAAAAAAAAAACCAAAACAAAAACAAAACAGTAATGACAGAGAgtcacaagctggtccc
------------------------------------------------

tcagtgactacccttcctccag
        3'-Splice     ▲
```

FIGURE 2. Nucleotide sequence of Intron B of the rGH
gene (13). The sequence contains a tandemly repeated DNA
sequence (195 bp); these repeats are 96% homologous. This
tandemly repeated structure is followed by a 73 bp fragment which is 66% homologous to the 5'-end of the repeated
units. The starting positions of these structures are indicated by asterisks. The repeated DNA is flanked by 18 bp
perfect direct repeats (indicated by the horizontal arrows).
Each of the tandemly duplicated DNA fragments contains an
open reading frame with the potential to code for similar
polypeptides containing 40 amino acids.

GH AND Prl GENE EXPRESSION

Cultured anterior pituitary cells (GH$_3$, GC and GH$_4$ sub-
lines) have been used extensively as a model system for studying the expression and hormonal regulation of the rGH and
rPrl genes (24-28). GH and Prl are synthesized in varying
amounts in the various cells; however, the regulation of GH
and Prl synthesis by thyroid and glucocorticoid hormones
appears to be similar in each subline.

Pre-mRNA Processing. The processing of the rGH-pre-mRNA
has been studied by hybridization with intron-specific probes
after size fractionation of RNA and transfer to nitrocellulose
(29). The primary transcript corresponds to a 2.3 kb fragment. In addition, prominent fragments of 2.1, 1.9 and 1.2
kb are observed. These fragments correspond to the predominant products that are formed during the processing (intron
removal) of the pre-mRNA. Analysis of these products with
intron-specific hybridization probes indicates that the pre-
mRNA is preferentially processed with intron removal in the
order: A, C, B and D. Other hybridizing fragments are also
observed, including a 1.7 kb fragment which lacks Exon I
and Exon II sequences and which appears to be initiated
within Intron B (discussed later). Hybridizing fragments
larger than the rGH pre-mRNA corresponding to transcripts of
3.1 and 4.5 kb are also found which lack Exons II and III.
The origin of these transcripts is unknown; however, it is
possible that they arise from a related gene. An rGH-
related gene which is homologous to the middle portion of
the rGH cDNA has been isolated from a rat DNA library (13).

Alternate Processing Pathways. In the case of hGH and
rPrl, alternate pre-mRNA processing pathways are involved
which can explain observed heterogeneities in hormone structure. Two predominant forms of hGH have been observed (30).
These include the normal hGH which contains 191 amino acids
(22K daltons) and accounts for 85% of the hGH produced in the
pituitary and a second form containing an internal deletion
of amino acids 32-46. The later form is designated the "20K
variant" and accounts for 15% of pituitary hGH. Analysis of
the hGH-1 gene structure indicates that Intron B separates
the codons for amino acids 32 and 33; however, an alternate
splice site present in Exon III also separates the codons
for amino acids 46 and 47 (19,31). Thus, the most likely
explanation for the generation of the two mRNAs is that there
is variable processing at the 3'-end of Intron B or within
Exon III. Splicing at the site within Exon III produces the
internally deleted form of hGH which lacks amino acids 33-46.
A similar splicing site variation in the rPrl mRNA accounts
for the finding of two different cDNAs (15) and may explain
amino terminal heterogeneity in prolactin (32).
     A second type of alternate mRNA processing occurs with
rGH mRNA. In cultured anterior pituitary cells (GH$_3$ and GC),
two distinct size forms of mature rGH mRNA are observed in
Northern blots (33). The production of these two forms can

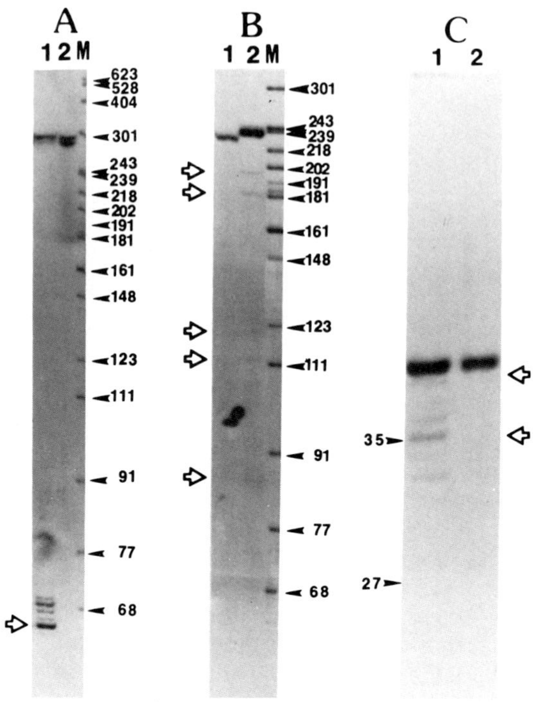

FIGURE 3. S1 nuclease mapping analysis of the rGH-related mRNA derived from mouse L cells transformed with the rGH Gene (L-rGH transformants). Mouse LTK⁻ cells were transformed with the HSV TK gene and a plasmid (prGHeh 5.8) containing the rGH gene (13). TK⁺ cells were selected in HAT medium and cloned. RNA was isolated from the cells grown in mass culture and S1 nuclease mapping was performed as described (35). In some cases pre-mRNA was selected by hybridization to filters containing rGH Intron C DNA. S1 nuclease profile of GC (5 µg, lane 1) and L-rGH-4 (150 µg lane 2) mRNA hybridized to a Bgl II/Pst I DNA fragment from the 5'-end of the rGH gene. The protected fragment (65 bp, lane 1) maps to the transcription initiation site of the rGH gene in GC cell mRNA (lane 1); no such fragment is present in L-rGH-4 mRNA (lane 2). (B) S1 endonuclease profile of L-rGH-4 mRNA (100 µg, lane 2) hybridized to a Pvu II/Kpn I endonuclease fragment derived from the rGH cDNA; the probe extends from Exon II to Exon IV. Lane 1: Labelled DNA probe plus S1 endonuclease (renaturing conditions). (C) S1 endonuclease profile of L-rGH-4 pre-mRNA hybridized to a genomic Dde I endonuclease DNA fragment extending from Exon II to 18 nucleotides 3' to the 18 bp direct repeat (5'-member, Fig. 2).

be varied by the addition of hormones (discussed later) and
is affected by the growth conditions of the GH_3 or GC cells.
The large rGH mRNA form predominates during the logarithmic
growth phase; however, when the cells become confluent the
smaller form appears. In addition, the smaller mRNA form is
the dominant one found in the rat pituitary.

The difference in the two mRNA forms has been investigated by primer extension of the mRNA and by S1 nuclease
mapping techniques (33). Both mRNA forms have identical 5'-
termini and S1 nuclease mapping studies suggest that the 3'-
termini are also identical. The poly A^- rGH RNA isolated
from cells expressing either the smaller or larger mRNA form
is the same size. In addition, as demonstrated by cell-free
translation, both mRNA forms direct the synthesis of the same
protein. Based on all of these observations, the difference
in the two mRNA forms appears to be due to the length of the
poly A segment.

rGH-Related Transcripts from L Cells Transformed with
the rGH Gene. Mouse L cells deficient in thymidine kinase
(TK⁻ cells) were cotransformed with the Herpes Simplex Virus
TK Gene and the rGH gene (35). Fifteen cell lines were
examined and eight of these expressed an rGH-related RNA product (0.8 kb) that was approximately 200-300 bp smaller than
the normal rGH mRNA. This RNA (0.8 kb) lacked sequences
corresponding to Exons I and II but contained sequences corresponding to Exons III-V. In addition, examination of the
pre-mRNA species indicated the presence of a 1.7 kb RNA
similar to that observed in GC cells which also lacked
sequences corresponding to Exons I and II.

As shown in Figure 3A S1 nuclease mapping analysis of
the pre-mRNA species using a probe which spans the 5'-end
of the gene indicated that the transcripts did not contain
these sequences. However S1 nuclease mapping of the pre-
mRNA (Fig. 3C) species using a DNA probe which originates
in Exon II and terminates just 3' to the 18 bp direct repeats (5'-member) which flanks the repetitive DNA within
Intron B (Fig. 2) suggested the presence of a transcription
initiation site at the beginning of the 18 bp direct repeat
(5'-member). Using a related DNA probe which extends 40 bp
further from the 18 bp repeat, a second putative initia-

(FIGURE 3 CONTINUED)

The labelled fragment (35 bp, lane 1) maps to the first
nucleotide immediately 5' to the 18 bp direct repeat.
Lane 2: Labelled probe plus S1 endonuclease (renaturing conditions).

tion site was detected which maps within the 18 bp repeat
(data not shown). These two transcription initiation
sites appear to be under the control of separate promoters. One of these putative initiation sites maps 14
bp 5' to an RNA polymerase III promoter and the other
appears to be an RNA polymerase II promoter located upstream
from the 18 bp direct repeat.

Interestingly, the larger DNA probe used for S1 nuclease
mapping (Fig. 3A) does not detect the transcription initiation
site which is proposed to be regulated by the RNA polymerase
II promoter, but detects instead the upstream site proposed
to be regulated by the RNA polymerase II promoter. This
suggests the possibility that a splice site exists in the
proximal 5'-portion of the first repetitive DNA element. In
addition, S1 nuclease mapping of the mRNA (0.8 kb) using a
cDNA probe which originates in Exon II and ends in Exon IV
reveals the presence of several distinct RNA species (Fig. 3B,
arrows). These bands map to positions within Exon III which
correspond to potential 3'-splice junctions. Consequently,
we hypothesize that the truncated rGH transcripts observed
in L cells transformed with the rGH gene initiate near the
5'-member of the 18 bp direct repeats flanking the repetitive
DNA within Intron B and that this RNA is subsequently
spliced to different sites within Exon III. As mentioned
previously some of these truncated transcripts might represent
RNA molecules which could code for rGH-related polypeptides;
however, there is no evidence that such proteins actually
occur.

HORMONAL REGULATION OF GH-RELATED GENES

The regulation of rGH gene expression by thyroid and
glucocorticoid hormones (24-28) and other factors such as
epidermal growth factor (3) has been studied in GH_3, GC and
GH_4 cells. The synthesis of GH is positively regulated
by thyroid (triiodothyronine, T_3) and glucocorticoid
(dexamethasone, Dex) hormones, whereas Prl is negatively
regulated by glucocorticoids. The regulation of GH appears
to be predominantly at the level of transcription; however,
glucocorticoids may also influence the stability of the rGH
mRNA. Currently, there is no information about the hormonal
regulation of the CS gene.

Hormonal Regulation of Transcription. The hormonal
influences on rGH gene transcription have been measured
utilizing an in vitro transcription system which depends on
in vivo initiation of transcription (28). This technique

(polymerase run-off) measures the number of polymerase
molecules which are actively transcribing the rGH gene in
control and hormone-treated cells. As shown in Fig. 4,

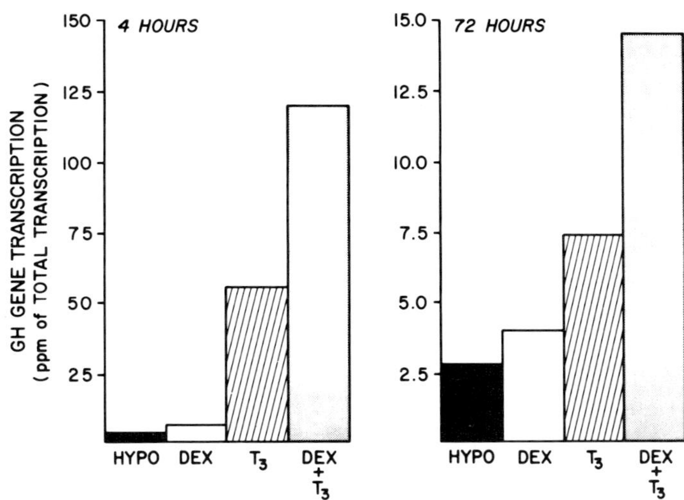

FIGURE 4.   In vitro transcription of the rGH gene.
Nuclei were isolated (28) from GC cells maintained in medium
supplemented with serum from a thyroidectomized calf for
120 hours. Prior to isolation of nuclei the cells were treated for 4 or 72 hours with no hormones, 1 μM Dex, 10 nM T_3 or
10 nM T_3 plus 1 μM Dex. Transcription (initiated in vivo)
was allowed to proceed in the presence of ^{32}P-labelled
ribonucleotide triphosphates; the RNA was isolated and
hybridized to nitrocellulose filters containing single-stranded DNA complementary to the rGH mRNA as described (28).
Transcription is represented as the fraction (ppm) of radioactivity hybridizing to the filters from the total radioactivity incorporated into RNA.

T_3-treatment produces a rapid (4 hr) and pronounced increase
in rGH transcription in GC cells. Dex-treatment alone has
little or no effect on transcription; however, in the
presence of T_3, Dex also increases synergistically the transcription of the rGH gene. After 72 hr of hormone treatment
the general pattern of induction is similar but the absolute
levels of transcription are decreased. Thus it is possible
that some factor(s) are induced by the hormones which

generally down regulated rGH gene transcription. Prior treatment of the cells with cycloheximide does not block the hormone effects, providing evidence that the hormones directly affect gene transcription. In addition to the effects on transcription, the data suggest that the hormones may influence post-transcriptional processes, possibly mRNA stability, since the observed increases in transcription were always less than the observed increases in mRNA levels.

Analysis of rGH gene transcription (polymerase run-off) in anterior pituitary tissue from thyroidectomized and adrenalectomized rats injected with T_3, Dex or Dex plus T_3 provides further evidence that these hormones regulate rGH gene transcription (36). In this case the control and induced levels of rGH gene transcription were approximately 10-fold higher than in GC cells. In contrast to the case with GC cells, Dex-treatment alone produced a significant increase in rGH gene transcription. Since the animals may have retained low levels of thyroid hormones, it is possible that thyroid hormones are required for the expression of the glucocorticoid response.

<u>Hormonal Regulation of rGH pre-mRNA</u>. The influences of thyroid and glucocorticoid hormones on the rGH pre-mRNA species were analyzed by hybridization analysis of RNA isolated from control and hormone-treated GC cells after size fractionation of the RNA and transfer to nitrocellulose. In the absence of both hormones no pre-mRNA forms were detected; however, the 1.7 kb RNA fragment which hybridizes to an Intron C probe but which lacks sequences corresponding to Exons I and II was observed. Treatment of GC cells with T_3 and Dex for 4 hr produced a dramatic increase in the primary rGH pre-mRNA transcript (2.3 kb) but did not affect the 1.7 kb RNA species. These data indicate that T_3 and Dex can selectively stimulate the transcription of the rGH pre-mRNA and suggest that the 2.3 and 1.7 kb RNA species originate from separate promoters. These findings further imply that the hormones exert local actions on the upstream rGH gene promoter.

<u>Hormonal Regulation of Multiple rGH mRNA Species</u>. The alternate rGH mRNA forms which appear to differ in the lengths of their poly A segments (discussed previously) are also regulated by thyroid and glucocorticoid hormones. For example, when GH₃ cells are maintained under depleted media conditions (maintenance of cells in the same media for 48-72 hours), both mRNA forms are observed. Under these conditions, addition of dexamethasone results in increased production of the larger mRNA form. However, addition of triiodothyronine

(T_3) to GH_3 cells maintained in depleted media results in production of the smaller rGH mRNA. These results suggest that the hormones can affect RNA processing (probably poly- adenylation); however, the mechanism(s) by which hormones influence mRNA processing are unknown.

Interestingly, the two rGH mRNA forms are not translated with equal efficiency in a cell-free translation system from reticulocytes (33). The translation of this larger mRNA proceeds much less efficiently than the smaller mRNA and suggests that rGH mRNA polyadenylation may have functional significance. For example, this represents a possible mechanism whereby hormones can regulate RNA processing which in turn could regulate rGH production at the translational level.

Hormonal Regulation of rGH in Transformed Cells. The influence of thyroid and glucocorticoid hormones on the rGH- related mRNAs derived from the mouse L cells transformed with the rGH gene (discussed previously) was examined. These cells contained truncated mRNA species (0.8 kb) and a putative pre-mRNA species (1.7 kb). Glucocorticoids were found to increase the concentration of both the 0.8 and 1.7 kb RNA fragments approximately 3-5 fold. Treatment of the transformed L cells with T_3 did not increase the concentrat- ion of either RNA species. Since the L cell nuclei con- tained low levels of a T_3-binding activity with an affinity which was 10-fold lower than the normal intranuclear T_3 receptor, this result suggests that the absence of bona fide receptors accounts for the lack of rGH regulation by T_3. Additionally, the glucocorticoid-mediated stimulation of the rGH mRNA levels in transformed L cells was shown not to be an effect on transcription (S.H. Mellon, unpublished observations). The data indicate that the normal rGH gene promoter is not very active in transformed L cells. Never- theless, a putative promoter (discussed previously) within Intron B can account for the production of a truncated, abberrant rGH transcript which like the case with GC cells is not under transcriptional control by glucocorticoids. The levels of the abberrant transcripts resulting from this promoter are nonetheless regulated by the steroid, presumably by affecting mRNA stability. Thus, the data suggest that structure(s) within the mRNA contains information important for the glucocorticoid hormone-mediated influence on mRNA stability.

ACKNOWLEDGMENTS

The authors express their appreciation to Susan Bromley for assistance in preparing the manuscript and Nancy Malich, Fran DeNoto, Deborah Glaister, Tuan Nguyen and Illana Mittman for technical assistance. Dr. Baxter is an Investigator and Dr. Eberhardt is a Senior Associate of the Howard Hughes Medical Institute.

REFERENCES

1. Niall, H.D., Hogan, M.L., Saya, R., Rosenblum, T.Y., and Greenwood, F.C. (1971). Proc. Natl. Acad. Sci. USA 68, 866-869.
2. Niall, H.D., Hogan, M.L., Tregar, G.W., Segre, G.V., Hwang, P., and Friesen, H. (1973). Rec. Prog. Hormone Res. 29, 387-416.
3. Johnson, L.K., Baxter, J.D., Vlodavsky, I., and Gospodarowicz, D. (1980). Proc. Natl. Acad. Sci. USA 77, 394-398.
4. Seeburg, P.H., Shine, J., Martial, J.A., Baxter, J.D., and Goodman, H.M. (1977). Nature 270, 486-494.
5. Martial, J.A., Hallewell, R.A., Baxter, J.D., and Goodman, H.M. (1979). Science 205, 602-607.
6. Miller, W.L., Martial, J.A., and Baxter, J.D. (1980). J. Biol. Chem. 255, 7521-7524.
7. Cooke, N.E., Coit, D., Weiner, R.I., Baxter, J.D., and Martial, J.A. (1980). J. Biol. Chem. 255, 6502-6510.
8. Cooke, N.E., Coit, D., Shine, J., Baxter, J.D., and Martial, J.A. (1981). J. Biol. Chem. 256, 4007-4016.
9. Miller, W.L., Coit, D., Baxter, J.D., and Martial, J.A. (1981). DNA 1, 37-50.
10. Shine, J., Seeburg, P.H., Martial, J.A., Baxter, J.D., and Goodman, H.M. (1977). Nature 270, 494-499.
11. Goodman, H.M., DeNoto, F., Fiddes, J.C., Hallewell, R.A., Page, G., Smith, S., and Tischer, E. (1980). In "Mobilization and Reassembly of Genetic Information" (W.A. Scott, R. Werner, D. Joseph, and J. Schultz, eds.), pp. 155-179. Academic Press, New York.
12. Seeburg, P.H. (1982). DNA (in press).
13. Barta, A., Richards, R., Baxter, J.D., and Shine, J. (1981). Proc. Natl. Acad. Sci. USA 78, 4867-4871.
14. Page, G.S., Smith, S., and Goodman, H.M. (1981). Nucl. Acids Res. 9, 2087-2104.
15. Cooke, N.E., and Baxter, J.D. (1982). Nature 297, 603-606.
16. Selby, M., Barta, A., Birnbaum, M., Baxter, J.D., Bell, G.I., and Eberhardt, N.L. (in preparation).

17. Owerbach, D., Martial, J.A., Baxter, J.D., Rutter, W.J.,
     and Shows, T.B. (1981). Science 212, 815-816.
18. Owerbach, D., Rutter, W.J., Cooke, N., Martial, J.A.,
     and Shows, T.B. (1981). Science 212, 815-816.
19. DeNoto, F.M., Moore, D.D., and Goodman, H.M. (1981).
     Nucl. Acids Res. 9, 3719-3730.
20. Pavlakis, G.M., Hizuka, N., Goden, P., Seeburg, P., and
     Hamer, D.H. (1981). Proc. Natl. Acad. Sci. USA 78,
     7398-7402.
21. Phillips, J.A., III, Hjelle, B.J., Seeburg, P.J., and
     Zachmann, M. (1981). Proc. Natl. Acad. Sci. USA 78,
     6372-6375.
22. Wurzel, J.M., Parks, J.S., Herd, J.E., and Nielsen, P.V.
     (1982) DNA (in press).
23. Cathala, G., Barta, A., Martial, J.A., Lan, N.C., and
     Baxter, J.D. (in preparation).
24. Martial, J.A., Baxter, J.D., Goodman, H.M., and Seeburg,
     P.H. (1977). Proc. Natl. Acad. Sci. USA 74, 1816-1820.
25. Martial, J.A., Seeburg, P.H., Guenzi, D., Goodman, H.M.,
     and Baxter, J.D. (1977). Proc. Natl. Acad. Sci. USA 74,
     4293-4295.
26. Ivarie, R.D., Baxter, J.D., and Morris, J.A. (1981).
     J. Biol. Chem. 256, 4520-4528.
27. Wegnez, M., Schachter, B.S., Baxter, J.D., and Martial,
     J.A. (1982). DNA 1, 145-153.
28. Spindler, S.R., Mellon, S.H., and Baxter, J.D. (1982).
     J. Biol. Chem. (in press).
29. Cathala, G., Savouret, J.-F., Lan, N.C., Gardner, D.,
     and Baxter, J.D. (in preparation).
30. Wallis, M. (1980). Nature 284, 512.
31. Fiddes, J.C., Seeburg, P.H., DeNoto, F.M., Hallewell,
     R.A., Baxter, J.D., and Goodman, H.M. (1979). Proc.
     Natl. Acad. Sci. USA 76, 4294-4298.
32. Catt, K.F., Moffatt, B., and Niall, H.D. (1967). Science
     157, 321.
33. Cathala, G., Savouret, J.-F., Barta, A., Gardner, D.,
     Lan, N.C., Martial, J.A., and Baxter, J.D. (in prepara-
     tion).
34. Cathala, G., Gardner, D., Lan, N.C., and Baxter, J.D.
     (in preparation).
35. Karin, M., Eberhardt, N.L., Richards, R.I., Barta, A.,
     Malich, N., Mellon, S.H., Martial, J.A., Baxter, J.D.,
     and Cathala, G. (in preparation).
36. Mellon, S.H., Baxter, J.D., and Spindler, S.R. (in
     preparation).

REGULATION OF THE HUMAN
GROWTH HORMONE GENE IN MOUSE CELLS[1]

Diane M. Robins
Inbok Paek
Richard Axel

Institute of Cancer Research
College of Physicians and Surgeons
of Columbia University
New York, N.Y.

Peter H. Seeburg

Genentech, Inc.
South San Francisco, Calif.

INTRODUCTION

A simple model of hormone action compatible with available data assumes that the interaction of steroid-receptor complex with appropriate DNA sequences enhances transcription. There is no direct evidence in support of this model. This problem has therefore been dissected into experimentally accessible questions: Are there specific sequences about inducible genes which render them sensitive to hormonal induction? Are there specific sequences about inducible genes that show high affinity for hormone-receptor complex? And finally, how does the interaction of receptor with DNA lead to transcription activation?

In this study, we have examined the expression of human growth hormone (hGH) genes introduced into the chromosome of murine fibroblasts. In vertebrates, growth hormone synthesis is restricted to the pituitary gland. In cultures of pituitary cells, either glucocorticoid or thyroid hormone generate a 3-fold increase in the level of GH mRNA; when added together, a 10-fold induction is observed (1,2). We have used cotransformation to introduce from 1 to 20 copies

---

[1]This work was supported by a Jane Coffin Childs
Postdoctoral Fellowship to DMR and by grants from the NIH.

of the human growth hormone (hGH) gene into thymidine kinase
deficient (tk⁻) murine fibroblast cells which express
functional glucocorticoid receptors (3). The administration
of hormone to these cotransformed cells results in a 3- to
5-fold induction of hGH mRNA and a similar induction in
secreted growth hormone protein. The DNA sequences
responsive to induction reside within 500 nucleotides of DNA
flanking the putative site of transcription initiation.
Fusion of this segment of DNA to the tk structural gene now
renders the tk gene responsive to hormone action.

RESULTS

THE REGULATED EXPRESSION OF HUMAN GROWTH
HORMONE GENES IN MOUSE CELLS

The human growth hormone gene is a member of a small
multi-gene family composed of at least two additional
hormones, chorionic somatomammotropin and prolactin (4,5).
Several different recombinant phage containing hGH sequences
have been isolated from a library of human DNA (P. Seeburg,
manuscript in preparation) constructed in the Charon phage
λ4A (6). Two different phage, designated λ20A and λ2C,
contain the entire growth hormone gene within a 2.6 kb
Eco RI fragment (Fig. 1). The complete nucleotide sequence
of this Eco RI fragment derived from λ20A indicates that
this fragment contains the entire hGH gene along with 500 5'
and 525 3' flanking nucleotides (P. Seeburg, manuscript in
preparation). The restriction map of the 2.6 kb Eco RI
fragment of λ2C is identical to that of λ20A. Partial
sequence analysis of λ2C from the 5' Eco RI site to the
Bam HI site adjacent to the first exon (Fig. 1) is identical
to that of λ20A. Both λ20A and λ2C, which presumably encode
the major form of growth hormone, were utilized in our
studies.
     Mouse Ltk⁻ cells express functional glucocorticoid
receptor (3). We have utilized a viral tk gene as a
selectable marker to introduce the 2.6 kb Eco RI fragment
into this cell line by DNA-mediated gene transfer (7,8).
Cotransformants were identified by blot hybridization (9)
and tested for the capacity to regulate the expression of
exogenous human growth hormone sequences. DNA was isolated
from eleven transformants obtained following cotransfer with
tk and hGH DNA, restricted with the enzyme Eco RI, and
analyzed by blot hybridization utilizing highly radioactive
hGH probes (Fig. 2). Nine of the eleven transformants inte-
grate at least one intact Eco RI fragment containing the hGH

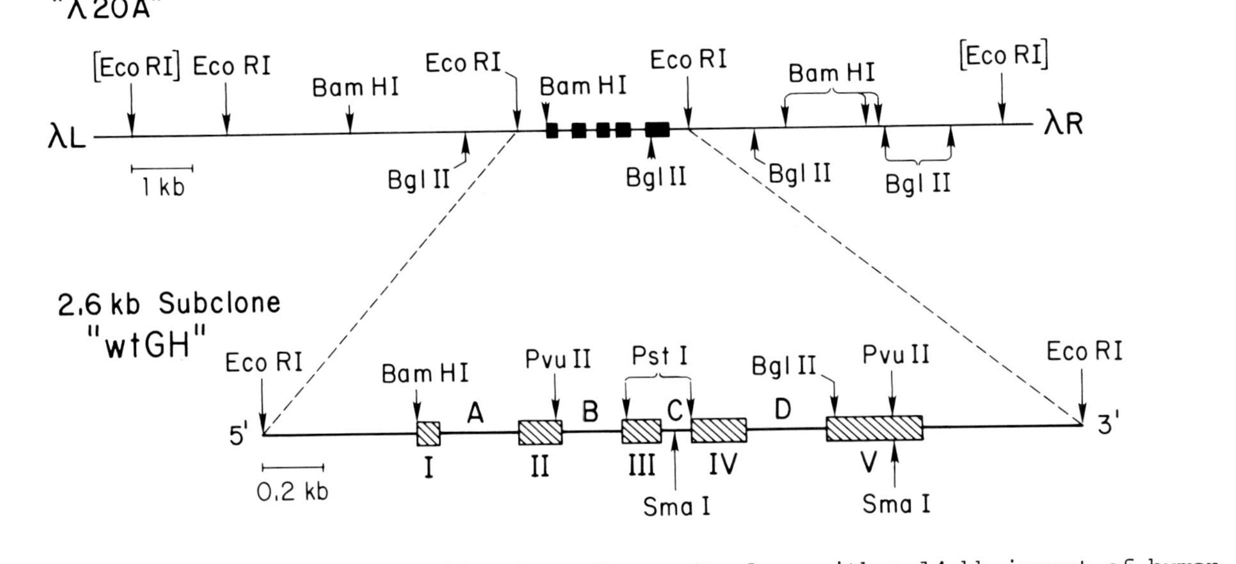

FIGURE 1. Recombinant hGH clones. λ20A is a Charon 4A clone with a 14 kb insert of human DNA. The hGH gene is contained in a 2.6 kb Eco RI fragment. Exons are shown by solid bars. A more detailed map of the sequenced 2.6 kb Eco RI fragment, "wtGH", shows the five exons (hatched bars) interrupted by introns A–D.

                    DIANE M. ROBINS *et al.*

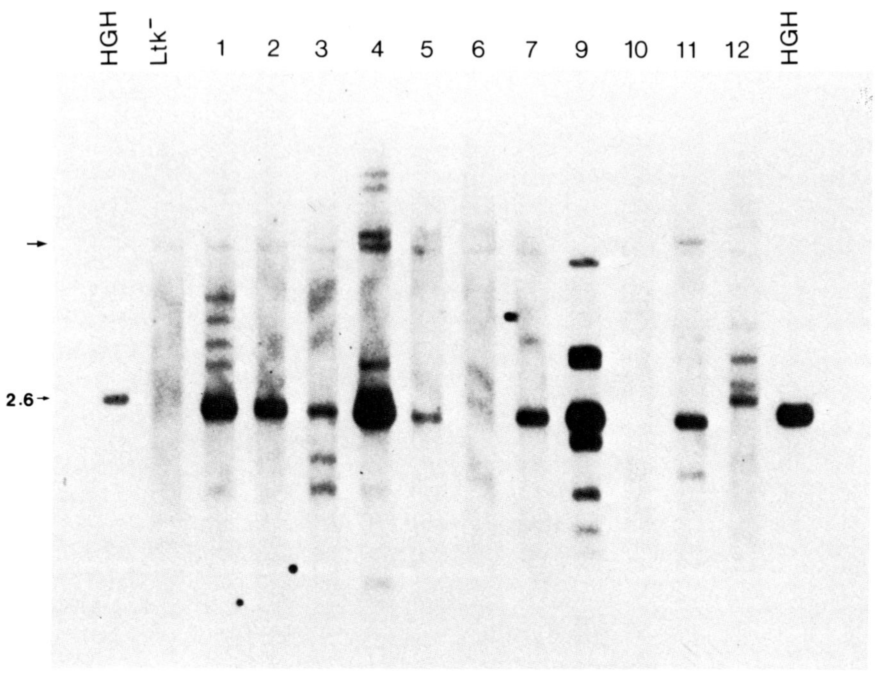

FIGURE 2. wtGH DNA in cotransformed L cells. Ltk⁻ cells
were cotransformed with 1 µg ptk, 1 µg wtGH and 20 µg Ltk⁻
DNA. Colonies surviving in HAT medium were picked and grown
into mass culture. High molecular weight DNA was digested
with Eco RI and blotted and hybridized. The hGH lane on the
left contains 20 pg of wtGH DNA and the hGH lane on the right
contains 100 pg of wtGH DNA, along with 20 µg of chicken DNA,
digested with Eco RI; these amounts are equivalent to approximately 1 and 5 hGH genes per genome, respectively. 20 µg
each of DNA digested with Eco RI from Ltk⁻ cells and 11 independently isolated tk⁺ transformants (1-12) are shown. An
arrow indicates the faint endogenous mouse GH gene present as
a high molecular weight Eco RI fragment in all the L cell
lanes.

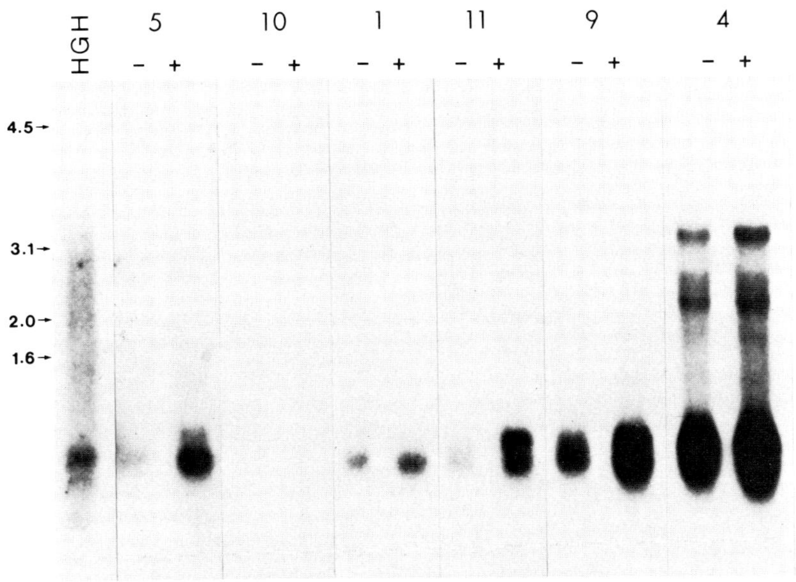

FIGURE 3. hGH RNA induction in L cells containing the
2.6 kb Eco RI hGH fragment. Cotransformant colonies (numbered as from Fig. 2) were grown for 72 hours in the absence
(-) or presence (+) of 10^{-6} M dexamethasone. Total cellular
poly A^+ RNA (12 µg/lane) was fractionated on an 0.8% formaldehyde gel and blotted onto nitrocellulose. 50 ng of human
pituitary RNA was run in the first lane, along with OD size
markers visualized by ethidium bromide staining. The nitrocellulose filter was hybridized to ^{32}P-labeled hGH DNA
(the 2.6 kb Eco RI fragment). As the 5 cotransformant cell
lines contain different amounts of hGH RNA, three autoradiographic exposures were examined. hGH, clone 5 and clone 10
(a tk^+ non-cotransformant) lanes were exposed 2 days, clones
1 and 11 were exposed 18 hours, and clones 9 and 4, 12 hours.

gene. The hGH copy number in these lines is variable. Some
lines contain 10-20 copies of hGH DNA.

These cotransformants were then tested for their ability
to regulate the expression of hGH RNA. Poly A$^+$ RNA was
prepared from five clones and examined by Northern blot
analysis (Fig. 3). All five lines synthesize an 825 bp hGH
mRNA which is inducible by the addition of glucocorticoid.
The levels of hGH mRNA, as well as the extent of induction,
differ among the different cell lines. Some lines show a
3-fold induction (lines 9 and 11), other lines show
intermediate levels of induction. A rough correlation
exists between the amount of mRNA produced and the number of
hGH copies integrated into transformed cell DNA. Thus, line
4 synthesizes the largest amount of mRNA and has integrated
the largest number of intact GH genes. Line 5 synthesizes
perhaps the least amount of mRNA and has probably integrated
only one or a few genes. Control transformants lacking an
hGH gene fail to express detectable human or mouse GH mRNA.
Similar data has emerged from an analysis of several
additional cotransformants obtained following exposure to
hGH λ phage (Table I). The frequency with which
cotransformants express elevated levels of GH mRNA upon
exposure to glucocorticoid, even if only a single intact hGH
gene is integrated into the chromosome, strongly suggests
that the information required for induction resides within
the 2.6 kb <u>Eco</u> RI fragment. Therefore, hormonal control is
not likely to result solely from the integration of hGH DNA
into pre-existing hormonally responsive sites in the
recipient cell. Further, hormonal control is not observed
for all cotransformed genes, but is specific for the hGH
gene. Analysis of the level of tk mRNA in these cell lines
by blot hybridization indicates that tk RNA levels remain
constant in the presence or absence of steroid hormone.

INDUCTION OF GROWTH HORMONE PROTEIN

We have asked whether the growth hormone mRNA sequences
present within our transformants direct the synthesis of a
secreted growth hormone polypeptide. In this manner, we
could determine whether the levels of induction of mRNA are
reflected by a proportionate induction in the level of
secreted protein. Control cells and transformants
containing hGH genes were grown either in the presence or
absence of hormone for five days without media change. The
media was then tested for hGH in radioimmune assays
utilizing antibody-coated filter disks. As shown in Table
I, cell lines which demonstrate significant levels of hGH
mRNA also secrete significant amounts of hGH into the tissue
culture medium. The maximum secretion of hGH is observed

TABLE I
QUANTITATION OF hGH mRNA AND
PROTEIN IN MOUSE CELLS

Cell line		cpm/μg A$^+$ [a]	mRNA copies/ [b] cell	Fold induction	Secreted GH (ng/ml) [c]
λGH-D	–	392	35	1.2	1.0 (below control)
	+	473	42		3.8
λGH-E	–	93	8	2.9	1.8 (below control)
	+	267	24		4.9
λGH-I	–	291	26	4.8	5.0
	+	1401	125		19.0
wtGH-1	–	58	5	2.5	N.D.
	+	146	13		
wtGH-4	–	1147	102	1.7	19.0
	+	2054	183		66.0
wtGH-9	–	487	43	3.0	18.4
	+	1455	129		66.0
wtGH-11	–	71	6	2.6	N.D.
	+	185	17		

[a] Determined from dot blots containing 3-5 concentrations of Poly A$^+$ RNA of each sample.

[b] There are 2×10^5 mRNA molecules per L cell, assuming 0.5 pg poly A$^+$ RNA/cell and an average mRNA size of 1 kb. From a standard curve derived by dot blotting human pituitary RNA in which approximately 20% of the mRNA is hGH, there are 450 cpm/100 ng of pituitary RNA, or 2.25 cpm/pg hGH mRNA. All cpm's have been normalized to the same standard curve.

[c] From radioimmune assay, in which the negative control is <2.5 ng/ml.

N.D. = Not Determined.

with cell line 9, which also exhibits a high level of mRNA in our series. This cell line synthesizes about 15 µg per liter in the absence of hormone and 60 µg of hGH per liter in the fully induced state. Thus, induction in the levels of GH mRNA result in a concomitant induction in the levels of secreted protein.

REGULATION OF AN hGH-tk FUSION GENE

We next asked whether the hGH sequence element responsive to hormonal induction can impart hormone sensitivity when fused to other structural genes. Expression of the 2.6 kb <u>Eco</u> RI fragment is hormonally regulated in mouse cells. In contrast, cotransformed tk gene expression is not regulated by glucocorticoids. We therefore constructed a fusion gene consisting of the 5' flanking sequences of hGH and the structural sequences encoding tk to discern whether the tk gene in this configuration is responsive to glucocorticoid induction. The 5' flanking region of hGH DNA extending from the Eco RI site to the Bam HI site, 3 nucleotides into the 5' untranslated region, was ligated to a fragment of tk DNA beginning in the 5' untranslated region of the tk message 60 nucleotides upstream from the translation start site (Fig. 4). This tk fragment contains the entire structural gene but lacks all essential promoter elements (10). This fusion gene was introduced into Ltk^- cells, and poly A^+ was then isolated from resulting tk^+ transformants. The patterns of the tk RNAs on Northern blot analysis are complex.

In control cells containing a wild-type tk gene, two transcripts, 1.3 and 0.9 kb, are observed. The 1.3 kb RNA is always expressed in transformants containing an intact tk gene and encodes the wild-type enzyme. The 0.9 kb RNA is observed in tk^+ transformants at a low level. This transcript initiates internal to the tk gene, consists solely of tk structural gene sequences, and must encode a truncated protein (11). RNA from one transformant containing the wild-type tk gene (line A), demonstrating these two transcripts, is shown in the Northern blot in Figure 5. RNA from three tk^+ transformants containing the hGH-tk fusion gene reveals two species of tk RNA. The larger species is the size expected (1.25 - 1.3 kb) if the growth hormone promoter initiates appropriately in hGH sequences to generate fusion mRNA. The 0.9 kb RNA presumably represents the aberrant tk transcript. In two of three transformants containing the fusion gene, we observe that the larger transcript is inducible upon glucocorticoid administration. The 0.9 kb transcript generated from an internal tk promoter provides a fortuitous internal control.

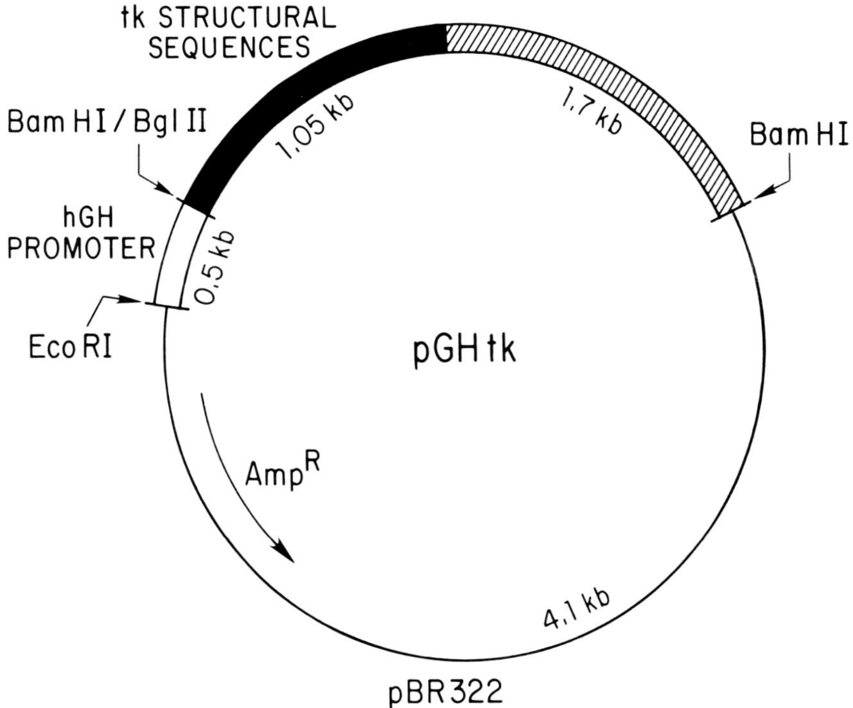

FIGURE 4.   Recombinant plasmid pGHtk contains and hGH-tk
fusion gene. Plasmid pGHtk contains an 0.5 kb <u>Eco</u> RI/<u>Bam</u> HI
fragment of hGH (☐) (from the 5' end of the 2.6 kb <u>Eco</u> RI
fragment) inserted at the <u>Bgl</u> II site of ptk, replacing the
tk promoter. Tk information includes the entire coding
sequence of the tk gene (■) as well as 1.7 kb of 3'
flanking DNA sequences (▨). The initiator AUG is located
50 nucleotides 3' to the <u>Bgl</u> II site. The <u>Bam</u> HI site of the
hGH fragment, including the putative promoter, is 3 nucleotides beyond the transcription initiation site of hGH (5).

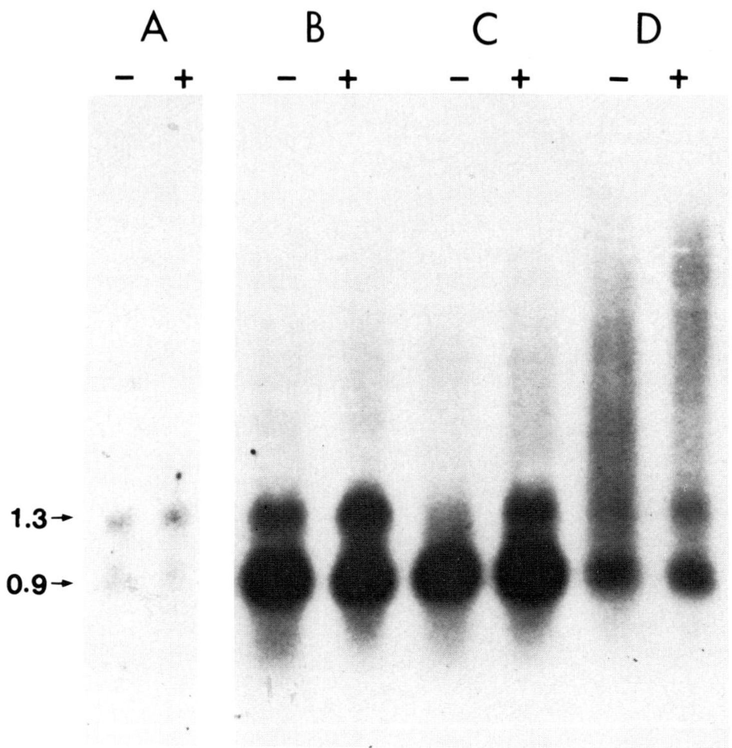

FIGURE 5.   Induction by glucocorticoid of a tk gene fused
to hGH DNA.   pGHtk was donated to Ltk⁻ cells and tk⁺ colonies
selected in HAT medium.   Several colonies were grown in the
absence (−) or presence (+) of glucocorticoid for 48 hours,
poly A⁺ RNA was isolated, and subjected to Northern blotting.
10 μg of poly A⁺ RNA was loaded in each lane.   Cell line A
is a wild-type tk⁺ transformant; lines B-D contain the pGHtk
fusion gene.   The filter was hybridized with ³²P-labeled tk
DNA (the 3.6 kb <u>Bam</u> HI fragment of ptk) and exposed to X-ray
film for 1 day.

Induction of RNA levels is observed only for the fusion transcript; the 0.9 kb RNA remains unresponsive to glucocorticoid administration. It is apparent from these studies that sequences responsive to induction reside within the 500 nucleotides flanking the 5' terminus of the hGH gene. Fusion of this element to other structural genes now renders these sequences hormonally responsive.

DISCUSSION

The introduction of recombinant clones containing the human growth hormone gene into mouse fibroblasts results in the regulated expression of hGH mRNA. Our results suggest that the induction we observe results from transcriptional activation rather than RNA stabilization. First, the hGH-tk fusion gene generates an inducible mRNA presumably with very few 5' hGH nucleotides; the remainder of the RNA encodes tk enzyme. In control cells, wild-type tk mRNA is not inducible by hormone. We consider it highly unlikely that the short segment of hGH can confer stability on the remaining length of mRNA independent of sequence or source. Second, we have identified hGH transformants synthesizing mature hGH mRNA constitutively. If induction results solely from stabilization, we would expect all transformants which synthesize hGH RNA to express enhanced levels in the presence of hormone. These results do not exclude mRNA stability as a contributing factor in the inductive process, nor do they argue that transcription level control is the sole determinant of induction. Our data do indicate that an element present in 500 bp of 5' flanking DNA is sufficient to render a gene responsive to hormone and that this element most likely operates to enhance transcription.

Evidence is accumulating for several genes that regulatory elements controlling the rate of transcription reside in 5' DNA quite close to the structural gene. Thus, MMTV (12), α-2u globulin (D. Kurtz, personal communication), hGH, tk (10), mouse globin (13), metallothionein (14) and Drosophila heat shock (15) genes remain responsive to widely differing inducing agents with 1 kb or less of 5' flanking DNA. It is possible that such elements exert their effects locally and do not "transduce" information over long distances. The hGH-tk fusion gene generates two mRNAs: a 1.25 - 1.3 kb RNA presumably initiated in GH sequences and a second 0.9 kb species initiating in the tk structural gene about 300 nucleotides downstream. Only the larger RNA is hormonally inducible, suggesting that the hGH regulatory element acts locally to activate transcription and has little or no effect upon the frequency of close, but downstream initiations. This argument must be tempered by the fact that cell lines expressing the fusion gene have integrated multiple copies of this gene. It is therefore possible that the smaller transcript derives solely from genes which remain unresponsive to hormone action.

One striking observation is that newly introduced hGH genes express significant levels of mRNA and protein while the endogenous murine GH gene in fibroblasts remains inactive either in the absence or presence of glucocorticoid. Since GH synthesis is restricted to the pituitary and is never expressed physiologically by fibroblasts, we are obliged to consider why newly introduced GH genes function in the recipient cell. It should be noted that although our control fibroblasts do not synthesize significant levels of endogenous GH mRNA, we do not know whether the murine gene is active in transformants expressing exogenous GH genes. However, in an analogous system, hormonal induction of exogenous rat α-2u globulin genes in a mouse fibroblast is not associated with activation of the endogenous mouse genes (16).

A pattern is emerging from gene transfer experiments which suggests that the mere introduction of exogenous genes into cells is sufficient to assure their expression. Thus, several genes, including globin (13,17), hGH and MMTV (18,19), when introduced into cells, synthesize significant quantities of RNA whereas their endogenous counterpart genes remain transcriptionally silent. Further, if appropriate control signals exist in the recipient cell, expression of the newly introduced genes may be properly regulated.

Studies of gene transfer together with studies of gene expression during normal development therefore define at least three states of genetic activity: "off", "on" and "regulated". The mere presence of a glucocorticoid receptor complex is inadequate to activate genes in the "off" state as is apparent for the endogenous GH gene in non-pituitary cells. Maintenance of this state perhaps reflects the chromosomal location of the endogenous gene or alternatively may result from prior developmental events about this gene which are self-perpetuating through cell division. Whatever mechanism is responsible for maintenance of the off state, it appears to be "cis" acting; a single exogenous gene introduced into a cell in which the endogenous gene is off can function in a regulated manner. Transformed genes may therefore escape the developmental history of the cell and immediately conform to the "on" state, a state accessible to appropriate regulators. In this state, genes may be regulated if appropriate controlling elements exist within the cell. The regulated state may therefore involve "trans" acting factors such as steroid hormone receptor complexes.

ACKNOWLEDGMENTS

We wish to thank Tom Livelli and John Fleming for technical assistance and Sandra Hayenga and Pam Ross for preparing the manuscript.

REFERENCES

1. Martial, J., Baxter, J., Goodman, H., and Seeburg, P. H., Proc. Natl. Acad. Sci. USA 74, 1816 (1977).
2. Tushinski, R., Sussman, P., Yu, L., and Bancroft, F. C., Proc. Natl. Acad. Sci. USA 74, 2357 (1977).
3. Lippman, M. E., and Thompson, E. B., J. Biol. Chem. 249, 2483 (1974).
4. Niall, H. D., Hogan, M. L., Sayer, R., Rosenblum, I. Y., and Creenwood, R. C., Proc. Natl. Acad. Sci. USA 68, 866 (1971).
5. DeNoto, F. M., Moore, D. D., and Goodman, H. M., Nucl. Acids Res. 9, 3719 (1981).
6. Maniatis, T., Hardison, R. C., Lacy, E., Lauer, J., O'Connell, C., Quon, D., Sim, G. K., and Efstratiadis, A., Cell 15, 687 (1978).
7. Graham, F. L., and van der Eb, A. J., Virology 52, 456 (1973).
8. Wigler, M., Sweet, R., Sim, G. K., Wold, B., Pellicer, A., Lacy, E., Maniatis, T., Silverstein, S., and Axel, R., Cell 16, 777 (1979).
9. Southern, E. M., J. Mol. Biol. 98, 503 (1975).
10. McKnight, S. L. Gavis, E. R., Kingsbury, R., and Axel, R., Cell 25, 385 (1981).
11. Roberts, J. M., and Axel, R., Cell, in press.
12. Lee, F., Mulligan, R., Berg, P., and Ringold, G., Nature 294, 228 (1981).
13. Chao, M., Mellon, P., Wold, B., Maniatis, T., and Axel, R., In: "Proc. of Symp. on Hemoglobin Synthesis" (E. Goldwasser, ed.) Elsevier, New York, in press.
14. Brinster, R. L., Chen, H. Y., Trumbauer, M., Senear, A. W., Warren, R., and Palmiter, R., Cell 27, 223 (1981).
15. Corces, V., Pellicer, A., Axel, R., and Meselson, M., Proc. Natl. Acad. Sci. USA 78, 7038 (1981).
16. Kurtz, D. T., Nature 291, 629 (1981).
17. Mantei, N., Boll, W., and Weissmann, C., Nature 281, 40 (1979).
18. Hynes, N. E., Kennedy, N., Rahmsdorf, U., and Groner, B., Proc. Natl. Acad. Sci. USA 78, 2038 (1981).
19. Buetti, E., and Diggelmann, H., Cell 23, 335 (1981).

MATERNAL AND EMBRYONIC TRANSCRIPTS OF THE SEA URCHIN EMBRYO[1]

Constantin N. Flytzanis, James W. Posakony,[2] Carlos V. Cabrera
Terry L. Thomas, Roy J. Britten and Eric H. Davidson

Division of Biology, California Institute of Technology
Pasadena, California 91125

Our view of the structure and function of the maternal RNA
stored in the egg of the sea urchin is changing as new data
accumulate. It is no longer convincing to consider the RNA
stored in the unfertilized egg simply as mature but inactive
messenger RNA. Maternal message, in the strict sense of the
term, is mRNA that is synthesized during oogenesis and is
utilized to support protein synthesis during early embryonic
development. Recent observations show that in addition to
bona fide maternal mRNA that is ready to be loaded on polysomes, a structurally more complex class of maternal transcript is also stored in the cytoplasm of both sea urchin and
amphibian eggs. The first indication of this was the observation of Hough-Evans _et al_. (1) that the complexity of the mRNA
loaded on early embryo polysomes is only about 70% of that of
the total egg RNA, although the egg RNA complexity is 1/5 of
that of the nuclear RNA in oocytes or embryos. That is, though
unusually complex for a cytoplasmic RNA population, the egg
RNA clearly cannot be regarded as an accumulation of total
germinal vesicle transcripts. Costantini _et al_. (2,3) more
recently reported that the poly(A) RNA of unfertilized
Strongylocentrotus purpuratus eggs includes transcripts of
several hundred different repetitive sequence families. At
least 70% of the mass of the poly(A) RNA contains repeat
sequence transcripts, and these are interspersed with single
copy sequence transcripts in an organization much like that of
the typical genomic DNA of this species. If these egg poly(A)
RNA molecules are allowed to renature, the complementary
repeat sequences form partially duplexed, multi-molecular
structures such as are shown in Figure 1. Evidently more than
one repeat sequence element exists in many of the egg poly(A)
RNA molecules, and the repeats are frequently located in
internal positions.

[1]This work was supported by NIH Grant HD-05753. C.N.F.
is supported by a Gosney Fellowship, and J.W.P was supported by
an NIH Predoctoral Training Grant GM-07616, and C.V.C is
supported by an NIH International Fellowship TW-02996.
[2]Present address: Department of Biochemistry and Molecular Biology, Harvard University.

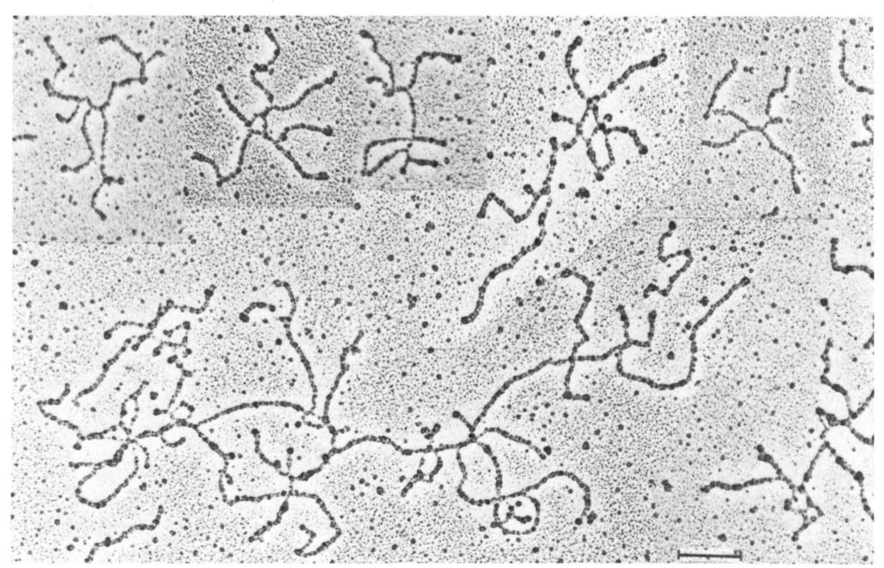

FIGURE 1. Electron   micrograph   of   multimolecular
structures formed by renaturation of sea urchin egg poly(A)
RNA.   Total egg poly(A) RNA was denatured and allowed to
renature to a C_0t of 600 M.S. The structures are held together
by short repetitive sequence duplexes that account for about
10% of the total length of the RNA molecules. The bar
represents a single-strand RNA length of 1000 nucleotides.
(Reprinted by permission from Nature, 287, p. 111. Copyright (c)
1980 Macmillan Journals Limited.)

Interspersed sequence organization of the oocyte poly(A)
RNA has been also shown for Xenopus laevis by Anderson et al.
(4). This investigation showed that as in sea urchin, about
70% of the mass of the Xenopus oocyte poly(A) RNA contains
interspersed repeat sequence transcripts. Studies carried out
with RNA extracted from manually enucleated oocytes showed
that the interspersed RNAs are located in the cytoplasm. Thus
interspersed poly(A) RNAs are a major part of the maternal
transcripts inherited by the early embryo in both species.
An unexpected finding (3) was that probes representing
both complements of genomic repetitive sequences are always
represented in sea urchin egg poly(A) RNA. This can be
demonstrated by the RNA gel blot method, as shown in Figure 2.
In this experiment (5) maternal poly(A) RNA was reacted with
strand separated probes from three different cloned repetitive
DNA families. It is evident that the band patterns obtained
differ for each strand of the same repetitive sequence. This
implies that different transcripts sharing a given repeat
are transcribed from different regions of the genome. To

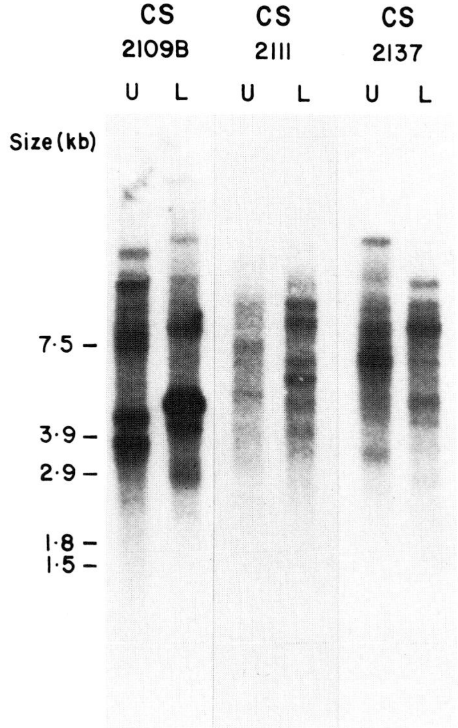

FIGURE 2. RNA blot hybridizations with cloned repetitive
sequences.   The cloned   probes 2109B, 2111 and 2137 were
described previously in detail by Posakony et al. (6).   All
lanes contain the same amount of egg poly(A) RNA. The DNA was
end-labeled and strand separated.   (U-upper) and (L-lower)
represent the slow and fast migrating strands respectively.
Molecular weight standards are given in kilobases.   (From
Posakony, J.W. (1982) Ph.D. Thesis, California Institute of
Technology).

demonstrate this directly Posakony cloned the poly(A) RNA of
sea urchin eggs using random primers, so that all regions of
the transcripts would be represented in the resulting cDNA
clone bank.   The sequence organization of several such clones
was studied in detail.   These clones were selected by screening
with the same cloned repetitive sequences as used in experiments such as those in Figure 2.   It was found that sets of
cDNA clones sharing the same repeat sequence indeed represent
specific transcripts coming from different regions of the
genome, since the single copy regions they contain do not cross
react.   RNA gel blot hybridizations carried out with such
single copy probe fragments each display a different sized

transcript, and the prevalence of these transcripts varies greatly even within a given family of cloned maternal transcripts (i.e., a set sharing a homologous repeat sequence). Only one strand of the single copy portions of these cDNA clones ever hybridizes to the maternal RNA. Transcription of the interspersed maternal poly(A) RNAs is thus clearly an asymmetric process.

Sequence determinations carried out on 8 of 9 different cloned repeats of genomic families represented in egg RNA (6) and also on the repetitive regions of some egg cDNA clones (5), have revealed translation stop codons in all possible reading frames, in both directions. While the number average length of S. purpuratus egg poly(A) RNA is about 3,000 nucleotides (nt) (2), Posakony (5) found that the interspersed RNA molecules are much longer, ranging from about 3 to at least 15 kilobases (kb). Since this is also much larger than the average polysomal mRNA of sea urchin embryos, these facts suggest that there are long, nontranslated regions somewhere in the interspersed maternal poly(A) RNAs. One such large poly(A) transcript (10 kb long) that is stored in the egg of S. purpuratus and that displays interspersed sequence organization has been studied by Thomas et al. (7). This maternal poly(A) RNA derives from a sequence known as "gene 88", the function of which has yet to be determined. Gene 88 transcripts are loaded on polysomes of early cleavage stage embryos but are later restricted to the cell nucleus, where they are found exclusively both in post-gastrula stage embryos and in adult tissues (8).

The presence of interspersed repeats in long polyadenylated transcripts is of course reminiscent of the structure of nuclear RNA, though as already pointed out the total egg RNA complexity is only a small fraction of that of the total nuclear RNA of the oocyte. Conceivably, as in the case in many nuclear RNAs, these repeats reside in unprocessed intervening sequences, a suggestion that implies that a major fraction of the stored maternal RNA might be utilized only after processing reactions have occurred in the embryo cytoplasm. On the other hand, the repetitive sequences may exist primarily in 3' terminal regions of the transcripts, where they might not interfere with immediate translation. In either case the fate of these maternal RNAs, and the biological significance of the nontranslatable sequences they contain, both repetitive and single copy, could be of basic importance in understanding the role of maternal transcripts in development. There are a number of interesting possibilities, as discussed by Davidson et al. (9). For example, cytoplasmic processing in the embryo could represent a control point in the utilization of maternal information, and might be accomplished differently at various stages or in the various regions of the embryo. The maternal

transcripts might include sequences (perhaps the repeat elements themselves) recognized by other macromolecules, such as
proteins, small nuclear RNAs or other RNAs, and such complexes
might be involved in sequestration of maternal mRNA; or in
subsequent processing; or in determining their turnover rates;
or in localization of the transcripts within the embryo; or
even in regulation of structure or function within the blasto-
mere nuclei.

During the three days required for the embryo to complete
development and attain the pluteus stage ($\sim$ 1,500 cells), the
number of ribosomes and the mass of both total RNA and poly(A)
RNA remain constant, though of course the amount of DNA and of
nuclear RNA increase. Prior to feeding, embryogenesis is thus
a process of cell division, reorganization of maternal components, and the appearance of new transcripts and their
products, all occurring within a closed system that is not
significantly increasing in macromolecular mass. Studies of
transcript prevalence during development, by the cDNA colony
screening method, showed only small changes for most highly
abundant cloned sequences (10). It was concluded that almost
always the more prevalent late embryo sequences are already
abundant in the maternal RNA, and that major changes in the
level of representation of such sequences during embryonic
development occur infrequently. This result was in accord
with prior measurements carried out on rare polysomal mRNA
sequences of gastrula and pluteus stage embryos. These
sequences are also represented, at the same low concentrations, in egg RNA (1,11). In addition, the studies of Bedard
and Brandhorst (unpublished data) have demonstrated that only
a minority of the protein species visible at the resolution of
two-dimensional gels clearly display sharp developmental
changes in their rate of synthesis. These changes occur mainly
in the blastula-gastrula period. However, most prominent
species of protein are synthesized throughout development in
sea urchin embryos and are coded initially by abundant maternal mRNAs (12). Recently we reported further developmental
prevalence measurements on about 200 cDNA clones represented
by abundant and moderately abundant embryo poly(A) RNAs (13).
Figure 3 shows graphically that in the typical case there are
at gastrula stage severalfold less copies of each transcript
species in the cytoplasm per embryo than there were originally
in the egg. However, by pluteus stage the number of transcripts of each species has risen to a level close to or
slightly greater than that at which they were originally
present in the maternal RNA. Only a few prevalent sequences
describe courses other than the typical pattern.

The prevalence of specific transcripts in sea urchin
embryos is determined by four parameters. These are the amount
of maternal transcript, the decay rate of the maternal

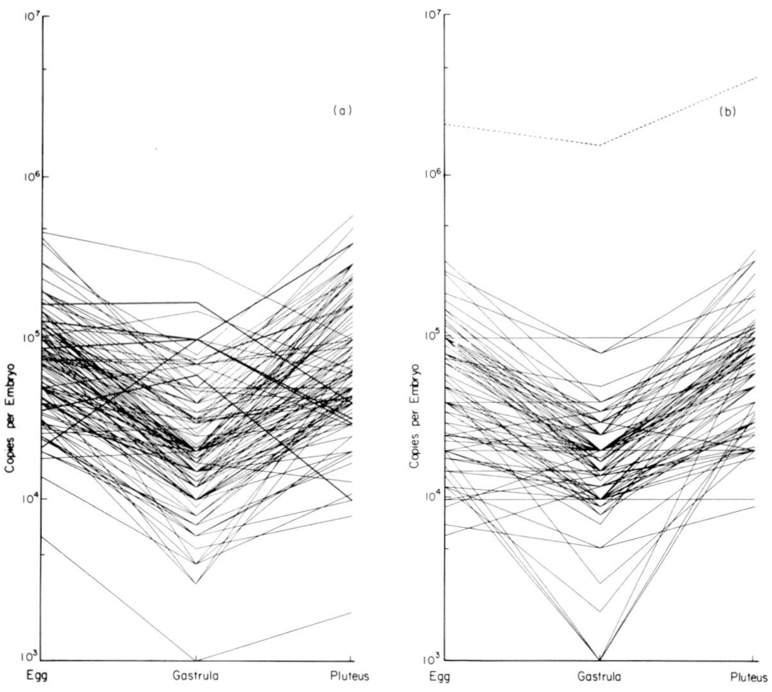

FIGURE 3.   Patterns of prevalence change for abundant cytoplasmic transcripts.   Light solid lines represent those clones displaying the "typical" developmental pattern and the heavier lines represent the few clones that display signifi- cantly different patterns. (a) 116 clones from a two-cell cDNA clone sublibrary; (b) 85 clones from a pluteus cDNA clone sublibrary. The dashed line represents the behavior of a 16S rRNA mitochondrial cDNA clone, which was used as a prevalence standard in these experiments. All of the other cDNA clones in the figure represent genomic rather than mitochondrial tran- scripts. (From Flytzanis et al. Develop. Biol. 91, 27, 1982).

transcript, the rate of flow of newly synthesized transcripts of that species into the cytoplasm and their decay rate. Cabrera et al. (14) have found that the newly synthesized transcripts belonging to high abundance classes characteristi- cally display low decay rates. Poly(A) RNAs of this category begin to flow into the cytoplasm at different times in develop- ment and at different rates. Thus the point at which the accumulation of these stable, newly synthesized transcripts can account for the total quantity of transcripts in the embryo differs for each sequence studied.   For some abundant sequences this point occurs in the blastula stage, and for most by the gastrula stage, but for other species the rate of synthesis is never high enough to account for more than a small fraction of the transcripts in the embryo. Thus some maternal poly(A) RNAs remain present even at the pluteus stage. The shape of the dominant representation patterns shown in Figure 3 thus reflect (a) the initial prevalence of the maternal transcripts; (b) a developmental decline in prevalence that is apparently due to turnover of the maternal transcripts; and (c) a late developmental increase in prevalence due to accumulation of stable, newly synthesized transcripts. It is interesting that the final result of this process is the reestablishment of approximately the same transcript sequence concentration (e.g., per ribosome) as in the unfertilized egg. This suggests that the sequence concentration of the abundant cytoplasmic poly(A) RNA is physiologically important.

Rare transcripts cannot easily be followed during development by the cloned cDNA colony screening procedure (10), but rare transcripts that become relatively prevalent at one or another embryonic stage can be detected. Flytzanis et al. (13) studied a set of cDNA clones that are extremely rare or are not represented at all in maternal RNA, and that become significantly more abundant during early development. This set of clones represents about 3% of the sample of cDNA clones screened, and it is safe to assume that < 10% of genes active in the embryo fall into the class of sequences that is not represented in the maternal RNA and later displays a dramatic developmental activation. Such transcripts, however, never approach the prevalence of the very high abundance embryo transcripts. Since only moderate numbers of transcripts (of the order of 10^4 /embryo) of these regulated genes are apparently required during embryogenesis, they can be supplied by new synthesis by the time there are only a few hundred embryo nuclei. By expressing these sequences the embryo may significantly alter its macromolecular constitution in a temporally and perhaps spatially specific way. From the same cDNA libraries as served for the experiment shown in Figure 3, Bruskin et al. (15) isolated several cDNA clones that represent poly(A) RNAs localized to the ectoderm of the late embryo. These sequences are also represented at very low levels if at all in the maternal poly(A) RNA, but they increase in prevalence rapidly during later development.

To summarize, we can distinguish several completely distinct classes of heterogeneous cytoplasmic embryo transcript. Among these are (a) bona fide maternal mRNAs that are translated on embryo polysomes following fertilization; (b) interspersed maternal poly(A) RNAs that contain extensive nontranslated regions, the function of which are yet unknown, even though they represent a major portion of the maternal poly(A) RNA mass; (c) embryo transcripts that replace homologous maternal transcripts, at more or less the same sequence concentration as originally in the unfertilized egg; (d) embryo transcripts that are absent from the maternal RNA and that appear only at given developmental stages. We shall have to learn a great deal about the structure, function, and regulation of all of these classes of gene product before the molecular biology of gene activity in the embryo begins to be understood.

REFERENCES

1. Hough-Evans, B. R., Wold, B. J., Ernst, S. G., Britten, R. J., and Davidson, E. H. (1977). Develop. Biol. 60, 258.
2. Costantini, F. D., Scheller, R. H., Britten, R. J., and Davidson, E. H. (1978). Cell 15, 173.
3. Costantini, F. D., Britten, R. J., and Davidson, E. H. (1980). Nature 287, 111.
4. Anderson, D. M., Richter, J. D., Chamberlin, M. E., Price, D. H., Britten, R. J., Smith, L. D., and Davidson, E. H. (1982). J. Mol. Biol. 155, 287.
5. Posakony, J. W. (1982). Ph.D. Thesis, California Institute of Technology.
6. Posakony, J. W., Scheller, R. H., Anderson, D. M., Britten, R. J., and Davidson, E. H. (1981). J. Mol. Biol. 149, 41.
7. Thomas, T. L., Britten, R. J., and Davidson, E. H. (1982). Devel. Biol., submitted for publication.
8. Lev, Z., Thomas, T. L., Lee, A. S., Angerer, R. C., Britten, R. J., and Davidson, E. H. (1980). Devel. Biol. 76, 322.
9. Davidson, E. H., Hough-Evans, B. R., and Britten, R. J. (1982). Science, in press.
10. Lasky, L. A., Lev, Z., Xin, J.-H., Britten, R. J., and Davidson, E. H. (1980). Proc. Natl. Acad. Sci. USA 77, 5317.
11. Galau, G. A., Klein, W. H., Davis, M. M., Wold, B. J., Britten, R. J., and Davidson, E. H. (1976). Cell 7, 487.
12. Brandhorst, B. P. (1976). Devel. Biol. 52, 310.
13. Flytzanis, C. N., Brandhorst, B. P., Britten, R. J., and Davidson, E. H. (1982). Devel. Biol. 91, 27.
14. Cabrera, C. V., Ellison, J. W., Moore, J. G., Britten, R. J., and Davidson, E. H. (1982). Submitted for publication.
15. Bruskin, A. M., Tyner, A. L., Wells, D. E., Showman, R. M., and Klein, W. H. (1981). Devel. Biol. 52, 308.

COMPLEX HORMONAL REGULATION OF MAMMARY GLAND MILK PROTEIN GENE EXPRESSION

Jeffrey M. Rosen[1], Andrew A. Hobbs[2], M.L. Johnson[3], John R. Rodgers[4], and Li Y. Yu-Lee[5]

Department of Cell Biology, Baylor College of Medicine, Houston, Texas 77030

ABSTRACT The control of milk protein gene expression by both peptide and steroid hormones is a complex inter- active process involving regulation at both the tran- scriptional and post-transcriptional levels. Both the rapidity and the large magnitude of induction observed in response to hormones in serum-free mammary explant cultures make this an ideal system for investigating the factors controlling specific mRNA accumulation in mammalian cells. The predominant milk proteins, the caseins, are a family of phosphoproteins, which appear to have evolved from a common ancestral gene. Consider- able sequence divergence has been observed among members of this gene family in both intra- and interspecies comparisons, with the exception of two conserved regions, the signal peptide sequences and the casein phosphoryla- tion sites. Members of the casein gene family are located on a single mouse chromosome, #5, while a fourth, abundant milk protein gene encoding a novel cysteine- rich, acidic whey protein is on a different chromosome. Analysis of the structure of genomic casein clones has revealed that the casein genes are large and complex, containing multiple intervening sequences resulting in a length of approximately 17 Kb for the γ-, and 25 Kb for the β-casein gene. These genes are, therefore, approxi- mately 18 and 22 times larger than their respective mRNAs. Cloned cDNAs have been used to measure the levels of these mRNAs during both normal mammary development

[1] This work was supported by a National Institutes of Health Grant CA-16303.
[2] Recipient of a Damon Runyon Fellowship DRG-482-F
[3] Recipient of a National Institutes of Health Fellowship HD-06157.
[4] Recipient of an NSF Graduate Research Fellowship.
[5] Recipient of a National Institutes of Health Fellowship CA-06645.

and in cultures exposed to the hormones, prolactin and
hydrocortisone. In organ culture both hormones were
found to be essential for the maximal induction of all
four mRNAs, with whey protein mRNA levels being particularly sensitive to the inductive effects of the steroid
hormone. The relative proportions of the α- and β-
compared to γ-casein mRNA changed from 300:35:1 in the
uninduced to 4:7.7:1 in the induced state, primarily
reflecting the very low basal levels of γ-casein mRNA.
Recent sequence analysis of γ-casein mRNA has revealed
the presence of an unusual poly(A) addition sequence,
AUUAAA, and a significantly greater proportion of γ-
casein mRNA was also observed in nuclear poly(A)-minus
RNA isolated from lactating glands. Changes in milk
protein mRNA concentrations may be explained by
differences in both the half-lives of these mRNAs, as
well as their rates of transcription in the presence
and absence of hormones.

INTRODUCTION

Milk protein gene expression is a developmentally and
hormonally regulated process (1). Milk protein mRNAs are
highly abundant in mammary tissue, comprising approximately
75% of the poly(A)RNA during mid-lactation. These hormone-
inducible mRNAs are, therefore, readily purified, and the
identification and isolation of specific cDNA probes are
straightforward (2,3). Specifically, three members of the
casein gene family, designated α-, β- and γ-casein, and a
fourth, whey acidic protein, have been utilized in our laboratory as markers of hormone action in the mammary gland (4).
Cloned cDNA probes for both mouse and rat milk protein mRNAs
have been generated permitting studies of the evolution of
these genes.
Milk protein gene expression is regulated by both
peptide and steroid hormones, principally prolactin and
hydrocortisone. Multihormonal regulation of gene expression
can be studied in mammary explant cultures using a serum-
free, chemically defined medium (5). Both the rapidity and
the large magnitude of induction observed in these cultures
make this an ideal system for investigating the factors
controlling specific mRNA accumulation in mammalian cells.
In particular it has been possible to study mRNA metabolism
using both steady state and pulse-chase methodologies (6).
Thus, it is possible to distinguish hormonal effects at both
the transcriptional and post-transcriptional levels. In the
following chapter we will describe briefly the results of our

studies of the structure, evolution, and hormonal regulation
of casein and whey protein gene expression.

THE CASEIN GENE FAMILY

The caseins are a family of milk phosphoproteins that
are secreted during lactation as large, calcium-dependent
aggregates termed micelles, in response to the lactogenic
hormones, prolactin and hydrocortisone (7). Caseins can be
isolated from skim milk either by acid precipitation at pH
4.0-4.6 or by centrifugation (8). Three major types of
caseins have been identified, which are present in constant
proportions in any one species (9): (1) α_S-caseins are precipitated by low concentrations of calcium, have the highest
phosphate content, little observable secondary structure,
and are the most poorly conserved caseins in terms of their
amino acid compositions and molecular weights; (2) β-caseins
are precipitated by moderate concentrations of calcium, have
some observable secondary structure, and are moderately
conserved; (3) κ-caseins are insensitive to calcium, have
few phosphate groups, are essential for micelle formation,
and are the most conserved of the casein proteins.

In the rat the three major milk proteins, the α-, β- and
γ-caseins, have apparent molecular weights of 42,000, 25,000
and 22,000, respectively (10), and comprise almost 80% of the
milk protein. These rat caseins appear to be analogous to
the bovine α_{S1}-, β-, and α_{S2}-caseins, respectively.
Cloned cDNA probes for each of the rat and mouse casein mRNAs
have been isolated in our laboratory (2,3).

Both heteroduplex and direct sequence analysis have indicated that there has been considerable sequence divergence
among the members of the casein gene family, such that even
under non-stringent hybridization conditions no cross-
hybridization was observed between members of the family in a
given species (2). Detailed sequence analysis of the rat
α-, β- and γ-casein mRNAs (11,12) has revealed two principal
regions of conservation among members of this gene family:
the signal peptide sequences (11,13) and the casein kinase
phosphorylation recognition sequences (11,12). As illustrated
in Figure 1, extensive homology is observed both at the
nucleotide (78%) and amino acid (73%) levels between the rat
β- and γ-casein signal peptide sequences. A similar 15 amino
acid signal peptide sequence has been observed in all
calcium-sensitive caseins in both intra- and interspecies
comparisons (14). For example, a 92% homology has been found
between the rat and ovine β-casein signal peptide sequences,
while the rest of the β-casein protein displays only a 38%

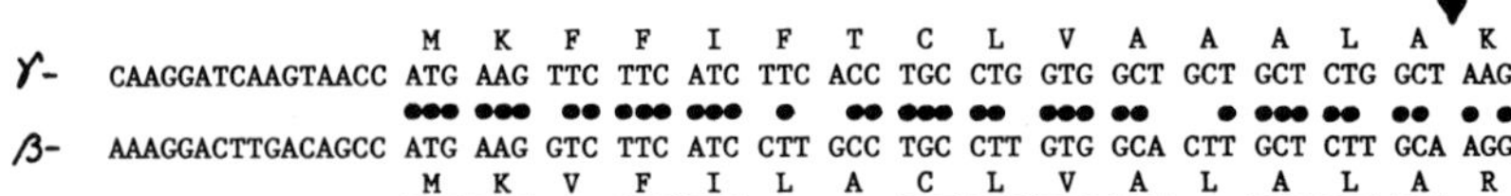

FIGURE 1.   Comparison of rat γ- and β-casein signal
peptide sequences. The single letter amino acid code is
shown. The arrow designates the amino termini of the
processed native caseins. Solid dots indicate conserved
nucleotides.

conservation (11). Thus, the casein signal peptide sequences
appear to have an unusual constraint on their evolution, suggesting a potentially important role for these sequences in
secretion.

The second region of conservation observed among members
of the casein gene family are the sites of serine phosphorylation and calcium binding, either -Ser-X-Glu or -Ser-X-Ser-P
(12). A comparison of the nucleotide sequences at the multiple phosphorylation sites in the α-, β- and γ-caseins indicates that similar serine and glutamic acid codons have been
utilized at these sites, while the remaining sequences in the
mature proteins have rapidly diverged. Thus, it is likely
that the caseins have evolved from a single primordial gene,
which most likely contained the site of phosphorylation.
Evolution of the signal peptide domain and gene duplication
may then have occurred prior to the mammalian radiation.
Rapid divergence of other regions of these proteins, whose
structure is probably dictated by their role as dietary
sources of essential amino acids, is, therefore, not
unexpected. Presumably, several essential regions of these
proteins involved in calcium binding, micelle formation and
casein secretion may evolve at a slower rate than the
remainder of the protein.

Additional support for this hypothesis has been obtained
recently by the determination of the chromosomal localization
of the members of the casein gene family. Mouse casein cDNA
clones homologous to the three rat α-, β- and γ-caseins, and
a fourth novel mouse cDNA clone encoding a protein similar to
these other calcium-sensitive caseins, were used to analyze a
panel of mouse : Chinese hamster somatic cell hybrids that
generated an 18-fold division of the mouse genome (15). All
four casein cDNA clones hybridized to DNA isolated from the
same two hybrid cell lines, which were the only hybrids
containing mouse chromosome #5 (3). The members of the casein
gene family are localized on a single chromosome, as might

be expected if they had originated by a relatively recent
duplication event from a common ancestral gene.

While genetic analysis has suggested that the caseins
may exist as a gene cluster on a single chromosome, little
information is available about the precise structure and
linkage arrangement of these genes. To obtain this information, a rat genomic DNA library was screened using a mixed
probe containing all three α-, β- and γ-casein cDNA clones.
Following plaque purification, each positive phage clone was
then screened with an individual casein cDNA probe. In no
instance was a single phage containing between 10 to 20 Kb
of rat DNA found to contain more than a single casein gene.
Based upon the analysis of approximately 100 Kb of DNA, it
now appears that the failure to isolate a single phage clone
containing more than a single casein gene, or even overlapping clones demonstrating the direct linkage of two of the
genes, is due to their large and complex nature as illustrated in Figure 2.

The structure of the γ-casein gene has been elucidated
by the analysis of two overlapping phage clones containing
approximately 28 Kb of rat DNA. Restriction mapping of these
clones and total rat DNA, as well as R-loop analysis, has
revealed that the single copy γ-casein gene is approximately
17 Kb in length or 18 times the size of the γ-casein mRNA
(16). The γ-casein gene is divided into at least nine short
exon sequences separated by large introns. An additional
characteristic of most eukaryotic genes also observed in the
casein genes is the presence of repeated sequence families
interspersed both in intervening and flanking sequences.
The presence of both very short exons and repeated sequences
in flanking and intervening sequences of the casein genes
presents a difficult technical problem in screening gene
libraries for members of this gene family.

The preliminary structure of the β-casein gene is also
shown in Figure 2. It is approximately 25 Kb in length or 22
times the size of β-casein mRNA. Although 40 Kb of genomic
DNA in this region have been analyzed, the precise determination of exon positions, as well as the complete structure of
this gene, has yet to be determined. If the members of the
casein gene family have originated by gene duplication, it
will be of interest to compare the structure of these genes
with respect to the positioning of introns and exons and the
location of putative functional domains within these proteins. These studies of casein gene structure are necessary
prerequisites for any experiments involving hormonal regulation of casein gene transcription.

Detailed restriction mapping has also allowed us to
investigate the methylation pattern of these genes. An

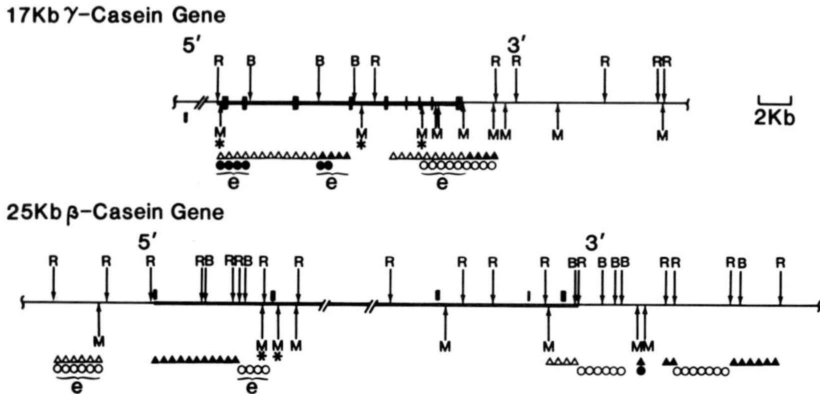

FIGURE 2. Structure of the rat γ- and β-casein genes.
The heavy solid line represents the gene regions containing
both introns (——) and exons (▮). The 5' and 3' flanking
sequences are depicted by the thinner solid lines. The
location of the following restriction enzyme sites are shown:
R = EcoRI, B = BamHI, M = MspI. The asterisks indicate MspI
sites that are methylated in non-expressing tissues such as
liver (see Figure 3). The solid and open triangles show the
positions of repeated DNA sequences present in both introns
and flanking sequences. The circles depict the position of
repeated DNA sequences homologous to a 2 Kb cloned rat repeat
sequence isolated from the 3' flanking region of the rat albumin gene (kindly provided by Dr. Tom Sargent). Open symbols represent weakly hybridizing repeat sequence families,
while solid symbols depict strongly hybridizing and possibly
more highly conserved repeated DNA sequences. The symbol
(e) shows the location of evolutionary conserved repeat
sequences, which are observed only when salmon sperm DNA is
not used in the hybridizations.

inverse correlation between tissue-specific patterns of
casein gene expression and methylation of certain CpG sites
has been observed (16). In lactating tissue in which the
casein genes are maximally expressed, an analysis of MspI/
HpaII sites has revealed that these sites are unmethylated
(Figure 3, lanes 3 and 4). However, in liver, a tissue in
which casein gene expression is not detected, three MspI/
HpaII sites are methylated (Figure 3, lanes 5 and 6). The
position of these MspI sites in the γ-and β-casein genes are
shown in Figure 2. In carcinogen-induced mammary carcinomas,
which retain their hormone-dependent growth properties, but
have greatly attenuated levels of casein gene expression

(16), a methylation pattern similar to liver is seen (Figure
3, lanes 1 and 2). In these tumors, a minor subpopulation
of cells has been shown to retain hormone-inducible casein
gene expression (17) and this may account for the faint
signals detected in the HpaII digests in Figure 3, lane 2.

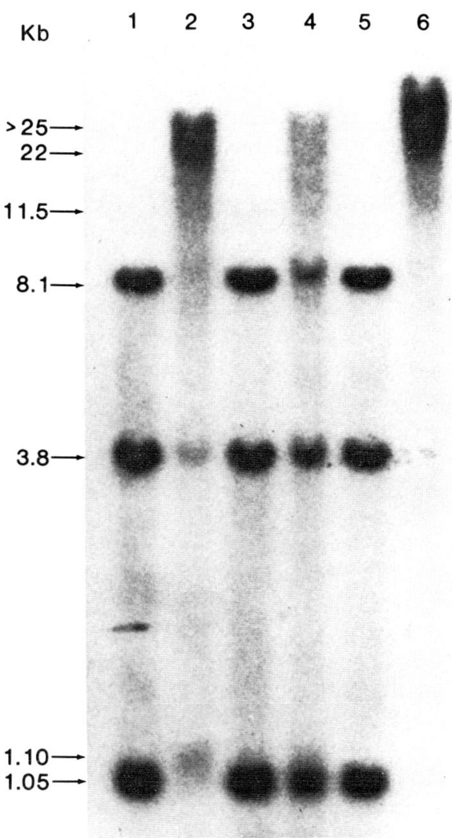

FIGURE 3. Analysis of methylation patterns in the γ-
casein gene. DNA (15 µg) was digested with either MspI
(lanes 1, 3 and 5) or HpaII (lanes 2, 4 and 6), electrophoresed on a 1% agarose gel and transferred to nitrocellulose.
Lane 1,2 = DMBA-induced mammary tumor DNA; lanes 3,4 = 8
day lactating rat DNA; lanes 5,6 = rat liver DNA. Hybridization was performed with a nick translated γ-casein cDNA
probe.

More extensive mapping studies have indicated that while,
in general, there is an inverse correlation between the
methylation pattern of the casein genes and the differentiation state of these tissues (or cell populations within
tissues), this is not an absolute "all or none" relationship.
Thus, both the positions and types of sites which are methylated may be important in the potential expressibility of
these genes. At certain HhaI sites methylated CpG sites are
present in the γ-casein gene both in lactating tissue and
liver, while certain other sites within these genes may be
unmethylated in all tissues. Furthermore, while methylation
may play a role in determining (or reflecting) the higher
order structure of these genes and their intrinsic ability
to be regulated by hormones, changes in DNA methylations
appear to have little mechanistic importance in hormonal
regulation of casein gene expression. We do not yet understand the origin of the observed differences in specific gene
methylation in transformed versus normal mammary cells.

A NOVEL MURINE WHEY ACIDIC PROTEIN

In addition to the caseins, which comprise the majority
of the milk protein, several whey proteins are secreted
during lactation. These include α-lactalbumin, and in murine
milk, a novel cysteine-rich protein designated whey acid protein (WAP, 18). Murine WAP has a molecular weight of about
14,000, an acidic isoelectric point, and may contain lipid.
While mouse WAP is not phosphorylated, rat WAP contains up to
three phosphate groups per molecule. A homologous whey protein has not yet been described from the much better characterized ruminant milks.
Cloned WAP cDNAs have been isolated and sequenced from
both the mouse and rat (18,19). Mouse and rat WAP (134 and
137 amino acids, respectively) contain an N-terminal signal
peptide of 19 amino acids. Most of the cysteines in WAP are
located in two domains containing six cysteines each,
arranged in an identical distribution (Figure 4). It is,
therefore, likely that the evolution of the murine WAP genes
involved an intragenic duplication of a single cysteine
domain. This distribution of cysteines resembles that of
several other small cysteine-rich proteins, such as the
neurophysins, snake venom neurotoxins, and wheat germ agglu-
tinins. However, a comparison of the two domains, at both
the nucleotide level by a dot matrix analysis, and at the
amino acid level, revealed no apparent homology between the
two domains either within WAP or between the murine WAP and
the neurophysins.

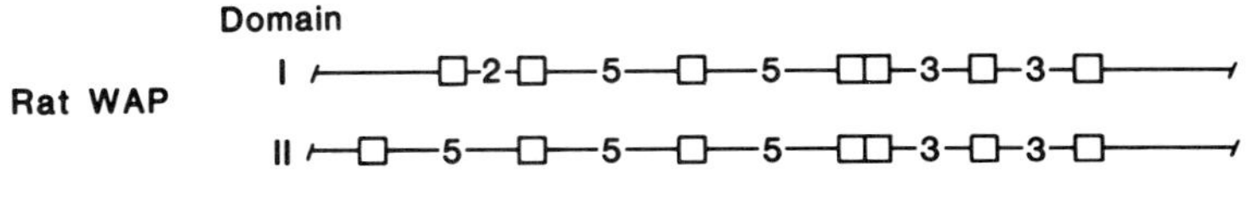

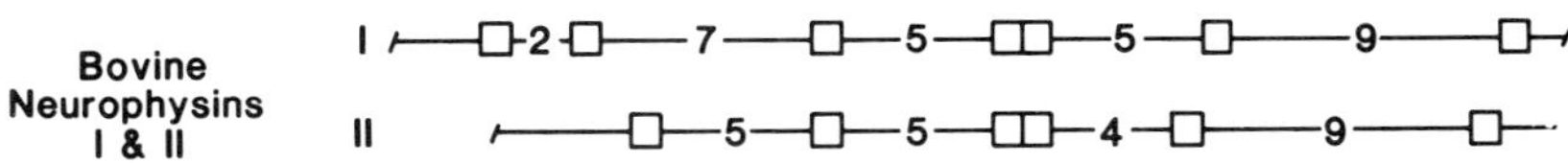

FIGURE 4. Two domain structure of murine WAP and its relationship to the bovine neurophysins. The cysteine residues are designated by the open squares and the numbers represent the amino acid residues between each cysteine.

Another unusual feature of the WAP mRNAs is the 91% conservation observed in the 3' non-coding regions of the mouse and rat WAP mRNAs. This is half the substitution rate of the coding region, and this low rate of sequence divergence in the 3' non-translated region of the mRNA may indicate a specialized function for this region. Finally, in the mouse, the WAP gene has been localized to a different chromosome from the casein gene family (3) and would, therefore, be expected to be regulated in <u>trans</u> by lactogenic hormones as compared to the caseins.

HORMONAL REGULATION OF MILK PROTEIN mRNA ACCUMULATION

Using single stranded cDNA probes specific for each of the three rat casein mRNAs and WAP mRNA, the accumulation of these mRNAs was quantitated during both normal rat mammary gland development and in mammary gland explant cultures, in response to the lactogenic hormones, prolactin and hydrocortisone (19). During mammary gland development from virgin to 8 days of lactation, the concentrations of α-, β-, and γ-casein and WAP mRNAS in total cellular RNA increased 8.6-, 23-, 140- and 300-fold, respectively (Figure 5). Since there is approximately a 10-fold increase in RNA content per g of tissue during this developmental period, these changes are even greater when expressed per mass of tissue. A comparison of the levels of each of these mRNAs at different stages of development indicated that the proportions of the three caseins remained relatively constant after the first few days of gestation and throughout lactation. However, their

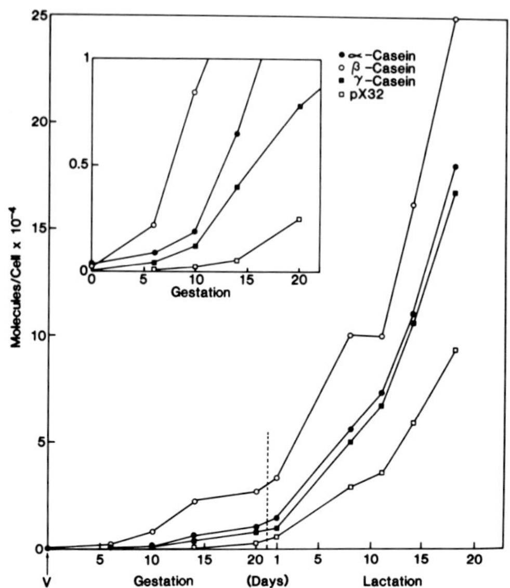

FIGURE 5.   Accumulation of mammary gland mRNAs during
normal development.   The concentrations of each of the four
mRNAs are shown as described in the figure and (19);   the
pX32 cDNA clone specifies the WAP mRNA.   The inset shows the
same results for the period of gestation, but with the y axis
magnified 100-fold.

relative concentrations were quite different in the virgin
gland, where there was approximately twice as much α- as β-
casein mRNA and the level of γ-casein mRNA was only 2% that
of α-casein mRNA.   By the 6th day of gestation, the proportions of each of the casein mRNAs was similar to that found
later in development.   By comparison, the concentration of
WAP mRNA was only one-third of the γ-casein mRNA in the
virgin gland, and the level of this mRNA increased much more
slowly than did any of the casein mRNAs during early and
mid-pregnancy.

   In order to investigate the mechanism of induction of
these milk protein mRNAs, further studies were carried out
using explant cultures obtained from 14 day pregnant animals.
Mammary gland explants were preincubated for 48 h in the
presence of insulin alone (I) or insulin plus hydrocortisone
(IF).   During the period of hormonal withdrawal, the levels
of casein mRNA decreased to less than 4% those observed in
the pregnant tissue.   The explants were then incubated for
24 h in fresh medium containing hormone combinations as

indicated, eg. I → IFM indicates preincubation with I alone followed by incubation with IF + prolactin (M) (Figure 6).

As illustrated in Figure 6, the markedly different basal levels of the three casein mRNAs observed in the virgin gland were also evident in cultures lacking prolactin, eg. I → I or IF → IF. Once again the basal levels of γ-casein mRNA were only 0.5-2% those of α-casein mRNA. In the presence of prolactin, γ-casein mRNA, therefore, displays the greatest fold induction (50- to 440-fold), although its maximally induced level is still only 30-40% of the α- and β-casein mRNAs.

The complex hormonal regulation of casein mRNA accumulation is also illustrated by the differential effects of hydrocortisone on the levels of the three casein mRNAs. Thus, the addition of hydrocortisone in the presence of prolactin has only a moderate effect on α-casein mRNA levels (1.6X), but increases γ- and β-casein mRNA levels 4.8- and 6.5-fold, respectively (I → IM vs. I → IFM). Hydrocortisone alone has little effect on the basal level of these three mRNAs (0.9-2.7X). Thus, this steroid hormone can differentially potentiate the effects of the peptide hormone on casein mRNA accumulation. While hydrocortisone has little effect on the basal level of the three casein mRNAs, its addition resulted in a 68-fold increase in the level of WAP mRNA. Thus, in contrast to casein gene expression, WAP gene expression appears to be regulated primarily by the steroid hormone. WAP mRNA levels increased only 2- to 3-fold following the addition of prolactin.

cDNA Probe	I ⟶ I	I ⟶ IM	I ⟶ IFM	IF ⟶ IF	IF ⟶ IFM
		(% of 8 day lactation)			
α	1.3	2.9(2.3x)	4.7(3.7x)	3.6(2.7x)	21.8(6.2x)
β	0.12	0.69(5.6x)	4.5(36x)	0.26(2.1x)	25.9(94x)
γ	0.021	0.21(10.3x)	1.0(50x)	0.019(0.9x)	8.5(440x)
pX32	0.048	0.11(2.3x)	0.45(9.5x)	3.2(68x)	9.8(3.0x)

FIGURE 6. Prolactin and hydrocortisone induction of casein and WAP mRNAs. The mRNA concentrations are expressed as percentages of the levels found in the 8-day lactating mammary gland.

POST-TRANSCRIPTIONAL REGULATION OF MILK PROTEIN
GENE EXPRESSION

From the preceding section it is apparent that the
control of milk protein gene expression by both peptide and
steroid hormones is a complex interactive process most likely
involving hormonal effects at both the transcriptional and
post-transcriptional levels. Previous studies from our
laboratory using a mixed cDNA probe representative of
primarily α- and β-casein mRNAs investigated the relative
contribution of transcriptional and post-transcriptional
processes on the hormonal induction of casein mRNA. Pulse-
labeling experiments performed using the mammary explant
cultures, and analyses of the rate of accumulation of casein
mRNA in the presence and absence of prolactin, both indicated
that there was a 2- to 4-fold increase in the rate of casein
gene transcription following prolactin addition (6). In
the absence of prolactin, i.e. in hormone withdrawn explants,
a rather high level of specific casein gene transcription
(approx. 1000 ppm) was observed, although there was little
net increase in casein mRNA levels (see Figure 6). Furthermore, a direct analysis of the half-lives of the casein
mRNAs revealed that there was between a 17- to 25-fold
increase in casein mRNA stability in prolactin-treated
cultures (6). This hormone-induced increase in casein mRNA
stability appeared to be selective, since only a small change
was observed in the turnover of total cellular poly(A)-
containing RNA. Subsequent studies of a number of other
viral (20,21) and eukaryotic mRNAs (22,23) have indicated
that changes in mRNA stability may play an important role
in regulating mRNA accumulation.

Although we understand very little about the mechanisms
involved in selectively stabilizing specific eukaryotic
mRNAs, recent structural studies of the milk protein mRNAs
have indicated a possible basis for the changes in their
relative levels in uninduced and induced mammary cells. In
particular, we have investigated the basis for the low
basal levels of γ-casein mRNA in the virgin mammary gland
and in explant cultures lacking prolactin. Sequence analysis
of the 3' ends of the four milk protein mRNAs, as shown in
Table 1, indicated the presence of an altered polyadenylation
site in the γ-casein mRNA. The polyadenylation signal AAUAAA
is found near the 3' end of almost all eukaryotic mRNAs and
has recently been shown to be involved in the polyadenylation
of SV40 late mRNA (24,25). Three of the milk protein mRNAs
have this usual polyadenylation signal present 16 to 23
nucleotides from their 3' ends. In contrast, γ-casein mRNA
had an altered polyadenylation site, AUUAAA, 12 to 17
nucleotides from its 3' terminus.

TABLE 1

POLY(A) ADDITION SITES OF RAT MILK PROTEIN mRNAs

	20	10	1

α U U C <u>A A U A A A</u> U C A C A U U U U A A G G C A C - (A_{52})

β U U C <u>A A U A A A</u> A U A A U C C U U U A C C G A U - (A_{40})

γ G C C A A G A C <u>A U U A A A</u> G U U U U A A A C G C - (A_{37})

WAP U G C U <u>A A U A A A</u> A A U C C A U U U G G C U U U - (A_{19})

The significance of this single base change in the
polyadenylation site of γ-casein mRNA is not apparent. In
lactating tissue and during hormonal induction in explant
cultures, the majority of all three casein mRNAs are found
in the poly(A)-containing mRNA fraction. The 20 to 30% of
apparent poly(A)-minus casein mRNA found in the fraction of
mRNA not retained on oligo-(dT)-cellulose appears to have
very short poly(A) tails, reflecting the distribution of
poly(A) tail lengths found in casein mRNA (26). However,
in recent experiments, an unusual distribution of γ-casein
mRNA sequences was observed in poly(A)-minus nuclear RNA
isolated from lactating tissue (Figure 7).

While the majority of both the α-casein and WAP mRNAs
are found in the poly(A)$^+$ nuclear RNA fraction, the γ-casein
mRNA was predominantly observed in poly(A)-minus nuclear RNA.
As expected for a sequence lacking poly(A), the predominant
band of γ-casein mRNA observed in the poly(A)-minus nuclear
RNA fraction was approximately 200 nucleotides shorter than
the corresponding poly(A)-containing γ-casein mRNA. When
radioautography was performed for 72 instead of 8 h, a very
small amount of WAP mRNA was also detected in the poly(A)-
minus nuclear RNA fraction. However, the predominance of
the γ-casein mRNA in the poly(A)-minus nuclear RNA fraction
was in marked contrast to the levels of the α-casein and
WAP mRNAs observed in the same RNA preparations.

These results suggested that the altered poly(A) signal
in the γ-casein gene may result in the rather inefficient
polyadenylation of this RNA in the nucleus. Presumably in
lactating tissue or during hormonal induction in culture, the
majority of the γ-casein mRNA sequences exiting the nucleus

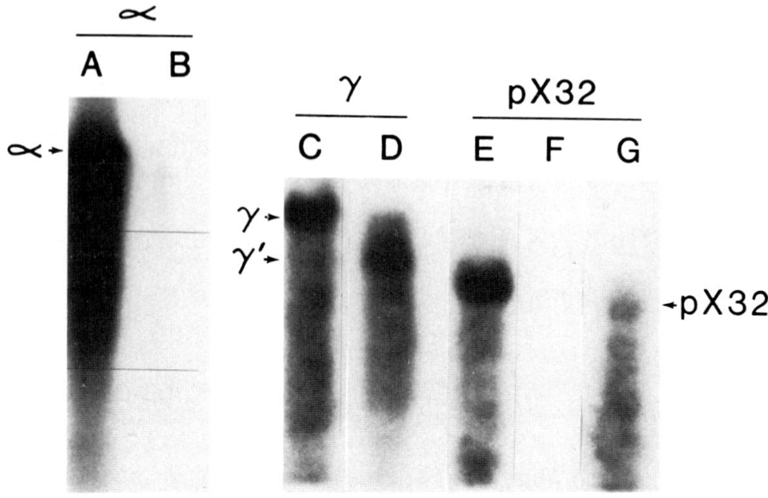

FIGURE 7.   Predominance of γ-casein mRNA in poly(A)-minus nuclear RNA.   Nuclear RNA was extracted from 10 day lactating tissue and fractionated by two passages over oligo-(dT)-cellulose into poly(A)$^+$ and poly(A)$^-$ RNA species.   The following amounts of RNA were electrophoresed on a 1-5% agarose gel containing 7.5 mM CH_3HgOH and transferred to DBM-cellulose.   poly(A)$^+$ RNA: lane A (10 μg), lanes C,E (2.3 μg); poly(A)$^-$ RNA: lane B (10 μg), lanes D,F,G (32 μg).   Hybridization was performed using nick translated recombinant cDNA clones pCαIII16 (α) lanes A,B; pCγ41 (γ) lanes C,D; and pX32 (WAP cDNA clone) lines E,F and G.   Radioautography was performed as follows: A-F were exposed for 8 h, lane G for 72 h.

are polyadenylated and, therefore, relatively stable. However, in the uninduced cultures or virgin tissue, the very low basal levels of γ-casein mRNA may reflect its lack of polyadenylation. Support for this hypothesis and the role of poly(A) in regulating mRNA stability comes from the studies of Zeevi et al. (27) using the inhibitor of poly-adenylation, cordycepin. Their results suggest that in the absence of polyadenylation, adenovirus nuclear RNA molecules are properly spliced and exit the nucleus as poly(A)-minus RNAs, which associate with polysomes. However, these mRNAs lacking poly(A) appear to have a much shorter half-life than their poly(A)-containing counterparts. Thus, poly(A) may play a critical role in increasing the half-life of cyto-plasmic mRNA. It is not clear how this process may be regu-lated by hormones, and at present this remains a speculative

hypothesis. Furthermore, while polyadenylation may play a
role in the accumulation of γ-casein mRNA, hormonal effects
on the stability of the other casein mRNAs may be regulated
in an independent manner. Further studies will be required
to elucidate the mechanisms by which peptide and steroid
hormones interact to regulate milk protein gene expression.
The preceding experiments serve merely to illustrate the
complexity of the regulation of mRNA metabolism in eukaryotic
cells.

ACKNOWLEDGMENTS

The authors would like to thank Mrs. Patricia Kettlewell
for assistance in the preparation of this manuscript.

REFERENCES

1. Topper, Y.J. (1970). Recent Prog. Horm. Res. 26, 287.
2. Richards, D.A., Rodgers, J.R., Supowit, S.C., and Rosen,
    J.M. (1981). J. Biol. Chem. 256,526.
3. Gupta, P.G., Rosen, J.M., D'Eustachio, P., and Ruddle,
    F.H. (1982). J. Cell Biol. 93, 199.
4. Hobbs, A.A., Richards, D.A., Kessler, D.J., and Rosen,
    J.M. (1982). J. Biol. Chem. 257, 3598.
5. Matusik, R.J., and Rosen, J.M. (1978). J. Biol. Chem.
    253, 2343.
6. Guyette, W.A., Matusik, R.J., and Rosen, J.M. (1979).
    Cell 17, 1013.
7. Waugh, D.F. (1971). In "Milk Proteins" (H.A. McKenzie,
    ed.), Vol. II, pp. 3-85. Academic Press, New York.
8. McKenzie, H.A., ed. (1971). Milk Proteins, Vol. I and
    II. Academic Press, New York.
9. Jenness, R. (1974). In "Lactation, A Comprehensive
    Treatise" (B.L. Larson and V.R. Smith, eds.), Vol. 3,
    pp. 3-96. Academic Press, New York.
10. Rosen, J.M., Woo, S.L.C., and Comstock, J.P. (1975).
    Biochemistry 14, 2895.
11. Blackburn, D.E., Hobbs, A.A., and Rosen, J.M. (1982).
    Nucl. Acids Res. 10, 2295.
12. Hobbs, A.A., and Rosen, J.M., unpublished observations.
13. Rosen, J.M., and Shields, D. (1980). In "Testicular
    Development, Structure and Function" (A. Steinberger
    and E. Steinberger, eds.), pp. 343-349. Raven Press,
    New York.
14. Mercier, J-C., and Gaye, P. (1980). Ann. N.Y. Acad. Sci.
    343, 232.

15. Swan, D., D'Eustachio, P., Leinwand, L., Seidman, J.,
     Keithley, D., and Ruddle, F. (1979). Proc. Natl. Acad.
     Sci. USA 76, 2735.
16. Rosen, J.M., Supowit, S.C., Gupta, P., Yu, L-Y., and
     Hobbs, A.A. (1981). In "Banbury Report 8: Hormones and
     Breast Cancer", pp. 397-424. Cold Spring Harbor
     Laboratory.
17. Supowit, S.C., and Rosen, J.M. (1982). Cancer Res. 42,
     1355.
18. Hennighausen, L.G., Sippel, A.E., Hobbs, A.A., and
     Rosen, J.M. (1982). Nucleic Acids Res. 10, in press.
19. Hennighausen, L.G., and Sippel, A.E. (1982). Nucleic
     Acids Res. 10, 2677.
20. Wilson, M.C., and Darnell, J.E. Jr. (1981). J. Mol.
     Biol. 148, 231.
21. Babich, A., and Nevins, J.R. (1981) Cell 26, 371.
22. Hynes, N.E., Groner, B., Sippel, A.E., Jeep, S., Wurtz,
     T., Nguyen-Huu, M.C., Giesecke, K., and Schütz, G.
     (1979). Biochemistry 18, 616.
23. Chang, S., Landfear, S.M., Blumberg, D.D., Cohen, N.S.,
     and Lodish, H.F. (1981). Cell 24, 785.
24. Proudfoot, N.J., and Brownlee, G.G. (1976). Nature 263,
     211.
25. Fitzgerald, M., and Shenk, T. (1981) Cell 24, 251.
26. Rosen, J.M. (1976) Biochemistry 15, 5263.
27. Zeevi, M., Nevins, J.R., and Darnell, J.E. (1982). Mol.
     Cell. Biol. 2, 517.

SELECTIVE GENE EXPRESSION
WITHIN HORMONALLY ACTIVATED MULTIGENE FAMILIES

Jamshed R. Tata

National Institute for Medical Research, Mill Hill, London NW7

ABSTRACT   Many hormonally regulated genes are members of small multigene families of 3-20 genes which are expressed in a highly tissue-specific manner. These include genes coding for chicken ovalbumin, Xenopus vitellogenin, rat α_{2u}-globulin and mouse α-amylase. Not all members of these multigene families are transcribed equally efficiently when a hormone activates specific genes. Particular emphasis will be placed in this report on the selective activation by estrogen of the vitellogenin multigene family in male Xenopus liver. The hormone induces de novo the transcription of at least 4 vitellogenin genes which can be distinguished, by virtue of a 20% divergence in coding sequence, into 2 groups (A and B) of two genes each. There may be additional vitellogenin sequences which are not transcribed. We have found, both in whole animals and in a cell culture system, that the A and B groups of vitellogenin genes are not simultaneously activated and that accumulation of individual mRNAs does not proceed at identical rates at the onset of primary stimulation in male Xenopus hepatocytes. During secondary hormonal stimulation in male or primary stimulation of female hepatocytes, all 4 vitellogenin genes were activated with identical kinetics and to the same extent. Differential hormonal effects on conformation (as judged by DNase I sensitivity) and transcription of individual vitellogenin genes, measured simultaneously in isolated nuclei, confirmed that the differential transcription of the A and B groups of vitellogenin genes were paralleled by a similar selective sensitization of these two groups of genes to DNase I. Selective expression or differential rates of transcription of individual hormonally activated genes within a multigene family poses new questions concerning the regulation of their expression and the sites of interaction between the genes and hormone receptors.

INTRODUCTION

Several papers in this Symposium refer to small multigene
families coding for proteins. The multigene families covered
in this report are characterized by small families of 3-20
genes (1). These include (a) developmentally and irreversibly
regulated genes such as globin, actin and myosin, and (b) genes
belonging to small multigene families reversibly inducible in
terminally differentiated cells. A salient feature of all
genes reversibly induced by external inducers is the high
degree of tissue specificity. Among the different inducers
or induction signals are hormones that regulate gene expression.

HORMONALLY REGULATED GENES

The study of hormonal regulation of genes has made a substantial contribution to the overall question of organization
of genes and the regulation of their transcription (2).
Perhaps the best example to illustrate this point is the
extensive knowledge gained from the analysis of regulation
by estrogen of genes coding for avian egg white proteins, in
particular chicken ovalbumin (3,4,5). With the widespread use
of gene cloning techniques a large number of genes can now be
listed whose transcription is hormonally regulated. Table 1
gives a partial list of some of these genes.

Among the important features emerging from Table 1 are
(a) tissue specificity, (b) hormone specificity, (c) species
specificity, and (d) gene multiplicity. Thus, for example,
estrogen activates genes coding for egg white or egg coat
proteins in the oviductal cells of all oviparous vertebrates
while it simultaneously induces egg yolk protein or vitellogenin genes in the hepatocytes of the same organism.
Vitellogenin genes are evolutionarily highly conserved and,
except in flies and mosquitoes, their activity is regulated
by juvenile hormone in the fat body of most insects (6). As
an example of hormonal specificity, the transcription of
α-amylase genes is modulated by insulin in the pancreas,
whereas by androgen in the salivary gland of rodents (7,8).
The hormones and genes listed in Table 1 can be grouped into
two categories. First, are those where the respective hormone
induces the activity of the gene de novo, i.e., in the absence
of the hormone, there is virtually no detectable transcript
of the gene regulated. This category includes genes coding
for egg proteins. In the other systems, the hormone modulates
(most often accelerating) the rate of transcription of a given
gene but that a low level of transcription can be detected in
the absence of the hormone. In all cases, however, the hormone
has to be present continuously for gene activation and that its

TABLE 1

SOME HORMONALLY REGULATED GENES

Gene	Regulated by	Tissue	Species
Ovalbumin	Estrogen	Oviduct	All birds
Vitellogenin	Estrogen	Liver	All oviparous vertebrates
	Juvenile hormone	Fat body	Most insects and invertebrates
Prostatic binding protein	Androgen	Prostate	Rodents
α_{2u}-Globulin	Androgen	Liver	Rodents
Glue polypeptides	Ecdysone	Salivary gland	Drosophila
Egg chorion proteins	Juvenile hormone?	Follicle cells	Drosophila
Casein	Prolactin	Mammary gland	Mammals
Prolactin	Estrogen	Pituitary	Rat
-X-MMTV-Y- *	Glucocorticoids	Mammary gland	Mouse
Growth hormone	Thyroid hormone	Pituitary	Rat
α-Amylase	Insulin	Pancreas	Rat
	Androgen	Salivary gland	Rodents

* MMTV, mouse mammary tumor virus; X, Y, neighboring host genome sequences.

effect is reversed upon hormone withdrawal. This aspect of
reversibility is an important advantage of studying hormonally
regulated gene expression since it allows the repeated analysis
of both the induction and de-induction processes. Finally,
most of the genes listed in Table 1 are members of small
multigene families of 3-20 genes. The ease of manipulating
gene expression with hormones makes it possible to answer the
question of whether or not all members of such multigene
families of reversibly inducible genes are co-ordinately
regulated and to equal extent (9). This question is discussed
below.

HORMONAL REGULATION OF INDIVIDUAL GENES
WITHIN A MULTIGENE FAMILY

Steroid hormone action is most markedly characterized by
regulation of transcription (2,3,4), which may involve either
an overall enhancement of the synthesis of all cellular
proteins or the specific induction of a few, but abundant,
tissue specific proteins. The induction de novo of egg white
and yolk proteins during estrogen-regulated egg maturation
exemplifies the latter action of estrogen and the exploitation
of this system which has made a major contribution to
understanding not only hormonal control of gene expression but
also to that of the relationship between gene structure and
organization. In particular, a large volume of literature
has accumulated on the regulation of transcription of genes
encoding chicken egg white proteins (3,4,10,11). More
recently, increasing attention has also been focussed on how
the same hormone controls the expression of vitellogenin genes
in the liver of chicken, Xenopus and other vertebrates (6,12).
In birds, estrogen and progesterone regulate the synthesis
in the oviduct of the egg white proteins ovalbumin, conalbumin,
ovomucoid and lysozyme (3,4,11). Injection of estrogen or
progesterone to "withdrawn" chicks, primed earlier with
estrogen, causes a re-accumulation in the secretory cells of
the oviduct of mRNA for all these proteins. Using cDNA
probes corresponding to each of these different messengers,
the rates of accumulation of mRNAs and of transcription of the
different hormonally regulated genes were found to vary over a
very wide range (13,14). The initial lag period and the time-
course of activation of transcription following a single
injection of estrogen or progesterone were also quite different
for the different egg white protein genes. It is possible
that these different steroid-sensitive genes may be organized
in different conformations. There is also some indirect
evidence to suggest that the hormone receptor complex may
interact in different modes to activate the different genes.

Such a conclusion is not altogether unexpected, but what is of
particular interest is the possibility that estrogen could
activate to different extents the transcription of individual
members of the same multigene family.

Recent studies on the genomic organization of the chicken
ovalbumin gene have revealed the presence of two other related
genes, termed "X" and "Y" genes, which are contiguously located
5' to the "authentic" ovalbumin gene. When the level of
messenger RNA in estrogen stimulated chicken oviduct was
measured with cloned probes corresponding to non-homologous
regions of the three natural ovalbumin-like genes, the number
of transcripts of each of these genes was found to vary by
100-200 fold (3,15,16). This differential rate of accumulation
can largely be accounted for by very unequal rates of transcription of the three hormonally activated genes, the rate of
transcription of the ovalbumin gene being several times that
of the "X" and "Y" genes. Thus, not only is the transcription
of the different egg white protein genes regulated differently
by estrogen, but the individual members of the same ovalbumin
gene family are differentially regulated. It is intriguing to
note that differential nuclease sensitivities of these three
ovalbumin genes do not exactly correspond to these differences
in the rates of their transcription (3,17), a finding that has
an important bearing on the question of how hormone-receptor
interaction with the genome might modify the configuration and
function of individual genes of the same family.

SELECTIVE REGULATION OF EXPRESSION
WITHIN THE VITELLOGENIN MULTIGENE FAMILY

Important advantages are offered by the vitellogenin
system over that of avian egg white protein induction,
particularly: (a) the ability to induce vitellogenesis with
estrogen in tissue from male animals, (b) the reproduction
of the full physiological induction in tissue culture,
(c) the possibility of studying the primary induction, and
(d) the non-involvement of DNA synthesis in the initial response.
The activation of egg white proteins in birds has been studied
almost uniquely during secondary hormonal induction which is
preceded by a massive increase in cell population, which in
turn vitiates the analysis of the initial events leading to a
possible re-organization of genes upon primary hormonal
activation.

As illustrated in Figure 1, _Xenopus_ vitellogenin cDNA
clones could be divided into two major groups (A and B) with a
20% sequence divergence (estimated from kinetics of thermal
denaturation of cDNA-mRNA hybrids) each sub-divided into two
genes (A_1, A_2 and B_1, B_2) distinguishable by a 5% sequence

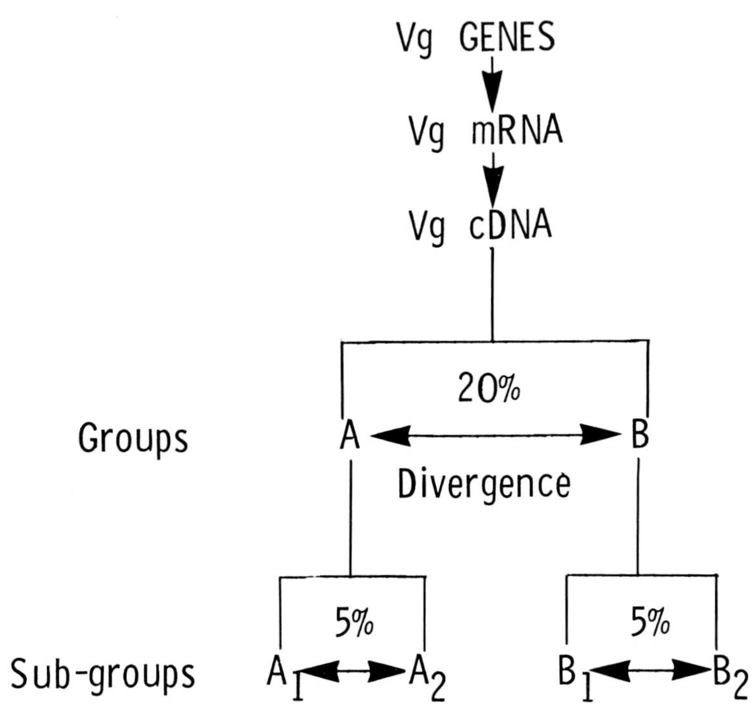

FIGURE 1.   Schematic representation of the four actively
expressed Xenopus vitellogenin genes and their classification
as 2 members each of the A and B groups of genes (18).

divergence (12,18).   Evidence for the distinctness of four
actively expressed vitellogenin genes was provided by
separation of four mRNAs by hybrid selection and translation
in a cell-free system to yield four distinct proteins (19).
Thus, Xenopus vitellogenin is encoded for by a small family of
genes, at least four of which are actively transcribed.   It
should be noted that the multiplicity of vitellogenin genes
may vary according to the species studied;   in chicken (20)
and locust (21) these may be represented in only two copies.
Are there additional gene copies in the Xenopus genome
that exist as pseudo- or "silent" genes and are not activated
by estrogen?   In our laboratory, we turned to the indirect
approach of differential nuclease digestion of chromatin to

answer this question (22). It is well established that DNase I
preferentially digested the ovalbumin gene in estrogen-
stimulated chick oviduct nuclei but not in control nuclei
(3,11,23). When we carried out similar studies on male Xenopus
liver nuclei, estrogen administration enhanced the DNase I
sensitivity of vitellogenin genes (24) but did not affect the
sensitivity of the albumin gene, which is constitutively
expressed in liver, or of the globin gene which is not
expressed. Similar results have also been reported with
cloned probes corresponding to each of the four actively
expressed Xenopus vitellogenin genes (25). The hormonal effect
of enhanced nuclease sensitivity persisted for nearly 30 days
although vitellogenin mRNA synthesis ceased before that.
A similar finding has been reported for the ovalbumin gene (11)
but the significance of this protracted nuclease sensitivity
is not known.

     A massive accumulation of vitellogenin mRNA in liver occurs
upon administration of estrogen (6,12,26). In all species
studied, the secondary response is both more rapid and of a
higher magnitude than when the hormone is administered for the
first time, especially in male animals. In order to understand
how the hormone activates the gene, it is essential to dissect
out the early stages of the transcriptional response as well
as the loss of response as the hormone is withdrawn. This is
more precisely achieved in tissue culture than in whole animals
because of the high variability between individual animals,
cellular heterogeneity of the liver, the complex dynamics of
hormone distribution, enterohepatic recycling of the hormone,
long-term storage and metabolism of estrogen, as well as the
difficulty of designing "pulse-chase" labelling experiments
in vivo. For these reasons, our laboratory has devoted much
effort in achieving both primary and secondary response
in vitro in primary cell cultures from a single pooled batch
of purified parenchymal cells (27,28). When the vitellogenin
mRNA levels were measured by a filter hybridization technique
using cloned vitellogenin cDNA probes, a more rapid
accumulation of mRNA was observed during secondary stimulation
as compared with the primary response. The removal of hormone
from the culture medium caused a very rapid loss of
vitellogenin mRNA levels, thus demonstrating that the hormone
is continuously required for a sustained transcription of the
vitellogenin genes. Since the half-life of Xenopus
vitellogenin mRNA is estimated to be 30-40 hours (6,26), the
rapid decay of the mRNA levels suggests that the cessation of
transcription is accompanied by an enhanced and selective
degradation of the vitellogenin mRNA. This finding is
consistent with those of Palmiter et al. on the stability of
mRNAs coding for egg white proteins in chicken oviduct upon
estrogen withdrawal (11). The rapid cessation of accumulation

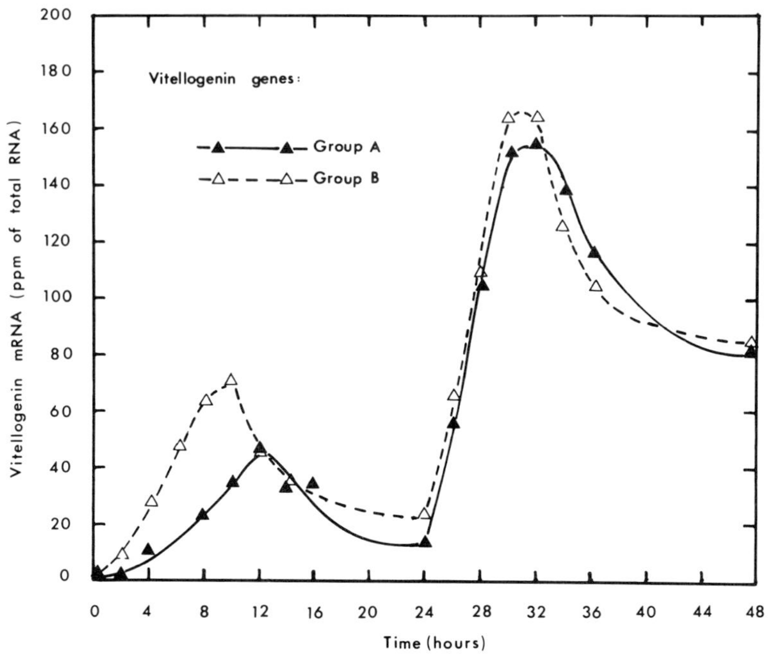

FIGURE 2. Non-coordinate and coordinate accumulation of
transcripts of A (▲) and B (△) groups of vitellogenin genes
induced in male _Xenopus_ hepatocyte cultures by the addition of
estradiol at times 0 and 24 hours (30).

of mRNA upon removal of the hormone from the culture medium is
also explained by the fact that _Xenopus_ parenchymal cells
metabolize estradiol extremely rapidly in culture with a half-
life of only 40 minutes (29). When the experiments were therefore designed in such a way that the hormone is replaced in
culture every 4 hours, a continuously increasing accumulation
of mRNA at rates comparable to those observed _in vivo_ was
obtained, thus establishing that the _Xenopus_ liver cell culture
system reproduces _in vitro_, both qualitatively and quantitatively, the full physiological process of vitellogenesis _in vivo_.
     It was important to determine if the genes of both the A
and B groups of the _Xenopus_ vitellogenin multigene family would
be activated simultaneously or to different extents. We therefore obtained RNA from cell cultures at very short time-
intervals after exposure to hormone and assayed vitellogenin
mRNA with cloned cDNA probes corresponding to both the A and B
groups of genes under hybridization conditions stringent enough
to prevent any significant cross-hybridization between mRNAs
and cDNAs corresponding to the two groups. Figure 2 presents

results showing that in cultured hepatocytes from male
Xenopus, not previously exposed to estradiol, the A and B
groups of vitellogenin genes were activated in a non-coordinate
fashion at the early stages of hormonal stimulation. More
precise studies (not shown here) have shown that at 2-3 hours
after the addition of the hormone, only the B group of genes
were transcribed. Figure 2 also shows that the accumulation
of mRNA ceases after about 8 hours, but that a second pulse
of hormone, 24 hours after the first, causes a fresh burst of
transcription of both genes. During this secondary response
the accumulation of mRNA corresponding to both the A and B
groups of vitellogenin genes occurred with identical kinetics
and to the same extent. The same result was obtained
irrespective of whether the hepatocytes were exposed to the
hormone in culture only 24 hours earlier (as in Figure 2) or
in vivo up to several weeks before removing the cells for
culture. When identical experiments were carried out with
cultures of hepatocytes from adult female Xenopus that
were vitellogenic, all genes were coordinately expressed (30).

    That the above hormonal responses observed in cell cultures
reflect a similar situation of coordinate and non-coordinate
activation of Xenopus vitellogenin genes in vivo was verified
by experiments carried out in whole animals. These experiments
were performed with nuclei isolated from livers of male and
female Xenopus following primary and secondary stimulation with
estradiol. All preparations of nuclei were split into two
batches. One batch was used to monitor the "run-off" transcription of vitellogenin genes by the endogenous RNA polymerase
II (31). The other batch was used to assess the sensitivity
to DNase I of vitellogenin genes by a technique of "nick
translation" of nuclei, whereby active gene sequences are
preferentially labelled with α-^{32}P-deoxyribonucleoside triphosphates after nicking with low levels of DNase I (32).
Thus, the configuration of a given gene as transcriptionally
active or inactive and its actual transcription were determined
simultaneously on the same sample of nuclei.

    Figure 3 shows that in nuclei isolated at early time-
intervals (up to 15 hours) after the first administration of
estradiol to male Xenopus, the B group of vitellogenin genes
were transcribed more readily than those of the A group (33).
However, 24 hours after the hormone was administered all
genes were transcribed at the same rate. Coordinate transcription of the A and B groups of vitellogenin genes was
observed at all times after secondary hormonal treatment of
males or upon the first treatment of vitellogenic females.
Thus, the results of differential transcription in isolated
nuclei of the vitellogenin multigene family corroborate those
of accumulation of vitellogenin mRNA in cultured cells
(Figure 2).

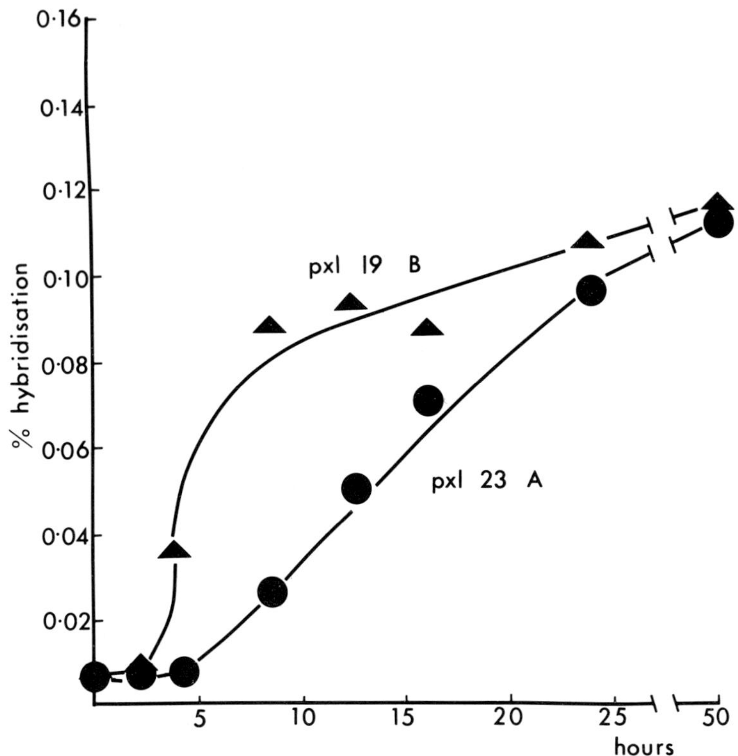

FIGURE 3. Differential activation of transcription of
A (●) and B (▲) groups of vitellogenin genes in isolated
nuclei from male Xenopus treated with estradiol. At the
different time-intervals indicated, after a single hormone
administration, nuclei were isolated and divided into two
groups, one aliquot was used for measuring "run-off" transcription for 30 minutes and the labelled transcripts assayed
for A and B group vitellogenin sequences by disc hybridization
to recombinant plasmids containing the respective cDNAs. The
other batch was used for monitoring sensitivity to DNase I (33).

Figure 4a shows that the increase in relative sensitivity
to DNase I of the A and B groups of vitellogenin genes nearly
paralleled their transcriptional activity in vitro in the
same samples of nuclei, as a function of time following primary
hormonal treatment of male Xenopus. On the other hand, the
high level of nuclease sensitivity was retained for six weeks
after the first injection of the hormone (time 0, Figure 4b),
at a time when the transcription of vitellogenin genes had

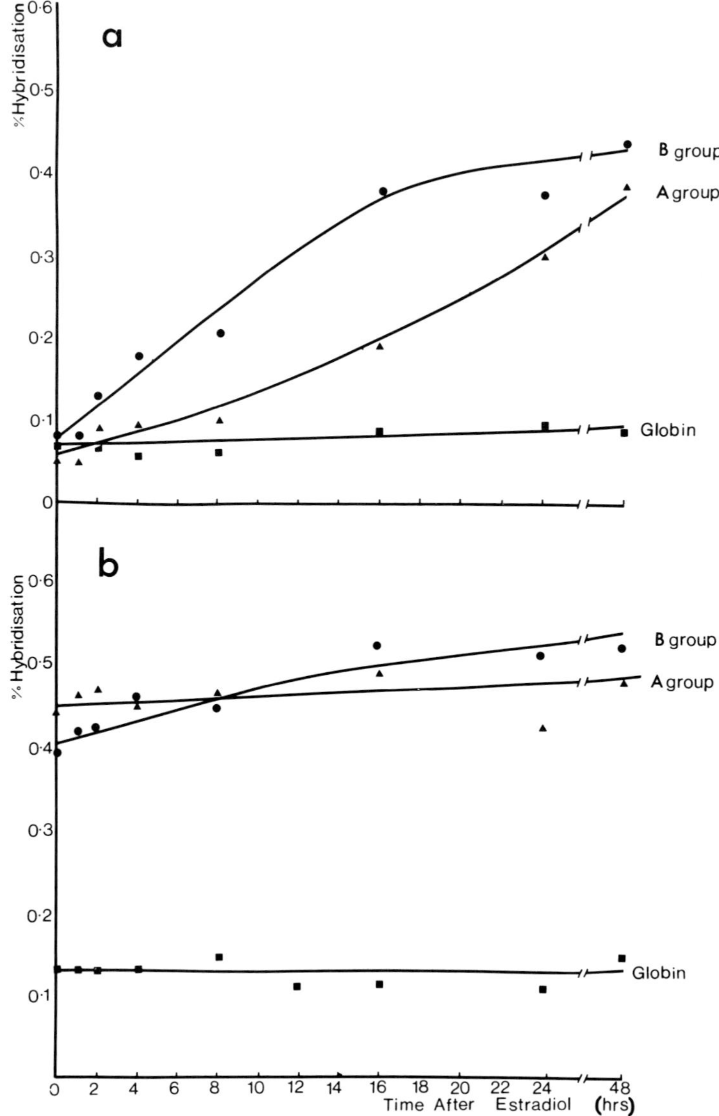

FIGURE 4. (a) Determination of DNase I sensitivity in an
aliquot of the same batches of nuclei used for transcription
studies (Figure 3) following primary estrogen stimulation.
             (b)   DNase I sensitivity following secondary
stimulation of male Xenopus. An injection of estradiol was
made at time 0 to animals that had been injected once (primary
stimulation) six weeks earlier (33). Vitellogenin group A
genes (▲); vitellogenin group B genes (●); plasmid (■)
containing Xenopus globin gene insert was used as a control.

ceased (results not shown). It can also be seen in Figure 4b
that the A and B groups of genes are equally sensitive to
DNase I and that this sensitivity does not significantly
change upon secondary hormonal stimulation of male animals.
Nuclei from hormonally treated or untreated female Xenopus
gave similar results.

       Thus, the principal finding of our combined studies on
cell cultures and in whole animals is that in hepatocytes from
male Xenopus the two genes of the A and B groups of the
vitellogenin multigene family are not in an equally
expressible configuration when they are activated by estrogen.
However, within a few hours of primary stimulation all genes
become equally sensitive and are coordinately transcribed, as
found in adult, vitellogenic females. The reason for this
shift from non-coordinate to coordinate expression of genes
within the same multigene family is not known. We are
currently extending our studies to immature females and
during early development in order to obtain further clues to
resolving this intriguing problem.

DISCUSSION

       Table 2 lists a few multigene families for which there
is evidence of unequal or differential expression of the
individual gene members. Besides the avian ovalbumin and
amphibian vitellogenin multigene families where individual
genes are activated and transcribed to different extents, as
seen above, it is worth referring to three other families,
all in some ways being hormonally controlled. The α-amylase
genes are an interesting example of a situation whereby a
given family of genes is expressed in different tissues, but
that the same individual gene may not be expressed in different tissues (7).   The accumulation of α-amylase mRNA in the
mouse pancreas, salivary gland and liver varies over a range
of three orders of magnitude. Only one of 5-9 genes is
predominantly transcribed in each tissue and it is also known
that the gene expressed in the pancreas is not the same as
that in liver and salivary gland (7,8). It is also known
that not all the members of the α_{2u}-globulin gene family are
expressed but it is not known whether those that are hormonally
regulated are transcribed coordinately or to the same extent
(34). The genes coding for larval serum protein in Drosophila
represent another type of differential gene regulation within
a family (35). Ecdysone induces the synthesis of a special
protein, termed larval serum protein, in the fat body of both
the male and female larva but the female synthesizes
considerably more protein. This sex-based difference has
been explained by a low-level (3-5 fold) amplification of

TABLE 2

DIFFERENTIAL GENE EXPRESSION WITHIN A MULTIGENE FAMILY

Multigene family	Species	Copy no.	Expressed in	No. genes transcribed (relative rates [ ])
Ovalbumin	Chicken	3	Oviduct	1[90], 1[10], 1[1]
Vitellogenin	Xenopus	>4	Liver	4[∿25], 2[0-25], 2[25-100]
α_{2u}-Globulin	Rat	∿19	Liver	<3 [?]
α-Amylase	Rodent	5-9	Salivary gland	1[>90], 1[<10]
			Liver	1[<10], 1[>90]
			Pancreas	1[100]
Larval serum protein	Drosophila	2-10	Fat body	2-6 [100]

genes in the female but not the male fat body and that most,
but not all, genes are equally actively transcribed. It is
not difficult to predict that more such examples of differential gene expression within multigene families will emerge in
the near future. The few examples listed in Table 2 already
indicate that a variety of mechanisms underlie the selective
regulation.

The selective hormonal regulation of expression of
individual genes within a multigene family raises new questions
and offers fresh approaches to understanding the different
levels at which gene transcription is controlled. According
to the model of steroid hormone action (3,6,36) the nuclear
receptor is depicted as interacting, either directly or via
an "acceptor" site, with a single hormone-regulated gene in
its target tissue. However, the fact that many steroidally
regulated genes (see Table 1) are members of multigene families
and that not all of them are equally activated by the hormonal
stimulus (see Table 2) now modifies the central question of
the role of the hormone receptor in regulating transcription.
Is the selective activation a property of the hormone receptor,
or is it independent of the receptor but merely a reflection
of the differential organization of the expressible and
inexpressible vitellogenin genes? Answers to such questions
require new experimental approaches.

Many steroid hormone regulated genes are not only members
of multigene families but are expressed in a highly tissue-
specific manner.   Thus, in two different tissues of a given
organism which have receptors for the same hormone the
hormone may activate or modulate the transcription of different genes. We have already seen that estrogen activates the
transcription of vitellogenin genes in the liver but that of
ovalbumin genes in the oviduct, both sets of genes
constituting small multigene families. It would be important
to establish the "leakiness" of transcription of these genes
in the hormone's target cells in which the proteins these
genes encode are not synthesized. It would also be important
to compare in the same tissue the organization and transcription of genes hormonally activated de novo with those that are
constitutively expressed but whose transcription is enhanced
by the hormone and those whose transcription is unaffected by
the hormone. Such a comparison would at least be more
revealing than that between expressed and non-expressed genes
in the same cells. Evidence available so far suggests that
steroid hormone receptors in different tissues, or even in
different species, are identical (10), so that the tissue
specificity of hormonal activation or modulation of gene
transcription must lie in some differences in the manner in
which the same gene is organized in different tissues. What

is the mechanism during cellular differentiation that pre-determines the availability of a gene for expression by the same hormonal or developmental signal in one tissue but not in another? Does the fact that many of these genes are members of small multigene families have any relevance to the tissue specificity of their inducibility? Answers to these questions may not establish the molecular basis of hormonal regulation of gene expression but the hormonal manipulation of regulation of transcription within multigene families will greatly add to our knowledge of how gene expression is controlled.

ACKNOWLEDGMENTS

I wish to thank Mr. A.P. Wolffe and Mr. J.L. Williams for the use of unpublished data and Mrs. Ena Heather for the preparation of the manuscript. I am also grateful to Professor W. Wahli, University of Lausanne, for the cloned cDNAs to _Xenopus_ vitellogenin used in this work.

REFERENCES

1. Ohta, T. (1980). "Evolution and Variation of Multigene Families." Springer-Verlag, Berlin.
2. Tata, J. R. (1980). Biol. Rev. 55, 285.
3. O'Malley, B.W., Roop, D. R., Lai, E. C., Nordstrom, J. L., Catterall, J. F., Swaneck, G. E., Colbert, D. A., Tsai, M.-J., Dugaiczyk, A.,and Woo, S. L. C. (1979). In "Recent Progress in Hormone Research" (R. O. Greep, ed.), Vol. 35, pp. 1-46. Academic Press, New York.
4. Schutz, G., Nguyen-Huu, M. C., Giesecke, K., Hynes, N. E., Groner, B., Wurtz, T.,and Sippel, A. E. (1978). Cold Spring Harbor Symp. Quant. Biol. 42, 617.
5. O'Malley, B. W., Stumph, W. E., Lawson, G. M., Tsai, M.-J., See this Symposium.
6. Tata, J. R.,and Smith, D. F. (1979). In "Recent Progress in Hormone Research" (R. O. Greep, ed.), Vol. 35, pp. 47-95. Academic Press, New York.
7. Schibler, U., Tosi, M., Pittet, A.-C., Fabiani, L., and Wellauer, P. K. (1980). J. Mol. Biol. 142, 93.
8. Hagenbuchle, O., Bovey, R., and Young, R. A. (1981). Cell 21, 179.
9. Tata, J. R. (1981). J. Steroid Biochem. 15, 87.
10. Roy, A. K., and Clark, J. H. (1980). "Gene Regulation by Steroid Hormones." Springer-Verlag, New York.
11. Palmiter, R. D., Mulvihill, E. R., McKnight, G. S., and Senear, A. W. (1978). Cold Spring Harbor Symp. Quant. Biol. 42, 639.

12. Wahli, W., Dawid, I. B., Ryffel, G. U., and Weber, R.
    (1981). Science 212, 298.
13. Hynes, N. E., Groner, B., Sippel, A. E., Jeep, S.,
    Wurtz, T., Nguyen-Huu, M. C., Giesecke, K., and Schutz, G.
    (1979). Biochemistry 18, 616.
14. McKnight, G. S., and Palmiter, R. D. (1979). J. Biol. Chem.
    254, 9050.
15. Heilig, R., Perrin, F., Gannon, F., Mandel, J.L., and
    Chambon, P. (1980). Cell 20, 625.
16. LeMeur, M., Glanville, N., Mandel, J. L., Gerlinger, P.,
    Palmiter, R., and Chambon, P. (1981). Cell 23, 561.
17. Lawson, G.M., Knoll, B.J., March, C.J., Woo, S. L. C.,
    Tsai, M.-J., and O'Malley, B. W. (1982). J. Biol. Chem.
    257, 1501.
18. Wahli, W., Dawid, I. B., Wyler, T., Jaggi, R. B., Weber, R.,
    and Ryffel, G. U. (1979). Cell 16, 535.
19. Felber, B. K., Maurhofer, S., Jaggi, R. B., Wyler, T.,
    Wahli, W., Ryffel, G. U., and Weber, R. (1980). Eur. J.
    Biochem. 105, 17.
20. Arnberg, A. C., Meijlink, F. C. P. W., Mullder, J.,
    van Bruggen, E. F. J., Gruber, M., and AB, G. (1981).
    Nucleic Acids Res. 9, 3271.
21. James, T. C., Bond, U. M., Maack, C. A., Applebaum, S. W.
    and Tata, J. R. Biochem. J. In press.
22. Botchan, M., and Watson, J. D. (eds.) (1978). Cold Spring
    Harbor Symp. Quant. Biol. 42.
23. Weintraub, H., and Groudine, M. (1976). Science 193, 848.
24. Dimitriadis, G., and Tata, J. R. (1982). Biochem. J. 202,
    491.
25. Gerber-Huber, S., Felber, B.K., Weber, R., and Ryffel,
    G. U. (1981). Nucl. Acids Res. 9, 2475.
26. Baker, H. J., and Shapiro, D. J. (1977). J. Biol. Chem.
    252, 8428.
27. Wangh, L. J., Osborne, J. A., Hentschel, C. C., and
    Tilly, R. (1979). Develop. Biol. 70, 479.
28. Searle, P. F., and Tata, J. R. (1981). Cell 23, 741.
29. Searle, P. F., Tenniswood, M. P. R., Wolffe, A. P., and
    Tata, J. R. Submitted to J. Biol. Chem.
30. Wolffe, A. P., and Tata, J. R. Unpublished results.
31. Weil, P. A., Luse, D. S., Segall, J., and Roeder, R. G.
    (1979). Cell 18, 469.
32. Levitt, A., Axel, R., and Cedar, H. (1979). Develop. Biol.
    69, 496.
33. Williams, J. L., and Tata, J. R. Unpublished results.
34. Kurtz, D. T. (1981). Nature 291, 629.
35. Roberts, D. B., and Brock, H. W. (1981). Experientia 37,103.
36. O'Malley, B. W., Vedeckis, W. V., Birnbaumer, M.-E., and
    Schrader, W. T. (1977). In "Molecular Endocrinology" (I.
    MacIntyre and M. Szelke, eds.), pp. 135-150, Elsevier,A'dam.

Multiple Roles for Calmodulin in the Regulation of Eukaryotic
Cell Metabolism

Anthony R. Means, James G. Chafouleas, Lisette Lagace,
Eugene Lai, and Joseph P. Stein[1]

Department of Cell Biology, Baylor College of Medicine,
Houston, Texas 77030; [1]Department of Endocrinology, Univ.
of Texas Health Science Center, Houston, Texas 77030

Abstract

Over the last five years the importance of calmodulin to
the eukaryotic cell has become more evident. Initially be-
lieved to be an activator for only one enzyme, cyclic nucleotide phosphodiesterase, this protein is now known to either
directly or indirectly mediate the Ca^{++} regulation of 15 key
enzymatic or structural systems. These facts taken together
with its amazingly conserved nature and ubiquitous distribution in all eukaryotic cells clearly qualify calmodulin as
the eukaryotic intracellular Ca^{++} receptor. To date, the
majority of studies on calmodulin have been directed toward
an understanding of the systems regulated and the mechanisms
by which this regulation occurs. However, regardless of
which system is regulated or by what mechanism, the ultimate
control mediated by calmodulin will be dependent upon its own
regulation. Utilization of the molecular probes for calmodulin will help unravel the complexity of the structural gene
as well as the genetic regulation of this protein.

Introduction

Calmodulin is a heat-stable, 17,000 M_r multifunctional
Ca^{++} binding protein which meets all the criteria set forth
for a Ca^{++} receptor (Means and Dedman, 1980). It contains
four equivalent Ca^{++} binding sites with a K_d of 2.4×10^{-6} M
which do not bind Mg^{++} under physiological conditions (Dedman
et al., 1977). Ca^{++} binding induces a more α-helical conformation of the protein which preceeds activation of the calmodulin-dependent phosphodiesterase (Ho et al., 1975; Klee,
1977; Dedman et al., 1977). In further investigating this
conformational transition accompanied by Ca^{++} binding, Richman (1979) has examined the microenvironment of the two tyrosine residues of the protein. In the absence of free Ca^{++},
both tyrosine residues 99 and 138 are accessible to acety-
lation by N-acetylimidazole whereas in the presence of Ca^{++},
only residue 99 can be acetylated. These findings suggest

that Ca^{++} binding alters the microenvironment of tyrosine
residue 138 and could play a role in the biological activity
of the protein.

This protein has been shown to mediate the calcium regulation of a large number of fundamental intracellular enzyme
systems as reviewed by Means et al. (1982). Calmodulin has
also been reported to be the calcium-binding protein regulating calcium transport in the sarcoplasmic reticulum (Katz
and Remtulla, 1978; LePeuch et al., 1979) and autophosphory-
lation of membrane proteins (Schulman and Greengard, 1977,
1978; DeLorenzo et al., 1979). In addition, immunofluorescence studies on a variety of cultured cells have demonstrated that calmodulin is localized on the actomyosin-con-
taining stress fibers in interphase and is a dynamic component of the mitotic apparatus (Welsh et al., 1978; Dedman et
al., 1978, 1979; Anderson et al., 1978a). In this regard,
this protein has been shown to regulate the calcium-dependent
assembly-disassembly of microtubules <u>in vitro</u> (Marcum et
al., 1978). Calmodulin has been demonstrated to be ubiquitous in eukaryotes (Kakiuchi et al., 1974; Waisman et al.,
1975; Chafouleas et al., 1979). The highly conserved nature
of this protein has been suggested by studies which demonstrate that the amino acid sequence of calmodulin is invariant in structure in the cow and rat (Watterson et al., 1980;
Dedman et al., 1978) and the sea pansy, <u>R</u>. <u>reniformis</u>
(Jamieson, et al., 1980), differing by no more than seven
conservative amino acid substitutions. Moreover, each of
these proteins contain four internally homologous calcium
binding domains, and only one of the substitutions (in <u>R</u>.
<u>reniformis</u>) occurs in these highly conserved regions. The
extent to which this protein is conserved in eukaryotes was
demonstrated by Chafouleas et al. (1979). Employing a radioimmunoassay for the protein they demonstrated that calmodulin
from representative species of primitive algae, slime mold
and coelenterate to the more advanced plants and mammals
exhibited immunological identity. Taken together these data
suggest that calmodulin is one of the most highly conserved
as well as widely dispersed proteins studied.

The Molecular Biology of Calmodulin

The calmodulin antibody has recently been used in an
immunoprecipitation assay to monitor the purification of
calmodulin mRNA (Chafouleas et al., 1979). Total poly(A)$^{+}$
RNA was isolated from the electroplax tissue of the electric eel and fractionated by sucrose gradient centrifugation
(Munjaal et al., 1981). One fraction was demonstrated to

contain approximately 39% calmodulin mRNA by _in vitro_ translation and hybridization analysis. This partially purified
mRNA was transcribed into double-stranded cDNA by reverse
transcriptase and cloned in the _Pst_ I site of pBR322 using
E. coli RRI as the host. A structural gene probe (pCM109)
was identified and shown by sequence analysis to contain the
nucleotides coding for amino acids 93-148 of calmodulin as
well as 180 nucleotides from the 3' nontranslated region
including a AAUAAA polyadenylation signal.

pCM109 was shown to hybridize to DNA isolated from
diverse animal and plant species suggesting that the calmodulin gene may also be highly conserved during evolution.
Analysis of the DNA hybridization data revealed that the calmodulin gene was represented less than 3 times in both eel
and chicken. The eel cDNA was also utilized to select and
characterize a full-length cDNA complementary to electroplax
calmodulin mRNA. DNA sequence of this recombinant molecule,
pCM116, has shown it to contain 26 nucleotides of 5'-non-
translated sequence, the initiator codon AUG, the entire
coding region, the terminator codon UGA, 409 nucleotides of
3'-nontranslated DNA and a poly(A) tail. Translation of
the coding region revealed only a single conservative amino
acid substitution compared to the published calmodulin sequence from bovine and human material. Hybridization of
^{32}P-labeled pCM116 to poly(A)$^{+}$ mRNA from electroplax distributed on a polyacrylamide gel and then transfered to
nitrocellulose paper yielded 3 bands of approximately equal
intensity at 820, 1100 and 2000 nucleotides, respectively.
These data were obtained whether the RNA was isolated from
total cell, cytoplasm or polyribosomes. However when nuclear
RNA was similarly analyzed an additional RNA species 5500
nucleotides in length was also observed. pCM116 was found
to contain an AAUAAA sequence at 580 nucleotides and another
at 873 nucleotides. The sequence of the 3' nontranslated
region of pCM109 was identical to the same portion of pCM116
up to the first polyadenylation site. A DNA fragment isolated 3' from the first AAUAAA of pCM116 hybridized to the
5500 nuclear RNA species as well as to the 2000 and 1100
nucleotide cytoplasmic mRNA but not to the 820 nucleotide
RNA. Together these data suggest that the 3 cytoplasmic
mRNAs differ only in the 3' nontranslated region and could
be derived from the 5500 nuclear molecule by differential
addition of poly(A). It also appears that pCM109 may have
been generated from the 820 nucleotide mRNA whereas the 1100
nucleotide species gave rise to pCM116. Multiple species of
calmodulin mRNA have been identified in poly(A)$^{+}$ RNA from
rat, cow, chicken, baboon and human arguing for a common
mechanism for processing of calmodulin mRNA.

The calmodulin cDNA probes have also been employed to screen a chicken DNA library. Two calmodulin (or calmodulin -like) genes have been isolated and characterized. The first gene was isolated as a 10.6Kb fragment and the calmodulin DNA was found to be localized to a 1.4Kb fragment at the extreme 5'-end. DNA sequence analysis has shown this fragment to contain all but 33 nucleotides complementary to the coding region of calmodulin mRNA (410 nucleotides) plus 481 nucleotides of 3'-nontranslated DNA, a polyadenylation site (AAUAAA), and a stretch of poly T residues. Derivation of the amino acids coded for by the structural segment illustrates that this genomic DNA begins with amino acid 12 (phe). Sixteen differences exist between this amino acid sequence and that derived from the eel cDNA (pCM116). Nine of the substitutions are nonconservative with two of the most striking being the presence of cys at positions 26 and 130. When this DNA was used as a hybridization probe in Northern analyses, only skeletal muscle demonstrated a positive signal (at 800 nucleotides) out of seven chicken tissues examined. Together these data suggest the possibility that this DNA represents either a processed calmodulin gene or a pseudogene.

The second chicken DNA was cloned as a 13.5Kb fragment. EcoRI cleaved this DNA into a 5' 7.0Kb and a 3' 6.5Kb fragment. Sequence analysis of the 6.5Kb fragment (5'$\longrightarrow$3') revealed that this DNA began with nucleotide 34 of the coding region of the calmodulin gene. Nucleotides 34-168 are identical to those of pCM116. However restriction hybridization analysis has shown the calmodulin DNA to encompass approximately 4Kb and to be interrupted by at least two introns; one in the structural sequence and one in the 3'-nontranslated region. Whether additional introns exist in the 5'-nontranslated region remains to be established. A mRNA was isolated from chicken brain poly(A)$^+$ RNA using pCM116 as a hybridization probe. This mRNA has been cloned and sequenced. It is now apparent that this mRNA is encoded by the calmoduin gene containing intervening sequences. Indeed the sequence of the chick brain cDNA is identical to pCM116 within the entire coding segment (444 nucleotides). Poly(A)$^+$ RNA from all 8 chicken tissues examined (including skeletal muscle) contain species complementary to the second chicken gene. In all cases the predominant mRNA is approximately 1800 nucleotides. Finally Southern analysis under stringent hybridization conditions suggest that this calmodulin gene is unique.

Mechanism of Action of Calmodulin

Although CaM is involved in the regulation of a variety of enzyme systems, it appears to mediate this regulation in most cases by activating specific protein kinases. Calmodulin plays a major role in mediating the Ca^{++} regulation of glycolysis as the calcium binding subunit of phosphorylase kinase. Of extreme interest is that this regulation occurs in two ways, for not only does phosphorylase kinase activate phosphorylase through phosphorylation in the presence of Ca^{++}, it also inactivates the antagonistic enzyme glycogen synthase through the same mechanism. As reported by Blackmore et al. (1981), these data explain the manner by which metabolism in hepatocytes in the absence of cAMP metabolism.

The possibility that calmodulin mediated phosphorylation may be a major mechanism for intracellular regulation is suggested by the work of Yamauchi and Fujisawa (1979), who have demonstrated that most of the Ca^{++}-dependent endogenous phosphorylation of rat brain cytosolic proteins requires calmodulin. More specifically, CaM-directed protein phosphorylation may be a major regulatory mechanism in stimulus-secretion coupling. This indeed has been demonstrated in the release of neurotransmitter in brain. DeLorenzo and Freedman (1978) observed that Ca^{++} stimulation of endogenous release of norepinephrine from isolated synaptic vesicles was associated with a rapid increase in phosphorylation of specific vesicle proteins. Schulman and Greengard (1978) demonstrated that an endogenous heat stable cytosolic protein was required for the calcium-dependent phosphorylation of synaptosomal membrane fractions from rat cerebral cortex. They also observed that authentic calmodulin could substitute for the endogenous protein. The authors therefore suggested that the calcium-dependent phosphorylation was mediated through calmodulin. That such a heatstable protein (most likely calmodulin) regulates neurotransmitter release was demonstrated by DeLorenzo et al. (1979). They showed that removal of the endogenous heat-stable protein from purified synaptic vesicles caused these vesicles to become refractory to Ca^{++} stimulation. The Ca^{++} stimulation of norepinephrine release and membrane protein phosphorylation could be restored to these depleted vesicles in a dose-dependent manner by the addition of authentic calmodulin. Schulman and Greengard (1978) have demonstrated that the Ca^{++}dependent phoshory-lation of membrane proteins is not restricted to neuronal tissues, but also occurs in many non-neuronal tissues. However, while these membrane phosphorylations all required Ca^{++} and calmodulin, tissue-specific endogenous substrates for the kinases were observed.

The possibility of a coordinate regulation of brain membrane protein phosphorylation by both cAMP and Ca^{++} is suggested by the study of Sieghart et al. (1979). They observed that the phosphorylation of the brain-specific proteins Ia and Ib were regulated by both Ca^{++} and cAMP. However, while both kinases phosphorylate these two proteins, they do so at different amino acid residues. LePeuch et al. (1979) have demonstrated that this coordinated regulation is not limited to these two brain proteins. They have shown that the rate of Ca^{++} uptake by cardiac sarcoplasmic reticulum is regulated by cAMP and Ca^{++}-calmodulin-dependent phosphorylation of the membrane protein phospholamban. cAMP mediates its regulation through cAMP-dependent protein kinase, while the Ca^{++}-dependent phosphorylation is mediated through a membrane-bound protein kinase which requires both Ca^{++} and calmodulin for activity. Both kinases phosphorylate different sites on the protein and it is the Ca^{++}-dependent phosphorylation which is mandatory for Ca^{++} uptake. The cAMP-dependent phosphorylation, while incapable by itself to stimulate Ca^{++} uptake, does however amplify uptake when Ca^{++}-dependent phosphorylation has taken place. The possibility that this membrane-bound calmodulin-dependent protein kinase may be similar to phosphorylase kinase or in fact a form of phosphorylase kinase devoid of its δ-subunit (calmodulin) is suggested by the fact that it is capable of phosphorylating exogenous phosphorylase in the presence of calmodulin. Conversely, exogenous phosphorylase kinase is capable of phosphorylating phospholamban. These findings are intriguing when taken with the work of Browning et al. (1979). They have shown that the preferential phosphorylation of a 40,000 MW protein in synaptic-plasma membranes following electrical stimulation can be mimicked by exogenous phosphorylase kinase, but not by cAMP-dependent protein kinase. These studies would suggest that the calmodulin regulation of protein phosphorylation mediated through phosphorylase kinase may be more extensive than previously thought. Of equal interest is the coordinated regulation of protein phosphorylation exhibited by calmodulin, since calmodulin in fact is involved in both processes. While calmodulin plays a direct role in Ca^{++}-dependent phosphorylation, it also plays an indirect role in cAMP-dependent phosphorylation by regulating cAMP metabolism through activation of both adenylyl cyclase and cyclic nucleotide phosphodiesterase.

Whether calmodulin mediates its regulatory function solely through protein phosphorylation or through other mechanisms, its ability to regulate is entirely dependent on the intracellular free Ca^{++} concentration. Since calmodulin controlled systems would either be constitutively turned on or off in the presence of a constant intracellular free Ca^{++}

concentration, it is apparent that true regulation must be mediated through the control of these intracellular levels. This, in fact, may be the mechanism by which external signals are transduced into intracellular responses in hormone action. Indeed, the most characteristic aspect of hormone action is the occurrence of Ca^{++} fluxes (Berridge, 1975; Rassmussen et al., 1976).

Although Ca^{++} fluxes are observed as a consequence of hormone action in target cells, it may not be a true increase or decrease in the intracellular levels of Ca^{++}, but a mobilization and redistribution of membrane-bound Ca^{++} that is essential. This is suggested by studies employing chlorotetracycline to monitor the shifts in membrane-bound calcium. LeBreton et al. (1976) demonstrated that the ADP- or ionophore A23187-induced shape change in human platelets was temporarily associated with a decrease in membrane-bound calcium. A similar phenomenon was also observed in rabbit neutrophils following stimulation with chemotactic factors (Naccache et al., 1979). In fact, the redistribution of Ca^{++} is so rapid that the authors have suggested this mobilization may represent one of the initial molecular events following the binding of chemotactic factor to its membrane receptor. Recent studies by Chafouleas et al. (1980) and Conn et al. (1981) have suggested that this redistribution of intracellular Ca^{++} may in fact represent a redistribution of the intracellular Ca^{++}-CaM complex. Since Ca^{++} has been known to play a prominent role in stimulus secretion coupling, Chafouleas et al. (1980) investigated the ability of various hormones to induce changes in the intracellular levels of CaM in their respective target cells. In no circumstance was CaM demonstrated to be elevated whether the systems employed were stimulation of estrogen and progesterone on chick oviduct, FSH on Sertoli cells, TSH on thyroid slices or GnRH on pituitary cells. Since calmodulin levels were unaltered by hormones, it was decided to determine whether the peptide hormones resulted in a redistribution of calmodulin within the cell. It was found that GnRH did promote a redistribution of calmodulin in rat pituitary cells whether the cells were fractionated in the presence or absence of calcium. This change in distribution preceded the secretion of LH as assessed by a specific radioimmunoassay and was dependent on the dosage of GnRH used. Whereas no changes were seen in nuclei or mitochondrial fractions, GnRH promoted significant redistribution of calmodulin in plasma membrane fractions. These changes could be accounted for by reciprocal changes in either microsomal or cytoplasmic fractions. The data suggest the possibility that calmodulin-mediated enzymes may be regulated due to a hormone-dependent redistribution

of this calcium binding protein. It is likely that this
redistribution is promoted by alterations in the net flux
or distribution of calcium within the cell rather than by
changes in the cellular content of calmodulin.

<u>Immunologic Techniques for the Study of Calmodulin</u>

It became apparent early on that data derived by any
of the biological assays for calmodulin might be inaccurate
due to the multifunctional aspect of the protein. Since
each assay was dependent on the presence of Ca^{++}-dependent
binding by a "biologically active" calmodulin molecule, it
was probable, that such assays could yield lower values for
the protein due to competition by other calmodulin binding
proteins. No such restrictions would be imposed, however,
by immunological probes since these assays are predicated on
the highly specific antigen-antibody interaction and would
not require Ca^{++}-dependent binding.

Interestingly, those aspects which made calmodulin exciting to study -- its highly conserved and ubiquitious nature as well as low M_r and pI -- proved to be deterents in
production of antisera to the protein. Initial attempts to
produce antibody against native protein in rabbits were
negative. However, Dedman et al. (1978a) were able to elicit
antibody to the native protein in goat. Chafouleas et al.
(1979) subsequently reported antibody toward native CaM produced in sheep which was used in development of an RIA. Detection of these antibodies, however, required affinity chro-
matography on a CaM affinity column. While these two reports
are the only ones dealing with antibody against the native
protein, other investigators have reported on the production
of antisera to calmodulin modified by alum precipitation
(Anderson et al., 1978), dinitrophenylation (Wallace and
Cheung, 1979) or peroxidate oxidation (Van Eldik and Watterson, 1980).

<u>Intracellular localization</u>

One of the first applications for the antibody in our
laboratory was the evaluation of the intracellular localization of CaM by the technique of indirect immunofluorescence.
In the interphase tissue culture cell, calmodulin was found
associated with the stress fibers (Dedman et al., 1978),
which are the actin-containing microfilaments. These organ-
elles are also the location for other contractile proteins
in nonmuscle cells, such as myosin and tropomyosin. The
localization of calmodulin on these stress fibers was most
intriguing when the mechanism of contractility in nonskeletal

muscle cells was reviewed. Unlike skeletal muscle where regulation of contraction is mediated through the troponin system, activation of myosin ATPase activity by actin in smooth and nonmuscle cells requires phosphorylation of the light chain of myosin. This phosphorylation results in a conformational change so that actin can come into contact with the myosin head group stimulating myosin ATPase leading to the development of tension and presumably motility. The enzyme responsible for the phosphorylation of the light chain of myosin, myosin light chain kinase (MLCK) has been shown to be a calcium-dependent enzyme, in fact, the calcium regulation of this enzyme is mediated by calmodulin (Yagi et al., 1978; Dabrowska et al., 1977; Hathaway and Adelstein, 1979). Recently, indirect immunofluorescence localization of the MLCK has revealed this enzyme to also reside on the stress fibers of interphase cells (Guerriero et al., 1981).

An equally exciting pattern of CaM staining was observed in the mitotic cell (Welsh et al., 1978). In metaphase, calmodulin was observed in association with the mitotic spindle. In anaphase, calmodulin was found only in the region of the spindle between the centrioles and the chromosomes and was completely absent from the interzonal region. This initial study suggested a similarity between calmodulin localization and that of the microtubules. In order to further evaluate this, Welsh et al. (1979) compared the immunofluorescence localization of calmodulin and tubulin in several mammalian tissue culture cells throughout mitosis. Although calmodulin was always restricted to the half-spindles, tubulin was found throughout the mitotic apparatus.

The cytoskeletal component in the mitotic apparatus to which calmodulin was associated was evaluated by a series of drug studies (Welsh et al., 1979) Treatment with cytochalasin B, a drug known to disrupt microfilaments, caused no change in either the concentration or localization of calmodulin or tubulin. However, when cells were treated with agents known to disrupt microtubules (i.e., colcemid), spindle structure was altered and tubulin and calmodulin specific fluorescence were equally affected. Colcemid treatment completely disrupted the spindle as reflected by tubulin immunofluorescence and equally abolished calmodulin localization in the mitotic apparatus. Treatment of cells with nitrous oxide, which causes disorganization of the spindle, but does not cause disassembly of spindle microtubules, disrupted spindle structure as visualized by tubulin-specific fluorescence. However, calmodulin was still concentrated in the cell center in the same region as the microtubules of the disorganized spindle. Finally, Brinkley and Cartwright (1971) had shown that two types of microtubules exist in the mitotic apparatus. The microtubules from pole to chromosome are known to be stable to cold temperatures whereas lowering the temperature to 4°C disrupts those microtubules present in the interzone region that extend from pole to pole. In order to substantiate the premise that CaM was associated with the former, cells were subjected to 4°C and immunofluorescence staining performed. As expected, treatment of cells at 4°C resulted in the disruption of the pole-to-pole microtubules as visualized by antitubulin immunofluorescence. However, the pole-to-chromosome tubules were intact. No change in the distribution of calmodulin was found upon exposure of the cells to cold temperatures suggesting that if calmodulin is associated with microtubules during mitosis, the most likely components are the pole-to-chromosome microtubules which comprise the half spindles. Subsequent studies by Andersen et al. (1978) confirmed this localization pattern using an antibody that had been prepared against alum-precipitated calmodulin. In addition, Lin et al. (1980) as well as deMey and deBrabander (1980) have localized calmodulin in mitotic mammalian cells by an electron microscopic method. Again, the distribution pattern described above was shown to be the case. Thus, a gradient of calmodulin concentration was seen in which the greatest concentration of CaM was present at the poles and the least at the chromosomes.

Although the intracellular distribution of CaM appears to be very elaborate, one consistant theme does arise. During mitosis there is a striking correlation between the location of CaM staining and the area in which microtubules are undergoing depolymerization. These experiments suggested that the calcium lability of microtubules may be mediated by calmodulin. In order to test this hypothesis microtubules were isolated by four cycles of polymerization/depolymerization as described by Borisy et al. (1975). Microtubule formation was monitored by a change in absorbancy at 320 nanometers and polymerization was initiated by the addition of GTP (Marcum et al., 1978). 0.3 micromolar calcium had no effect on microtubule polymerization. Increasing the concentration to 11 micromolar caused only about a 10% reduction in the extent of microtubule polymerization. When calmodulin was added to the mixture in buffer containing 0.3 micromolar free Ca^{++}, little effect on polymerization was observed. However, elevating the Ca^{++} concentration to 11 micromolar calcium in the presence of calmodulin completely prevented microtubule polymerization. Similarly, addition of 11 micromolar calcium plus calmodulin to polymerized microtubules resulted in a complete depolymerization of these structures. Further examination of this effect revealed that calmodulin caused a shift in the concentration of calcium required for microtubule polymerization (Dedman et al., 1980). Thus, in the absence of calmodulin, the amount of calcium required for depolymerization was approximately 10^{-3} M. However, in the presence of calmodulin the concentration of calcium required for complete depolymerization of microtubules was 10^{-5} molar. Therefore, a 100-fold change in sensitivity of calcium was achieved by the addition of calmodulin.

While the experiments demonstrated the involvement of calmodulin in the calcium sensitivity of microtubule depolymerization, the molar ratios of CaM to tubulin required to elicit these effects was 6:1. Since CaM has been shown to interact with its other regulated systems in 1:1 stoichiometry, these experiments were less than ideal. In order to further substantiate these data in a more in situ situation, a detergent-permeabilized cell system was used as described by Brinkley et al. (1980). In this system, isolated tissue culture cells are incubated with colcemid to disrupt the cytoplasmic microtubule complex. The cells are then permeabilized by treatment with Triton X-100 and washed to remove depolymerized 6S tubulin and the colcemid. Visualization of these cells by indirect immunofluorescence microscopy using antitubulin reveals that the only fluorescent structure present in each interphase cell is the single organizing center associated with the centrosomal region. Addition of 6S tubulin to this preparation results in the specific nucleation of microtubule assembly from the single nucleating site. Incubation of these lysed cells with 11 micromolar free calcium produces no difference in the degree of microtubule polymerization. However, if calmodulin is added in the presence of 11 micromolar calcium, complete abolition of microtubule polymerization is achieved. Brinkley et al. (1980) established that the optimal concentration of tubulin for microtubule assembly was 1 milligram per ml. Assuming a molecular weight of 110,000 for the 6S dimer, this results in a tubulin concentration of approximately 10 micromolar. The optimal concentration of calmodulin in this system was 6 micromolar. Thus, the ratio of calmodulin to tubulin required for optimal effects was much closer to 1:1 than the concentrations required in the in vitro microtubule polymerization experiments.

Taken together, these experiments suggest that there should be an inverse relationship between the concentration of calmodulin in the cell and the number of polymerized microtubules. Several systems have been developed to examine this possibility. The first was achieved by first determining the distribution of anti-calmodulin in mitotic cells by

immunofluorescence microscopy and then serially sectioning
them for electron microscopy (Dedman et al., 1980). Each
of the sections was evaluated for the number of microtubule
profiles as described by Brinkley and Cartwright (1971). In
metaphase, when high concentrations of calmodulin were found
at the poles, no polymerized microtubules were counted in
those sections. However, as the gradient of calmodulin
decreased from the poles toward the chromosomes, the number
of polymerized microtubules increased, reaching the highest
numbers at the kinetochore plates of the chromosomes. On
the other side of the chromosomes an inverse relationship was
again observed where high concentrations were observed close
to the kinetochore followed by a decreasing gradient towards
the opposite pole. This relationship held throughout mitosis
even in late anaphase where calmodulin was found to be
transiently associated on both sides of the chromosomes. In
such cells, the only polymerized microtubules were found in
the middle of the cell corresponding to the position of the
developing cleavage furrow. Once again, therefore, there was
an inverse relationship between the concentration of calmodulin and the number of polymerized microtubules.

The second approach was to isolate mitotic spindles from
sea urchin eggs as described by Salmon and Segall (1980).
These spindles demonstrate micromolar sensitivity to calcium
and were found to contain calmodulin by radioimmunoassay
(Dedman et al., 1980). Staining of the metaphase mitotic
spindle by antitubulin demonstrated that the entire mitotic
spindle contained tubulin, including the asters at each
pole. Examination of the immunofluorescence pattern of calmodulin revealed localization only in the region of the centrioles. Serial sections of the mitotic spindles revealed that
no polymerized microtubules were found in association with
the poles. As the distance between the pole and chromosome
increased, the number of polymerized microtubules increased.
Again, in this system, there is an association between the
calmodulin concentration and the number of polymerized microtubules. Finally, Cande and Wolniak (1978) have developed
a eukaryotic mammalian cell system in which the rate of metaphase-anaphase chromosome movement can be studied. This
movement is markedly enhanced by the addition of micromolar
calcium. In collaborative experiments, we have recently
shown that this movement can be blocked by incubation of the
mitotic spindles with calmodulin antibody or the anticalmodulin agent, W13, described by Hidaka et al. (1980). It seems
reasonable to assume, therefore, that calmodulin may be
involved in the regulated depolymerization of microtubules
that occurs during metaphase-anaphase chromosome movement.
It should also be pointed out that in those systems where

calcium concentration has been measured, micromolar concentrations of calcium are not achieved in the spindle. Therefore, high concentrations of calmodulin as observed by immunofluorescence would be required in order to assure proper depolymerization of microtubules. However, the exact mechanism by which calmodulin affects microtubule depolymerization remains to be determined.

Radioimmunoassay

The next major use for the calmodulin antibody was in the development of a RIA for calmodulin. Since the antibody had been successfully prepared, the only major hurdle to cross was radiolabeling the tracer calmodulin. In that the sensitivity of the assay would be directly related to the specific activity achieved, the procedure of radiolabeling was very important. Calmodulin contains only two tyrosine residues, one of which is inaccessible to the exterior of the protein. In addition, calcium binding to calmodulin potentiates a dramatic change in the tyrosine fluorescence of the molecule. These observations, therefore, would suggest that labeling procedures which were directed to the tyrosine residues might result in low specific activity as well as reduced biological activity. Indeed, when the lactoperoxidase and chloramine T procedures were utilized to radioiodinate calmodulin, low specific radioactivity and significant loss in biological activity was observed (Chafouleas et al., 1979). Calmodulin does, however, contain 8 lysine residues and so the procedure of Bolton and Hunter (1975), in which the iodine is conjugated to the ε amino side groups of lysine was used. Calmodulin labeled by this procedure had high specific radioactivity (2400 Ci/mmol) and retained complete biological activity (Chafouleas et al., 1979). The resultant radioimmunoassay was shown to have an interassay variability of less than 5% and an intraassay variability of less than 3%. The statistical assay sensitivity was 150 pg whereas the limit of detection was 15 picograms (Chafouleas et al., 1979).

The radioimmunoassay was first used to confirm the highly conserved nature of calmodulin. When pure calmodulin isolated from bovine brain, rat testis, Renilla reniformis and the peanut plant (Arachis hypogea) were used to obtain standard dilution curves, it was found that all 4 proteins described the same curve (Chafouleas et al., 1979). On the other hand, troponin C from rabbit skeletal muscle required 660-fold greater protein concentration to achieve 50% competition and demonstrated a statistically different slope. Again, this emphasized the fact that although troponin C and calmodulin are homologous, they are not identical. Another small molecular weight calcium-binding protein, parvalbumin, demonstrated no cross reactivity even at 50,000-fold protein excess.

The radioimmunoassay was also used to evaluate calmodulin levels in heat-treated supernatant solutions from highly diverse sources. Cell samples tested, regardless of source, demonstrated immunological identity to the rat testis calmodulin standard (Chafouleas et al., 1979). Tissues and organisms examined to date include rat, eel, rabbit, bovine, human, amphibian, reptile, slime mold, plants, algae, coelenterate, paramecium, tetrahymena and amoeba. These data once again illustrate the fact that the immunologic nature of calmodulin is highly conserved between the most primitive unicellular organism and man.

The radioimmunoassay was utilized to illustrate another very interesting point regarding the nature of calmodulin's interaction with its various binding proteins. The most usual assays to determine the amount of calmodulin in a sample are based on its ability to stimulate a partially purified preparation of bovine brain cyclic AMP phosphodiesterase or chicken gizzard myosin light chain kinase. When the amount of calmodulin found in various tissues and species by radioimmunoassay was compared to those determined by the phosphodiesterase assay, it was found that in all instances, the radioimmunoassay yielded higher values (Chafouleas et al., 1979). In fact, in organisms such as <u>Dictyostelium</u> and <u>Chlamydomonas</u>, no calmodulin could be detected by the phosphodiesterase assay, whereas significant concentrations were found by radioimmunoassay. The discrepencies observed between the two types of assays can be explained when the procedures employed for the biological assay are compared. The assay for calmodulin by enzyme activation depends upon the ability of "biologically active" calmodulin to activate the enzyme through Ca^{++}-dependent binding. There are many calmodulin-binding proteins present in cells. These proteins can be found in both the heat-treated samples to be assayed as well as the enzyme preparation itself. Therefore, these calmodulin-binding proteins could interfere in the assay by competing for the calmodulin and effectively resulting in a lower measured level of this protein. Since the RIA is not dependent on Ca^{++}-dependent activation, the results from this assay would not be so influenced. Indeed, when purified calmodulin is quantitated by both assays, the values obtained are in close agreement. These observations would suggest that the radioimmunoasay is the assay of choice when evaluating calmodulin levels in tissue extracts.

Regulation of Calmodulin in Transformed Cells

The experiments performed using calmodulin in microtubule polymerization would suggest that there is a positive correlation between calmodulin content and microtubule depolymerization. Interestingly, Watterson et al. (1976) and LaPorte et al. (1979) had reported 2-fold elevations in calmodulin levels in chicken embryo fibroblasts transformed by Rous Sarcoma Virus when compared to the nontransformed cell. Since a diminished cytoplasmic microtubule network has been reported as one characteristic response to transformation (Brinkley et al., 1975), these reports would further substantiate the role of calmodulin in microtubule polymerization. In order to ascertain whether the elevation in calmodulin level observed in the transformed chicken embryo fibroblasts is a result of transformation in general, Chafouleas et al. (1981) compared calmodulin levels in Swiss mouse 3T3 and 3T3 cells transformed by the DNA type virus SV40, as well as normal rat kidney cells and those cells transformed by the RNA type virus, Rous Sarcoma Virus. Since the steady state concentration of a protein is a function of the rates of synthesis and degradation, these turnover parameters were determined in both the 3T3/SV3T3 and NRK/SNRK systems. In addition to calmodulin, similar determinations were performed for tubulin. In both cell systems, calmodulin levels were elevated at least 2-fold in the transformed cells compared to their appropriate nontransformed counterpart. This was in contrast to tubulin which retained the same levels in the normal and transformed cells. The rate of synthesis of calmodulin, tubulin and total protein were all increased in the transformed cells. While tubulin and total protein demonstrated about a 2-fold increase, calmodulin was synthesized three times as fast. In addition, the transformed cells also exhibited greater rates of degradation. Calmodulin and total protein exhibited similar changes. Tubulin, on the other hand, had a rate of degradation which was twice that present in the normal cell. Therefore, tubulin levels are unaltered in the transformed cells through reciprocal changes in the rates of degradation and synthesis. A net increase in the intracellular levels of calmodulin is achieved through a selective increase in the rate of synthesis which is twice as great as the change in degradation. These changes in the steady state concentrations of calmodulin and tubulin result in a 2-fold increase in the calmodulin-to-tubulin molar ratio in the transformed cells. Such an alteration may account for the apparent differences in not only the cytoskeleton, but other key metabolic activities in the transformed cell.

Regulation of Calmodulin during the Cell Cycle

Since the turnover studies were performed on asynchronous populations of cells, Chafouleas et al. (1982) investigated the time during the cell cycle in which calmodulin was synthesized. Because of their relatively short cell cycle and ease of handling, Chinese hamster ovary cells, CHO-K1, were selected for the synchrony experiments. Prior to synchronization, the cell cycle parameters of G_1, S, G_2 and M were ascertained by the pulse-labeled mitosis procedure and found to be 5, 8, 2 and 1 hr, respectively. All cells were synchronized by mitotic shake (Terasima and Tolmach, 1961) and calmodulin levels were determined by radioimmunoassay (Chafouleas et al., 1979) during the subsequent cell cycle. Calmodulin levels at mitosis were 160 ng/10^6 cells and fell by 50% in early G_1 coincident with separation of the daughter cells. During late G_1 calmodulin increases to the value determined in mitotic cells and remains at this concentration for the duration of the cell cycle. Subsequent experiments with cell populations that varied in the length of G_1 revealed that calmodulin always doubled in late G_1 or early S suggesting that this protein might be important in G_1/S transition. This possibility was investigated by using the anticalmodulin drug W13 and its less active dechlorinated homolog W12 (Hidaka et al., 1980). Addition of W13 to asynchronous cells resulted in a 52% reduction in cell number after 24 hr whereas W12 had no effect. Evaluation of the W13-treated cells by flow cytometry revealed a build-up of cells in G_1 at the expense of the S population. At the concentration utilized in these experiments W13 was not cytotoxic and the cell cycle block was readily reversible. The results of additional experiments showed that the inhibition of progression through the cell cycle was due to a single block at the G_1/S boundary. W13 had no effect on progression through either G_2 or M. These data reveal that calmodulin is synthesized entirely during late G_1/early S and suggest the possibility that the increased intracellular concentration of this protein may be important for DNA repair and/or synthesis.

The calmodulin concentration also seems to be important for the reentry of plateau (G_0) cells into the cell cycle. Calmodulin levels increase by 50% as cells leave G_1 and enter plateau and remain at this concentration for the duration of the G_0 phase. Upon release of the cells into the growth cycle, calmodulin decreases by 50% within the first hr and remains at the concentration for 4-5 hr. By 6 hr the intracellular levels have again doubled and remain at this level as cells pass through S, G_2 and M. This concentration is that normally achieved after the increase in calmodulin at the G_1/S boundary (Chafouleas et al., 1982). Addition of W13 at the time of treatment with fresh medium prevented entry into S phase but the changes in the intracellular concentration of calmodulin were unaltered. When cells were treated with W13 at various times following release from plateau, a direct correlation was observed between the percentage of cells entering S phase and the time of drug addition (r=0.99). The labeling index increased as the interval between drug treatment and G_0 release increased. However, although some cells entered S phase, no progression through this period was observed when W13 was added as late as 5 hr following addition of fresh medium. Removal of the drug resulted, after a 5 hr lag period, in progression of all cells through S phase in a synchronous fashion. These data strengthen the contention that calmodulin is important for the progression of cells through DNA synthesis.

Recent experimental results can be interpreted to suggest that the increase in calmodulin at G_1/S is important for optimal DNA repair prior to replicative DNA synthesis. Bleomycin is a drug known to cause DNA damage by strand scission. Concentrations of this agent can be readily found that result in potentially lethal damage to tissue culture cells. Under these conditions approximately 90% of the cells are killed. The remaining 10%, however, can recover from the drug by repairing DNA and will eventually repopulate the culture dish. If cells are selected for potentially lethal damage and released from Bleomycin into the presence of media containing W13, all the cells are killed within 3 hr. The most obvious explanation of these results is that calmodulin is required for DNA repair. When calmodulin is neutralized with W13, then the cells cannot repair the DNA damage which resulted from Bleomycin. This results in cell death during the subsequent replicative phase. If such a scenario is the case then some of the enzymes involved in DNA repair must be regulated by calmodulin. It is known that such enzymes are induced at the G_1/S boundary as is the case for calmodulin. Studies are currently underway to determine whether the DNA repair enzymes are calmodulin-binding proteins and whether the increased calmodulin synthesis at G_1/S is transcriptionally or post-transcriptionally regulated.

REFERENCES

Andersen, B., Osborn, M. and Weber, K. (1978). J. Cell Biol. 17,354.

Blackmore, P.F., El-Refai, M.F., Dehaye, J.-P., Strickland, W.G., Hughes, B.P. and Exton, J.H. (1981). FEBS Lett. 123,245.

Berridge, M.J. (1975) In: Advances in Cyclic Nucleotide Research, Vol. 6, edited by P. Greengard and G.A. Robison. Raven Press, New York, pp. 1-98.

Bolton, A.E. and Hunter, W.M. (1973). Biochem. J. 133,529.

Borisy, G.G., Marcum, J.M., Olmstead, J.B., Murphy, D.B. and Johnson, K.A. (1975). Ann. N.Y. Acad. Sci. 253,107.

Brinkley, B.R. and Cartwright, J. (1971), J. Cell Biol. 50, 416.

Brinkley, B.R., Pepper, D.A., Cox, S.M., Fistel, S., Brenner, S.L., Wible, L.J. and Pardue, R.L. (1980). In: Microtubules and Microtubule Inhibitors, edited by M. DeBrabander and J. DeMey. Amsterdam, Elsevier, pp. 281.

Brinkley, B.R., Fuller, G.M. and Highfield, D.P (1975). PNAS 72,4981.

Browning, M., Bennett, W., and Lynch, G. (1979). Nature 278, 273.

Cande, W.Z. and Wolniak, S.M. (1978). J. Cell Biol. 79,573.

Chafouleas, J.G., Bolton, W.E., Hidaka, H., Boyd, A.E. III and Means, A.R. (1982). Cell 28,41.

Chafouleas, J.G., Dedman, J.R., Munjaal, R.P. and Means, A.R. (1979). J. Biol. Chem. 254,10262.

Chafouleas, J.G., Pardue, R.L., Brinkley, B.R., Dedman, J.R. and Means, A.R. (1980) In: Calcium-Binding Proteins: Structure and Function, edited by F.L. Siegel, E. Carafoli, R.H. Kretsinger, D.H. MacLennan and R.H. Wasserman. Amsterdam, Elseiver, pp. 189-196.

Chafouleas, J.G., Pardue, R.L., Brinkley, B.R., Dedman, J.R. and Means, A.R. (1981). Proc. Natl. Acad. Sci. USA 78,996.

Conn, P.M., Chafouleas, J.G., Rogers, D. and Means, A.R. (1981). Nature 292,264.

Dabrowska, R., Sherry, J.M.F., Aromatorio, D.K. and Hartshorne, D.J. (1977). Biochemistry 17,253.

Dedman, J.R., Jackson, R.L., Schreiber, W.E. and Means, A.R. (1978). J. Biol. Chem. 253,343.

Dedman, J.R., Lin, T., Marcum, J.M., Brinkley, B.R. and Means, A.R. (1980). In: Calcium-Binding Proteins: Structure and Function, edited by F.L. Siegel, E. Carafoli, R.H. Kretsinger, D.H. MacLennan and R.H. Wasserman. Amsterdam, Elsevier, pp. 181-188.

Dedman, J.R., Potter, J.D., Jackson, R.L. and Means, A.R. (1977). J. Biol. Chem. 252,8415.

Dedman, J.R., Welsh, M.J. and Means, A.R. (1978a). J. Biol. Chem. 253,7515.

DeLorenzo, J.R. and Freeman, S.D. (1978). Biochem. Biophys. Res. Commun. 80,183.

DeLorenzo, J.R., Freeman, S.D., Yohe, W.B. and Maurer, S.C. (1979). Proc. Natl. Acad. Sci. USA 76,1838.

DeMey, J., Moeremans, M., Gevens, G., Muydens, R., VanBelle, H. and DeBrabander, M. (1980). In: Microtubules and Microtubule Inhibitors, ed. M. DeBrabander and J. DeMey. Amsterdam, Elsevier, pp. 227-241.

Guerriero, V., Jr., Rowley, D.R. and Means, A.R. (1981). Cell 27,449.

Hathaway, D.R. and Adelstein, R.S. (1979). Proc. Natl. Acad. Sci. USA 76,1653.

Hidaka, H., Naka, M. and Yamaki, T. (1980). Biochem. Biophys. Res. Commun. 90,694.

Ho, H.C., Dasai, R. and Wang, J.H. (1975). FEBS Lett. 50(3), 374.

Jamieson, G.A., Jr., Hayes, J., Blum, J. and Vanaman, T.C. (1980). IN: Calicum-Binding Proteins: Structure and function, edited by F.L. Siegel, E. Carafoli, R.H. Kretsinger, D.H. MacLennan and R.H. Wasserman. Amsterdam, Elsevier, pp. 165-172.

Kakuichi, S., Yamazaki, R., Techima, Y. and Miyamoto, E. (1974). Biochem. J. 146,109.

Katz, S. and Remtulla, M.A. (1978). Biochem. Biophys. Res. Commun. 83,1373. Klee, C.B. (1977). Biochemistry 16(5), 1017.

LaPorte, D.C., Gidwitz, S., Weber, M.J. and Storm, D.R. (1979). Biochem. Biophys. Res. Commun. 86,1169.

LeBreton, G.C., Dinerstein, R.J., Roth, L.J. and Feenberg, H. (1976). Biochem. Biophys. Res. Commun. 71,362.

LePeuch, C., Haiech, J. and Demaille, J.G. (1979). Biochemistry 18:5150.

Lin, C.T., Dedman, J.R., Brinkley, B.R. and Means, A.R. (1980). J. Cell Biol. 85,473.

Marcum, M., Dedman, J.R., Brinkley, B.R. and Means, A.R. (1978). Proc. Natl. Acad. Sci. USA 75,3771.

Means, A.R. and Dedman, J.R. (1980). Nature 285,73.

Means, A.R., Tash, J.S. and Chafouleas, J.G. (1982). Physiol. Rev. 62,1.

Munjaal, R.P., Chandra, T., Woo, S.L.C., Dedman, J.R. and Means, A.R. (1981). Proc. Natl. Acad. Sci. USA 78,2330.

Naccache, P.H., Volpi, M., Showell, J.H., Becker, E.L., Sha'afi, R.I. (1979). Science 203,461.

Rasmussen, H., Goodman, D.B.P., Friedman, N., Allen, J.E. and Kurvkawa, K. (1976). In: Handbook of Physiology-Endocrinology. Vol. VII. American Physiological Society, Washington, D.C., pp. 225-264.

Richman, P. (1978). Biochemistry 17,3001.

Salmon, E.D. and Segall, R.R. (1980). J. Cell Biol. 85,355.

Schulman, H. and Greengard, P. (1977). Nature 271,478.

Schulman, H. and Greengard, P. (1978). Proc. Natl. Acad. Sci. USA 75,5432.

Sieghart, W., Forn, J. and Greengard, P. (1979). Proc. Natl. Acad. Sci. USA 76,2475.

Van Eldik, L.J. and Watterson, D.M. (1981). J. Biol. Chem. 256,4205.

Wallace, R.W. and Cheung, W.Y. (1979). J. Biol. Chem. 254, 6564.

Watterson, D.M., Sharief, F. and Vanaman, T.C. (1980). J. Biol. Chem. 255,962.

Watterson, D.M., VanEldik, L.J., Smith, R.E. and Vanaman, T.C. (1976). Proc. Natl. Acad. Sci. USA 73,2711.

Welsh, M.J., Dedman, J.R., Brinkley, B.R. and Means, A.R. (1978). Proc. Natl. Acad. Sci. USA 75,1867.

Welsh, M.J., Dedman, J.R., Brinkley, B.R. and Means, A.R. (1979). J. Cell Biol. 81,624.

Yagi, K., Yazawa, M., Kakiuchi, S., Ohshima, M., Uenishi, K. (1978). J. Biol. Chem. 253,1338.

Yamauchi, T. and Fujisawa, H. (1979). Biochem. Biophys. Res. Commun. 909,1172.

INDUCTION OF MICROTUBULE ASSEMBLY _IN SITU_ FROM UNPOLYMERIZED TUBULIN POOLS IN SV40 TRANSFORMED 3T3 CELLS

William J. Deery and B. R. Brinkley

Department of Cell Biology, Baylor College of Medicine, Houston, Texas 77030

ABSTRACT Many types of cultured cells contain an elaborate network of cytoplasmic microtubules which associate with the centrosome and radiate out toward the cell periphery. This array of cytoskeletal filaments is referred to as the cytoplasmic microtubule complex (CMTC) and has been reported to be diminished in many transformed cells. Addition of cAMP has been shown to restore the normal cytoskeletal phenotype. Following lysis of SV40 transformed cells with the non-ionic detergent Brij-58, endogenous tubulin is retained within the cells and subsequent incubation of these cells in microtubule assembly buffer results in the formation of an enhanced CMTC. In this system, the presence of cAMP, nucleotide triphosphates and protein synthesis do not appear to be required for microtubule assembly. Ca^{++} (100 uM) inhibits assembly and cAMP plus ATP appears to produce an opposing effect on this inhibition. The implications of these results in understanding the mechanism of altered microtubule assembly in transformed cells are discussed and previous findings concerning the effects of transformation on cytoplasmic microtubules are reviewed.

INTRODUCTION

The cytoskeleton is a term given to a complex system of anastomosing intertwining fibrous structures throughout the cytoplasm of eukaryotic cells identified as microfilaments, intermediate filaments and microtubules through electron microscopic, immunocytochemical and biochemical studies (for review see Refs. 1-3). The components of the cytoskeleton are involved in many diverse functions some of which being cell morphology (4-6), cell spreading and adhesive properties (7-9), maintenance of contact inhibition and anchorage-dependent growth (10-13), and mobility of cell surface receptors (14 - 17). Recent studies have also implicated their involvement in DNA synthesis, gene transcription, and growth control. Thus, the major functions of cytoskeletal elements certainly appear to be related to and dependent upon their close association and interaction with the plasma membrane and discrete foci within the cytoplasm.

In many cells, the process of transformation _in vitro_ by viral or chemical agents is accompanied by alterations in the wide range of cellular activities mentioned above and this suggests that the cytoskeleton can be an intermediate in the expression of the transformed phenotype in malignant cells (for review, see Refs. 18, 19). Although strong evidence exists for this correlation, it should be mentioned that the morphology and other properties of some carcinoma cells can change very little upon transformation and that such cells do not appear to be grossly altered in microtubule and microfilament expression (20). This report will focus on one of the major cytoskeletal components, microtubules, and the factors which may affect their assembly _in vivo_, and describe specific changes that occur in these structures upon cellular transformation.

The fate of cytoplasmic microtubules in transformed cells as well as their possible involvement in the transformation process has been the subject of much controversy. Experiments from Puck's laboratory first suggested that microtubules may be involved in cellular transformation (21,22). It was found that after addition of cAMP, CHO cells were transformed from their rounded shapes to an elongated, flattened, normal morphology. Such "reverse transformation" could be prevented by the microtubule inhibitor colchicine suggesting that the microtubule system might be involved in the expression of the transformed phenotype. Obviously, cells undergoing dramatic morphological changes as observed in many malignant transformations suggests that the cytoplasmic microtubule complex (CMTC) should be altered or rearranged to some degree.

The extent to which the CMTC might be altered with transformation has been examined extensively by several investigators using a variety of techniques and protocols. These studies have yielded contrasting conclusions. Using immunofluorescence techniques, Brinkley and co-workers (23,24), Edelman and Yahara (25) and Weber et al. (26) observed an extensive CMTC in untransformed 3T3 cells. However, Brinkley and Edelman's groups found found that transformed cells appeared to contain a diminished CMTC, whereas others have concluded that microtubules are unchanged (27-30). Support for a diminished CMTC has come from several investigators utilizing transmission electron microscopy (31-33). Morphometric EM studies showed that normal rat kidney cells had twice the number of microtubules as their transformed counterparts (32) whereas 3T3 and transformed SV40-3T3 cells had similar numbers of microtubules in the centrosomal region but fewer microtubules extended into the cortical region of the cytoplasm (33). These morphological findings were in agreement with the hypothesis of Edelman (34,35) that "surface modulating assemblies" consisting of microtubules (and microfilaments) were responsible for the regulation of cell surface receptors and growth control, and that transformation might be accompanied by alterations in the cytoskeleton.

If microtubule networks are significantly altered in various transformed cells, one possible explanation for their diminution is that the active pool size of the microtubule subunit protein, tubulin, may be proportionately smaller. This could occur from either decreased tubulin synthesis (or increased degradation) or modifications resulting in defective tubulin molecules. The latter possibility is unlikely since tubulin isolated from transformed cells has been shown to be competent to polymerize in vitro (25,36,37). There is also strong evidence which indicates that the tubulin content in various transformed cells (25,32,37-39) as well as undifferentiated and differentiated cells (40) is not significantly different. These findings suggest therefore that suppression of tubulin assembly may reside in the components of the centrosome (a microtubule organizing center, MTOC) from which cytoplasmic microtubule assembly appears to be nucleated, and/ or factors in the cytoplasm which could associate with tubulin and regulate assembly.

In order to study the characteristics and possible differences in microtubule assembly from MTOCs in 3T3 and SV3T3 cells, a biochemical method using a Triton X-100 lysed cell system was developed by Brinkley and co-workers (41, 42). When exogenous tubulin dimer from bovine brain was added to tubulin-depleted lysed cells, microtubules reformed preferentially from the centrosomes (MTOCs) of both cells. However, under identical conditions, the microtubules in lysed 3T3 cells were two to three times longer and more numerous than those reassembled in SV3T3 cells. Since the two cell types were exposed to the same tubulin, the attenuated assembly in the SV3T3 cells indicated differences in the components of MTOCs and/or factors tightly associated with the cytoplasmic matrix.

Additional lysed cell studies revealed that microtubule growth characteristics in SV3T3 cells could be restored to those observed in 3T3 cells by the addition of cAMP plus ATP in the assembly buffer (42). Furthermore, when [^{32}P]ATP was used, tubulin as well as several other proteins were found to be phosphorylated. These findings were consistent with studies which showed that the pleiotropic effects of transformation could be reversed by increased intracellular levels of cAMP (22,43) and that cytoskeletons of CHO cells stimulated to undergo reverse transformation with cAMP contained nearly 40% more polymerized tubulin than those of their unstimulated controls (44). Recent studies by Lockwood et al. (45) using a lysed cell system have also demonstrated that phosphorylation-dephosphorylation of specific cytoskeletal and cytoplasmic proteins occurs during cAMP-induced reverse transformation. Thus, cAMP and the phosphorylation state of certain proteins appears to influence cellular activities and morphology while concomitantly stimulating the assembly of cytoskeletal elements which organize the cytoplasm.

Although cAMP levels and protein phosphorylation evidently play a major role in the transformation process, cytoplasmic Ca^{++} levels appear to be equally important. It was originally demonstrated by Weisenberg (46) that microtubule polymerization in vitro could be inhibited when excess Ca^{++} was present in the reassembly buffer, and since then several investigators have proposed that calcium ion flux may regulate microtubule assembly in vivo (47-49). Further experiments utilizing the lysed cell system to study the assembly of purified exogenous tubulin have provided strong evidence for the regulation of CMTC formation by calcium and the calcium-dependent regulatory protein, calmodulin, which increases calcium inhibition of microtubule assembly in vitro (50). Radioimmunoassay studies by Chafouleas et al. (39) demonstrated that transformed cells had an enhanced rate of calmodulin synthesis and a two- or three-fold increase in the calmodulin/tubulin ratio compared to untransformed cells. Following lysis, 95% of the cellular tubulin was extracted but calmodulin was retained. When lysed 3T3 cells were preincubated with concentrations of calmodulin equivalent to that which exists in SV3T3 cells, microtubule growth was reduced to that observed in SV3T3 cells (51,52). Conversely, microtubule assembly in SV3T3 cells could be restored to that observed in 3T3 cells by preincubation of lysed SV3T3 cells with calmodulin antibody (53). Thus, it is possible that the elevated levels of calmodulin in transformed cells may in some ways suppress the formation of a full CMTC in these cells.

In this report, we have utilized a Brij-58 detergent system which has been reported to provide greater preservation of cytoskeletal components (54,55). Unlike the Triton-lysed cell system, Brij-lysed cells retain endogenous tubulin (56) and this has allowed us to examine the effects of various molecules on microtubule assembly and disassembly in situ, from endogenous, unpurified microtubule protein. The results presented here indicate that unpolymerized pools of tubulin exist in transformed SV3T3 cells which can be readily induced to polymerize and enhance the preexisting CMTC. In this system, the presence of cAMP plus ATP appears capable of antagonizing the inhibitory effect of Ca^{++} on microtubules.

METHODS

Swiss mouse fibroblast (3T3) cells or their transformed SV40-3T3 counterparts were grown on glass coverslips for 24-48 hours to 90-95% confluency in plastic petri dishes containing Dulbecco's modified Eagle's medium supplemented with 10%

(3T3) or 5% (SV3T3) fetal calf serum at 10% CO_2. Just prior to lysis, cells on coverslips were removed from petri dishes and washed for 15 sec at room temperature in 0.08 M PIPES buffer pH 6.9 containing 10 mM EGTA, 1 mM $MgCl_2$ (extraction buffer). Cells were lysed in extraction buffer containing 0.15% Brij-58 for 2.5-3 min followed by a 30-sec wash in extraction buffer to remove detergent and 30 sec in reassembly buffer, (RB), (extraction buffer at 1 mM EGTA) to reduce EGTA concentration. Microtubule assembly or disassembly was then studied by inverting cells on coverslips over a drop of buffer which contained various molecules known to affect microtubules.

After cells were lysed and exposed to various experimental conditions, they were fixed for 30 min at 25°C in 3% formaldehyde made up in extraction buffer containing 2 mM EGTA and 1% DMSO and processed for indirect immunofluorescence microscopy using sheep antibody to tubulin and fluorescein-tagged goatanti-sheep IgG as described previously in detail (57). Cells were examined with a Leitz Orthoplan microscope equipped with epiillumination and photographs were recorded on Tri-X Pan film (Kodak).

RESULTS

Extent of Cytoplasmic Microtubular Changes in Transformed Cells. It has been reported by several investigators that virally or chemically induced transformation of cells is accompanied by conspicuous alterations of the cytoskeleton (23,31,35). In order to examine the CMTCs, two protocols were used to fix the microtubules and lyse the cells prior to indirect immunofluorescence with antitubulin. When nontransformed 3T3 cells were either fixed first with 3% formaldehyde-phosphate buffered saline and subsequently lysed in acetone for 7 min at -20°C (Fig. 1a), or lysed first with 0.05% Triton X-100 containing 4% polyethylene glycol (PEG) to stabilize the microtubules followed by fixation and immunofluorescence staining (Fig. 1c), an extensive lacy network of long microtubules was apparent. However, when transformed SV3T3 cells were fixed first then lysed, the CMTC appeared diminished and more diffusedly stained (Fig. 1b). Transformed cells which were extracted first with the Triton-stabilizing buffer for 4.5 min then fixed (Fig. 1d) were found to display a much more extensive network of microtubules and less diffuse staining, however, the CMTC still appeared diminished and contained fewer microtubules extending to the plasma membrane compared to untransformed cells.

Effect of Cyclic AMP on Transformed Cells. It is well known that cyclic AMP and its derivative, 3':5'-dibutyryl-cyclic AMP can stimulate cells to undergo microtubule-related morphological changes (21,22,43). Untreated transformed SV3T3 cells which were fixed then lysed displayed only diffuse staining

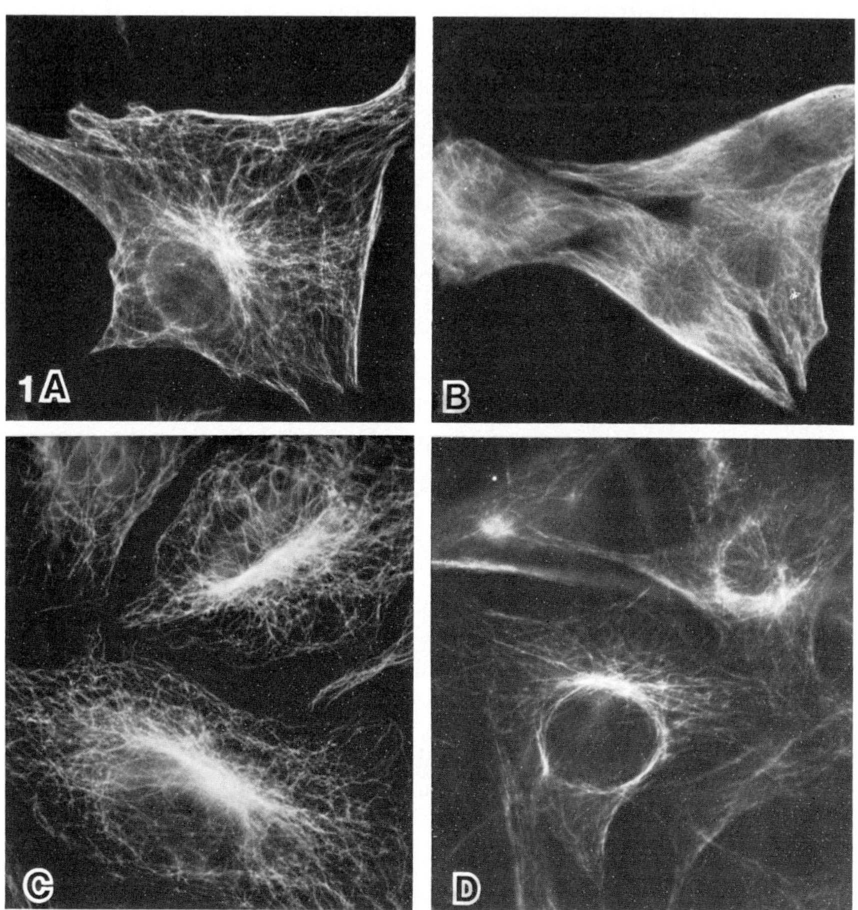

FIGURE 1.   Cytoplasmic microtubule complex (CMTC) in cultured cells.   3T3 cells were either fixed then lysed, (a), or lysed then fixed , (c).   SV3T3 cells were either fixed then lysed, (b), or lysed then fixed, (d).   This was followed by staining with antitubulin immunofluorescence.

with tubulin immunofluorescence (Fig. 2a).   When they were treated with 0.3 mM dibutyryl cAMP and 1 mM theophylline in culture, the cells became more flattened and fibroblastic in appearance.   Moreover, they exhibited a more organized, extensive CMTC (Fig. 2b) suggesting that the cyclic nucleotide stimulated microtubule assembly.

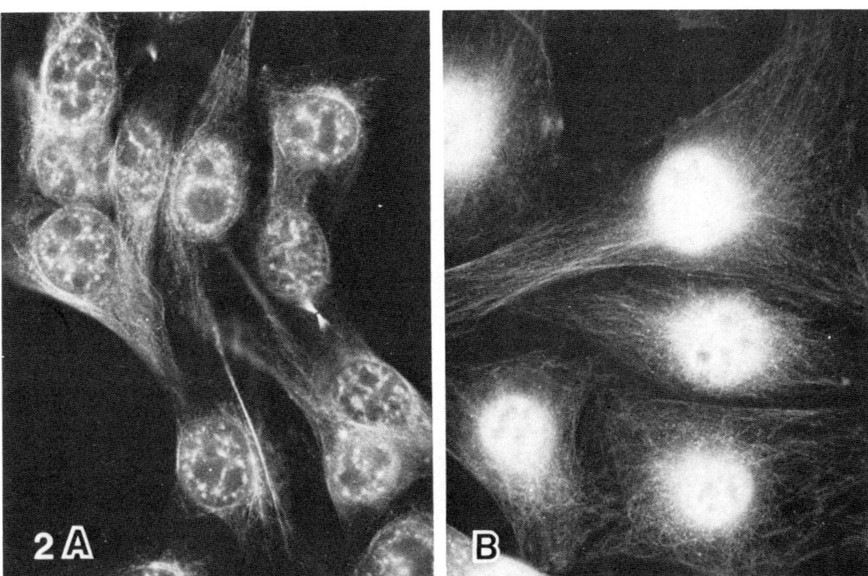

FIGURE 2.   Effects of dibutyryl cyclic AMP on cell morphology and CMTC.   SV3T3 cell before treatment with cAMP, (a).
After treatment with cAMP, the cells become more fibroblastic
and an extensive CMTC is more apparent, (b).

Comparison of Microtubule Initiation and Elongation in
Lysed 3T3 and SV3T3 Cells. Because of the apparent difference
in the expression of cytoplasmic microtubules between normal
and transformed cells, a Triton extracted cell system was used
to compare microtubules assembled from exogenous, pure tubulin
in 3T3 and SV3T3 cells (41,42).   When 3T3 or SV3T3 cells, which
were treated with colcemid and lysed in 0.05% Triton to depolymerize the CMTC and extract endogenous tubulin, were exposed
to pure tubulin (1 mg/ml in 0.5 mM GTP reassembly buffer) for
15 min at 37°C, microtubule assembly occurred in association
with the organizing centers (MTOCs) (Fig. 3a and b).   However,
the average length of microtubules assembled in the transformed
cells was two or three times less than in 3T3 cells (Fig. 3c).
In addition, the number of microtubules per MTOC in SV3T3
cells was 30-40% smaller (not shown).
     Since cAMP stimulated microtubule growth in cultured cells,
the possibility of cAMP and ATP derivatives to promote microtubule polymerization in the lysed cells was examined.   Addition of 10 μM cAMP or 8-bromo cAMP plus 0.5 mM ATP to tubulin
solutions and SV3T3 lysed cells stimulated microtubule growth
to a level observed in 3T3 cells (Fig. 3c).   These results suggested that cAMP-dependent phosphorylation of some protein(s)

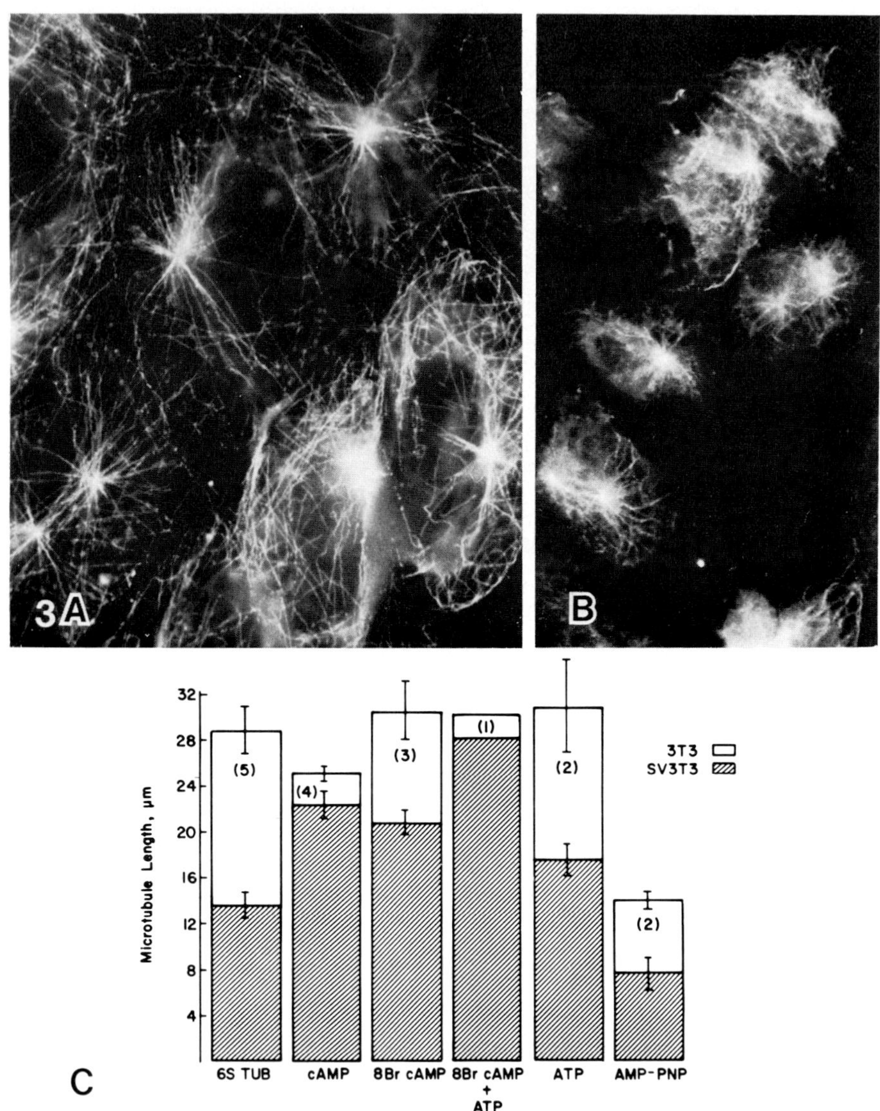

FIGURE 3.   Analysis of microtubule initiation and elongation in lysed 3T3 and SV3T3 cells exposed to bovine brain tubulin. (a), Immunofluorescence of regrowth in 3T3, or (b), SV3T3
lysed cells. X1000. (c), Effect of cAMP and ATP derivatives
on microtubule growth in 3T3 and SV3T3 lysed cells. Numbers
in parentheses represent the number of experiments performed
for each test group (at least 100 microtubules were measured
per test group per experiment). (From Tash et al., 1981).

is involved in stimulating microtubule assembly particularly since the nonhydrolyzable ATP analog, AMPPNP (1 mM), inhibited assembly and ATP alone failed to stimulate.

<u>Endogenous Cytoplasmic Microtubule Assembly and Disassembly in Lysed SV3T3 Cells.</u> We have previously reported that 3T3 cells lysed in Brij-58 could retain unpolymerized, endogenous tubulin and that this tubulin was competent to reassemble (56). These observations together with evidence by others that unpolymerized tubulin pools exist in transformed (32) and undifferentiated (40) cells led us to examine the effects of various molecules on SV3T3 microtubule assembly and disassembly using the Brij-lysed cell system.

By lysing SV3T3 cells in 0.05% Triton-reassembly buffer containing 4% polyethylene glycol prior to fixation (Fig. 4a), the existing CMTC can be preserved. This was used to compare and determine the extent of assembly or disassembly under various conditions in the Brij-lysed cells. When cells were lysed in Brij (as described in Methods) and subsequently incubated in GTP-RB for 20 min at 37°C, a more fully formed CMTC was observed with essentially all microtubules extending to the plasma membrane (Fig. 4b). It appeared that most assembly was via elongation of shorter microtubules since assembly was observed in GDP (Fig. 4c). This was also consistent with in vitro polymerization in the presence of GDP (58-60). The addition of 0.1 mM cycloheximide in the RB did not affect the enhancement of the CMTC indicating that synthesis of tubulin was not required. In the presence of 100 μM Ca^{++}, assembly was inhibited and the CMTC began to depolymerize, the extent varying from cell to cell (Fig. 4d). However, this inhibition could be overcome if ATP and cAMP were present (Fig. 4e) and in many cases the CMTC's under these conditions were actually indistinguishable from the enhanced CMTC's in the absence of Ca^{++} (Fig. 4f).

DISCUSSION

The issue of whether the CMTC plays an active or passive role in determining the transition between normal and transformed cell morphology remains unresolved. It is clear, however, that the CMTC conforms to the cell's shape be it elongated and flattened or rounded. There is good evidence that the extent of the CMTC is reduced upon cellular transformation and that unpolymerized tubulin pools exist. In the present report we show two conditions depicted in Fig. 5 which induce the assembly of this unpolymerized tubulin.

When living transformed SV3T3 cells are treated with cAMP or when Brij-lysed cells are incubated in microtubule assembly buffer, a more extensive CMTC is observed (Fig. 5). In the

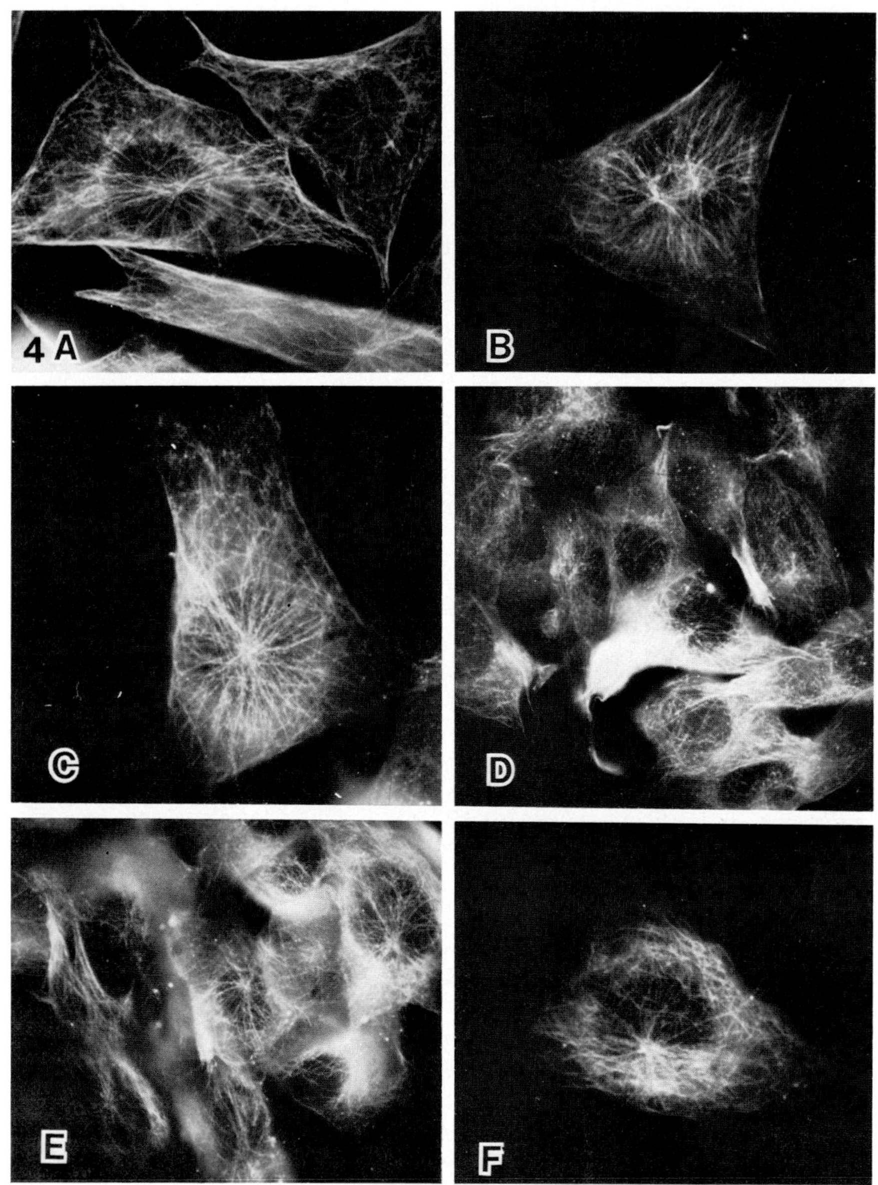

FIGURE 4. Cytoplasmic microtubule assembly in Brij-58 lysed SV3T3 cells from endogenous unpolymerized tubulin pools and the opposing effects of Ca^{++} vs. cAMP + ATP on assembly. (a), SV3T3 cells lysed in Triton-PEG. (b), SV3T3 cells lysed in Brij and incubated for 20 min at 37°C in 1 mM GTP-RB; (c), 1 mM GDP-RB; (d), 1 mM GTP-100 μM Ca^{++}-RB; (e), 1 mM ATP-0.3 mM GDP-0.01 mM 8-Bromo cAMP-100 μM Ca^{++}-RB; or (f), 1 mM ATP-0.3 mM GDP-0.01 mM 8-Bromo cAMP-RB.

former case, since the only known action of cAMP in eukaryotic cells is the activation of cAMP-dependent protein kinases, it is likely that its target of action is phosphorylation of proteins which associate with tubulin (e.g. microtubule associated proteins, MAPs) and directly influence its capacity to polymerize, and/or tubulin itself. Another possibility is the indirect influence of protein phosphorylation resulting in decreased intracellular levels of Ca^{++} (which inhibits microtubule assembly)and/or reduced synthesis of calmodulin, the protein which can increase Ca^{++} inhibition. In the case of the lysed cells, the apparent phosphorylation-independent induction of microtubule assembly could result from the chelation of Ca^{++} by EGTA or the extraction of an inhibitor which may be inactivated by phosphorylation in the living cell.

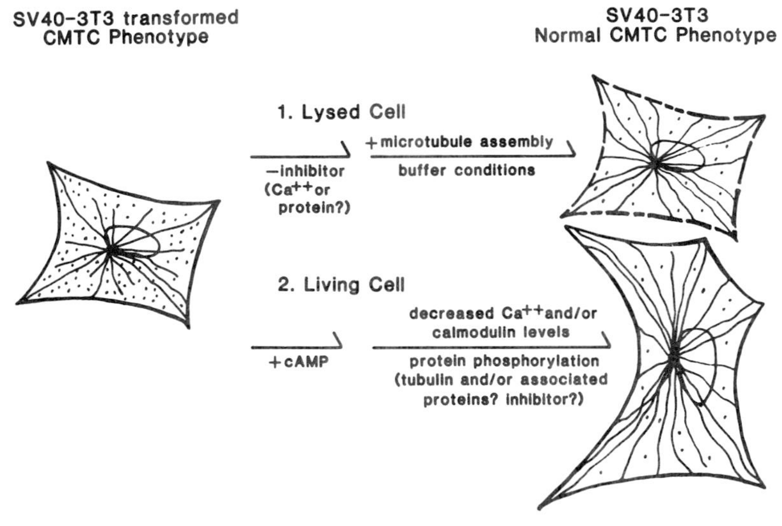

FIGURE 5.   Enhancement of the CMTC following either lysis of SV3T3 cells with Brij-58 detergent or addition of cAMP to cells in culture (reverse transformation).   The dots in the cytoplasm represent unpolymerized tubulin.

From our data and the work of others, we can speculate on the possible factors and intracellular sites at which the assembly of cytoplasmic microtubules in transformed cells might be regulated. This is summarized in Fig. 6. Since unpolymerized tubulin pools appear to exist and can be readily induced to assemble following cell lysis, it seems unlikely that the transformed CMTC phenotype is a direct consequence of impairment at the level of the genes, transcription, or translation. Both in vitro (50,61,62) and lysed cell assembly studies (51-53) show that Ca^{++} and calmodulin can decrease the fraction of tubulin participating in the polymerization reaction and therefore are attractive explanations for intracellular microtubule inhibition. In opposition to Ca^{++} inhibition, cAMP-dependent phosphorylation can stimulate the assembly reaction (Figs. 3 and 4) and actually appears capable of overriding the Ca^{++} effect. Thus, an interplay between transformation-related alterations of phosphorylation events and intracellular levels of Ca^{++}/calmodulin may be key factors determining the state of the CMTC.

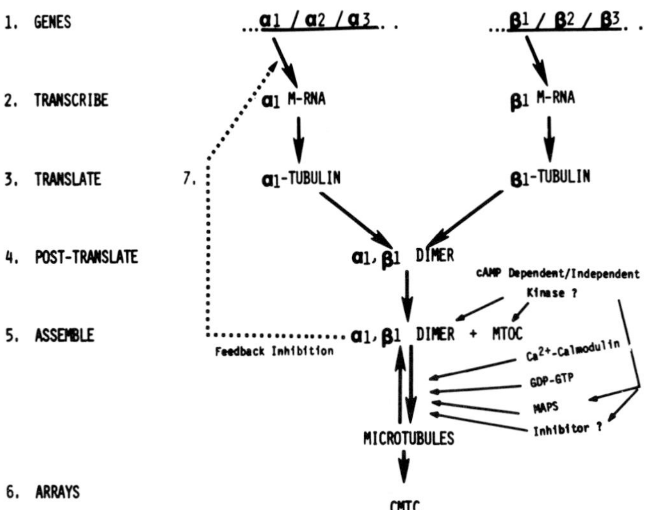

FIGURE 6. Possible sites of regulation for cytoplasmic microtubule assembly in cells.

ACKNOWLEDGMENTS

The authors gratefully acknowledge the thoughts and scientific experimentations of Drs. A.R. Means, J.G. Chafouleas, J. S. Tash, J.R. Dedman, R.L. Pardue, D.A. Pepper and S.L. Brenner. We also thank L.J. Wible and S.M. Cox for the excellent technical assistance and Ms. Pat Williams for typing the manuscript. This research was supported by NIH grant CA 23022.

REFERENCES

1. Goldman, R., Pollard, T., and Rosenbaum, J. (eds.) (1976) "Cell Motility. Book A, B. and C", Cold Spring Harbor Labatory.

2. Watson, J. D., and Albrecht-Buehler, G. (eds.) (1982) "Organization of the Cytoplasm", Cold Spring Harbor Symposium on Quantitative Biology, Vol. XLVI.

3. Wilson, L. (1982) "Methods in Cell Biology" Vol. 24 A,B, Academic Press Inc., New York.

4. Goldman, R. D. (1971) J. Cell Biol. 51, 752.

5. Goldman, R. D. (1972) Cold Spring Harbor Symp. Quant. Biol. 37, 523.

6. Porter, K. R. (1966) "Principles of Bimolecular Organization" Boston: Little, Brown & Co., pg. 308.

7. Bragine, E., Vaslieu, M., and Gelfand, I. (1976) Exp. Cell Res. 94, 241.

8. Goldman, R. D., and Follett, E. A. C. (1970) Science 169, 286.

9. Goldman, R. D., Yerna, M. J., and Schloss, J. A. (1976) J. Supramol. Struct. 5, 155.

10. McNutt, N. S., Culp, L. A., and Black, P. H. (1971) J. Cell Biol. 50, 691.

11. McNutt, N. S., Culp, L. A., and Black, P. H. (1973) J. Cell Biol. 56, 412.

12. Miller, C. L., Fuseler, J. W., and Brinkley, B. R. (1977) Cell 12, 319.

13. Pollack, R., and Rifkin, D. (1975) Cell 6, 495.

14. Ash, J. F., Louvard, D., and Singer, S. J. (1977) Proc. Natl. Acad. Sci., USA 74, 5584.

15. Yahara, I., and Edelman, G. M. (1975) Exp. Cell Res. 91, 125.

16. Nicolson, G. L. (1976) Biochim. Biophys. Acta 457, 57.

17. Yin, H.H., Ukena, T. E., and Berlin, R. D. (1972) Science 178, 867.

18. Brinkley, B. R., Miller, C. L., Fuseler, J. W., Pepper, D. A., and Wible, L. J. (1978) In "Cell Differentiation and Neoplasia" (G. F. Sanders, eds.) pp. 419-450. Raven Press, New York.

19. Brinkley, B. R. (1982) In "Chemical Carcinogenesis" (C. Nicolini, ed.) NATO Advanced Study Institute (in press).

20. Asch, B. R., Medina, D., and Brinkley, B. R. (1979) Cancer Res. 39, 893.

21. Hsie, A., and Puck, T. (1971) Proc. Natl. Acad. Sci. USA 77, 1311.

22. Puck, T., Waldren, C. A., and Hsie, A. W. (1972) Proc. Natl. Acad. Sci. USA 69, 1943.

23. Brinkley, B. R., Fuller, G. M., and Highfield, D. P. (1975) Proc. Natl. Acad. Sci. USA 72, 4981.

24. Fuller, G. M., and Brinkley, B. R. (1976) J. Supramol. Struc. 5, 497 (349).

25. Edelman, G. M., and Yahara, I. (1976) Proc. Natl. Acad. Sci., USA 73, 2047.

26. Weber, K., Bibring, T., and Osborn, M. (1975) Exp. Cell Res. 95, 111.

27. Osborn, M., and Weber, K. (1977) Cell 12, 561.

28. Tucker, R. W., Sanford, K. K., and Frankel, F. R. (1978) Cell 13, 629.

29. DeMey, J., Joniau, M., DeBrabander, M., Moens, W., and Geuens, J. (1978) Proc. Natl. Acad. Sci., USA 75, 1339.

30. Watt, F. A., Harris, H., Weber, K., and Osborn, M. (1978) J. Cell Sci. 32, 419.

31. Fonte, V., and Porter, K. R. (1974) Int. Congr. Electron Microsc., 8th Aust. Acad. Sci., Canberra pg. 334.

32. Rubin, R. W., and Warren, R. H. (1979) J. Cell. Biol. 82, 103.

33. Zimmer, D. B., Turner, D. S., Goldstein, M. A., and Brinkley, B. R. (1981) Cell Biol. Internatl. Repts. 5, 1115.

34. Yahara, I., and Edelman, G. M. (1975) Exp. Cell Res. 91, 125.

35. Edelman, G. M. (1976) Science 192, 218.

36. Fuller, G. M., Ellison, J., McGill, M., Sordahl, L., and Brinkley, B. R. (1975) In "Internatl. Symp. Microtubules and Microtubular Inhib. (M. Borgers and M. DeBrabander, eds.) pp. 379-390. Amer. Elsevier, New York.

37. Wiche, G., Lundblad, V. J., and Cole, R. D. (1977) J. Biol. Chem. 252, 794.

38. Hiller, G., and Weber, K. (1978) Cell 14, 795.

39. Chafouleas, J. G., Pardue, R. L., Brinkley, B. R., Dedman, J. R., and Means, A. R. (1981) Proc. Natl. Acad. Sci., USA 78, 996.

40. Olmsted, J. B. (1981) J. Cell Biol. 89, 418.

41. Brinkley, B. R., Cox, S. M., Pepper, D. A., Wible, L., Brenner, S. L., and Pardue, R. L. (1981) J. Cell Biol. 90, 554.

42. Tash, J. S., Lagace, L., Lynch, D. R., Cox, S. M., Brinley, B. R., and Means, A. R. (1981) Cold Spring Harbor Conf. on Cell Prolifer. 8, 1171.

43. Johnson, G. S., Friedman, R. M., and Pastan, I. (1971) Proc. Natl. Acad. Sci., USA 68, 425.

44. Bloom, G. S. (1978) J. Cell Biol. 79, 309a.

45. Lockwood, A. H., Trivette, D. D., and Pendergast, M. (1981) In "Symp. on Quant. Biol." Cold Spring Harbor Laboratory, Vol. XLVI pp. 909-919.

46. Weisenberg, R. C. (1972) Science 177, 1104.
47. Fuller, G. M., Artus, C. S., and Ellison, J.J. (1976) J. Cell Biol. 70, 68a.
48. Kirschner, M. W., Williams, R. C., Weingarten, M., and Gerhart, J. C. (1974) Proc. Natl. Acad. Sci., USA 71, 1159.
49. Solomon, F. (1976) In "Cell Motility" (R. Goldman, T. Pollard, and J. Rosenbaum, eds.) pp. 1139-1148. Cold Spring Harbor Laboratory, New York.
50. Marcum, J. M., Dedman, J. R., Brinkley, B. R., and Means, A. R. (1978) Proc. Natl. Acad. Sci., USA 75, 3771.
51. Means, A. R. (1981) Rec. Prog. Horm. Res. 37, 333.
52. Means, A. R., Tash, J. S., and Brinkley, B. R. (1981) In "Func. Correl. of Horm. Recep. in Reprod." (V. Mahesh, T. Muldoon, B. Saxena, and W. Sadler, eds.) 12, 187-204. Amsterdam, Elsevier.
53. Tash, J. S., Means, A. R., Brinkley, B. R., Dedman, J. R., and Cox, S. M. (1980) In "Microtubules and Microtubule Inhibitors" (M. DeBrabander and J. DeMey, eds.) 3, 269-279. Amsterdam, Elsevier.
54. Schliwa, M. (1980) Proc. 38th Mtg. Elec. Micro. Soc. Am. 814.
55. Cande, W. Z., McDonald, K., and Meeusen, R. L. (1981) J. Cell Biol. 88, 618.
56. Deery, W. J., and Brinkley, B. R. (1981) J. Cell Biol. 91, 337a.
57. Brinkley, B. R., Fistel, S. H., Marcum, M., and Pardue, R. L. (1980) Internatl. Rev. of Cytol. 63, 59.
58. Carlier, M. F., and Pantaloni, D. (1978) Biochem. 17, 1908.
59. Karr, T. L., Podrasky, A. E., and Purich, D. L. (1979) Proc. Natl. Acad. Sci, USA 76, 5475.
60. Zackroff, R. V., Weisenberg, R. C., and Deery, W. J. (1980) J. Mol. Biol. 139, 641.
61. Karr, T. L., Kristofferson, D., and Purich, D. L. (1980) J. Biol. Chem. 255, 11853.
62. Weisenberg, R. C., and Deery, W. J. (1981) Biochem. Biophys. Res. Commun. 102, 924.

MODULATION OF INTERMEDIATE FILAMENT ORGANIZATION BY CYCLIC NUCLEOTIDE ANALOGUES AND β-ADRENERGIC AGONISTS DURING MYOGENESIS IN VITRO[1]

Elias Lazarides
David L. Gard[2]

Division of Biology
California Institute of Technology
Pasadena, California

I. ABSTRACT

In chicken skeletal myogenic cells grown in tissue culture, desmin and vimentin containing intermediate filaments exist in the form of a dense filamentous network early after the onset of fusion. However, late in myogenesis (within 7 days) there is a redistribution of this network and both desmin and vimentin become associated with the Z-lines of assembling myofibrils. Here we show that the cyclic AMP analogue, 8-BrcAMP, or the β-adrenergic agonist isoproterenol, when applied to cultured myotubes for more than 16 h between days 4 and 8 in culture, inhibit the transition of the two intermediate filament proteins to the myofibril Z-lines, as assayed by immuno-fluorescence. Little or no effect is observed with either BrcAMP or isoproterenol in cells older than 8 days, when most of desmin and vimentin are in association with Z-lines. Similarly these two compounds have no observable effect in the distribution of the

[1]*This work was supported by grants from the National Institutes of Health, the National Science Foundation and the Muscular Dystrophy Association of America. D.L.G. was also supported by Predoctoral Training Grant No. GM 07616 and E.L. is the recipient of a Research Career Development Award, both from the National Institutes of Health.*

[2]*Present address: University of California, San Francisco, California.*

intermediate filament network prior to 5 days in culture. The nucleotide analogues, BrAdenosine, BrAMP, and BrGMP when applied to 8-
day-old cultures have little or no effect in the redistribution of
desmin and vimentin to the Z-line, while BrcGMP at the same concentration as BrcAMP has only a slight effect. Double immunofluorescence with desmin and α-actinin antibodies of BrcAMP treated
mature myogenic cultures indicate that this nucleotide analogue has
no apparent effect on the appearance of α-actinin containing Z-line
striations under conditions when the association of desmin with Z-
lines is inhibited. The effect that BrcAMP and isoproterenol have on
the association of desmin with the Z-line appears to be reversible,
since 8-day-old cultures, which exhibit little or no association of
desmin and vimentin with Z-lines after a 24 h exposure to either of
these compounds, will exhibit normal levels of association of these
two proteins with Z-lines when the cultures are incubated for 48 h in
normal growth medium without either of these two compounds.
These results, in conjunction with earlier observations showing that
phosphorylation of desmin and vimentin is stimulated 2-3 fold by
treatment with BrcAMP or isoproterenol only in mature (8 days or
older) myogenic cells, suggest that changes in the level of desmin and
vimentin phosphorylation mediated by cyclic AMP play a crucial role
in the regulation of the association of these two intermediate
filament proteins with the assembling myofibril Z-line.

II.   INTRODUCTION

Avian embryonic skeletal myotubes grown in vitro express two
intermediate filament (IF) proteins: Desmin, the IF protein first
identified in smooth muscle and subsequently identified as the major
subunit of intermediate filaments in adult skeletal and cardiac
muscle; and vimentin, the IF protein found in most mesenchymal cells
and all cells in culture (see reference 1, for review). Early in
myogenesis both filament proteins are found in an extensive network
of cytoplasmic filaments (2, 3). Late in myogenesis, however, both
proteins become associated with the Z lines of myofibril bundles
where they surround the Z discs and may function to interlink Z discs
of adjacent myofibrils (3-5). The mechanisms which regulate this
redistribution of intermediate filaments during myogenesis are not
yet understood.

We have previously observed that both desmin and vimentin are
phosphorylated in cultured skeletal muscle myotubes, and that in
mature myotubes such phosphorylation can be stimulated by the
addition of β-adrenergic hormones or cAMP analogues to the culture
medium (6, 7). Sensitivity to cAMP stimulated increases in IF protein
phosphorylation is dependent upon the state of differentiation of the
myotubes. The onset of sensitivity appears to coincide chrono-

logically with the redistribution of intermediate filaments to the Z line during myogenesis (7). This result suggested that a change in the phosphorylation of intermediate filament proteins is intimately associated with the filament distribution.

In this paper we demonstrate that while 8-BrcAMP or isoproterenol have no visible effect on intermediate filament distribution in early myotubes (3-4 days), exposure of mature cells to 8-BrcAMP or isoproterenol results in a marked decrease in the proportion of cells exhibiting filament redistribution to the Z lines. The possible correlation between the cAMP effects on filament redistribution and phosphorylation are discussed.

III. METHODS

Cultures of avian embryonic myotubes were prepared as previously described (3). Cells were grown on collagen-coated 60 mm petri plates containing 2-5 collagen-coated coverslips. Cultures were treated with 10 µM cytosine arabinofuranoside (days 4-7) to prevent overgrowth of residual fibroblasts. Cyclic nucleotide derivatives (8-BrcAMP, 8-BrcGMP, 8-Br Adenosine, 8-BrAMP, 8-BrGMP; Sigma) and isoproterenol (Sigma) were prepared as stock solutions in Earle's balanced salt solution (EBSS) and added to the indicated final concentrations.

Cells were fixed and processed for immunofluorescence as previously described (3). Characterization of antibodies to desmin and vimentin have been presented elsewhere (4, 5).

The percentage of multinucleate myotubes showing Z-line striations with antibodies to desmin or vimentin in immunofluorescence was determined by counting cells during a grid or random scan of each coverslip. Due to the asynchronous process of myogenesis, most myotubes exhibit both filamentous and Z line IF proteins. These cells were counted as having Z-line-associated proteins. The extensive network of branches formed by many myotubes made it difficult to prevent multiple counts of a single cell. The actual percentage of cells showing striations may thus be larger than the estimates presented.

Incubation of cultures with $^{32}PO_4$ and subsequent analysis of the myogenic cells by 2D IEF/SDS-PAGE was as previously described (7).

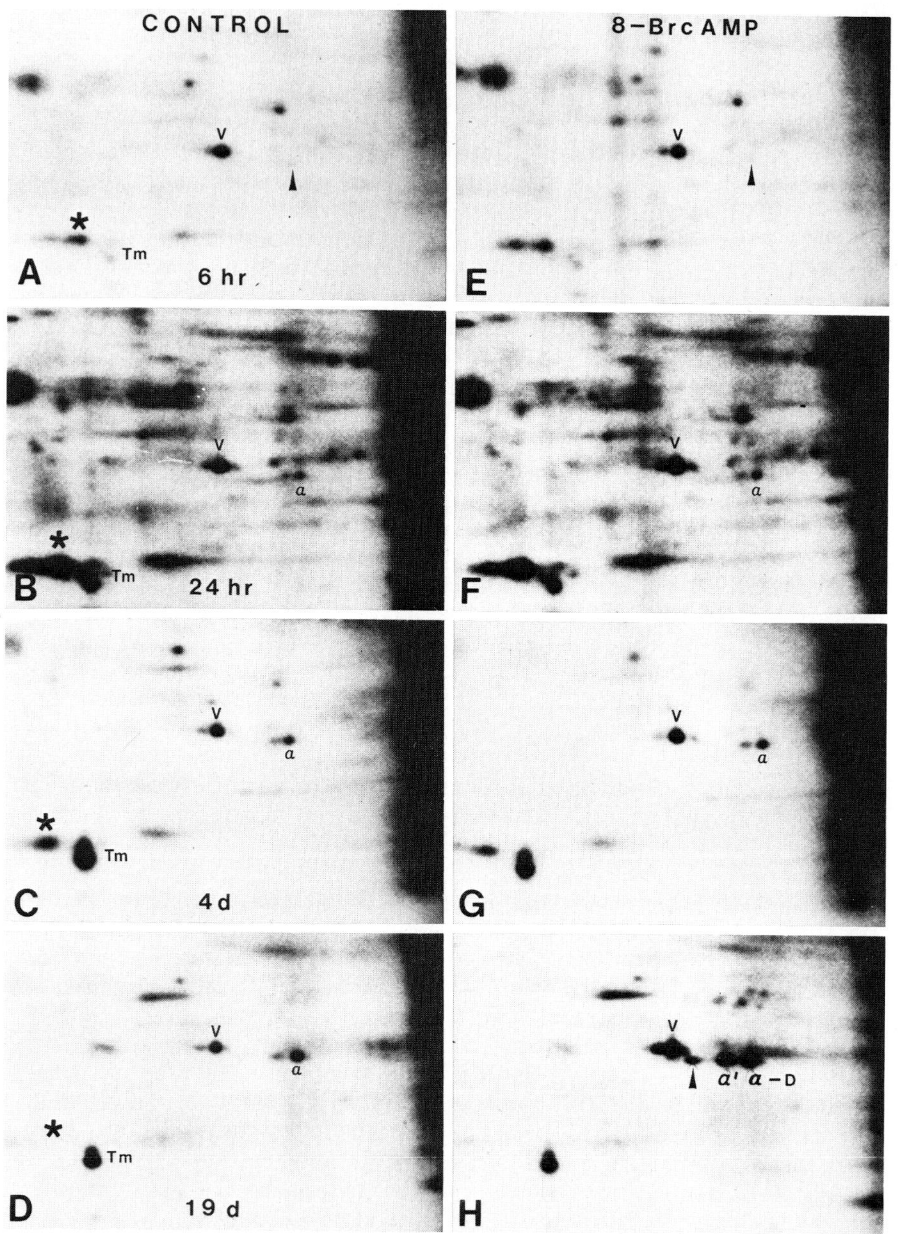
CONTROL
8-Brc AMP
V
*
Tm
A
6 hr
E
V
a
*
Tm
B
24 hr
F
a
V
a
*
Tm
C
4 d
G
V
a
V
a
*
Tm
D
19 d
V
a' a-D
H

IV. RESULTS

A. *Sensitivity of Desmin and Vimentin Phosphorylation to 8-BrcAMP During Myogenesis*

As we have previously shown, incorporation of radioactive phosphate into both desmin and vimentin is observed during all stages of myogenesis <u>in</u> <u>vitro</u> (7). However, sensitivity of IF protein phosphorylation to the addition of BrcAMP or isoproterenol varies during differentiation. In secondary myogenic cultures 6 h after plating, which consist predominantly of unfused replicating myoblasts, vimentin is the major intermediate filament protein while desmin is a minor protein species restricted to the few fusing myoblasts or young myotubes (3). Though $^{32}PO_4$ incorporation is greatest in vimentin (Fig. 1A,B), a $^{32}PO_4$-labeled species identifiable as α-desmin is observed (Fig. 1A,B, arrowheads). By 24 hr after plating phosphorylated desmin is prevalent, with both α- and α'-desmin identifiable. At later stages (4, 7 or 19 days) desmin represents one of the major cellular phosphoproteins (along with vimentin and tropomyosin). The observed increase in phosphorylated desmin presumably parallels the increase in synthesis of desmin and

FIGURE 1, opposite page. Phosphorylation during myogenesis. Myogenic cultures were labeled for 4 h in $^{32}PO_4$. 8-BrcAMP and theophylline were added during the last 45 min of the labeling period, to final concentrations of 10^{-3} M in E, F, G and 5 x 10^{-4} M in H. A,E. Control and 8-BrcAMP-treated 6 h myoblasts. B,F, Control and 8-BrcAMP-treated 24 h myoblast/myotube. C,G. Control and 8-BrcAMP-treated 4-d myotubes. D,H. Control and 8-BrcAMP-treated 19-d myotubes. Phosphorylation of desmin (α) and vimentin (V) is apparent in control cultures at all times examined. In 6 h myoblasts (A) desmin (arrowhead) is a very minor component (see text), however, the amount of phosphorylated desmin and muscle tropomyosins (Tm) increase dramatically during myogenesis. A phosphoprotein prominent in early myogenic cultures (asterisks) is lost in mature myotubes (19-d; see text). 8-BrcAMP (E-H) stimulated phosphorylation increases are restricted to mature myotubes. No response is observed in myoblasts (E,F) or 4-d myotubes (G), but is found in myotubes older than 7-d (see reference 7) and continues through at least 19-d in culture (H). (A, B, E, F autoradiographs exposed 48 h; C, D, G, H exposed 24 h.)

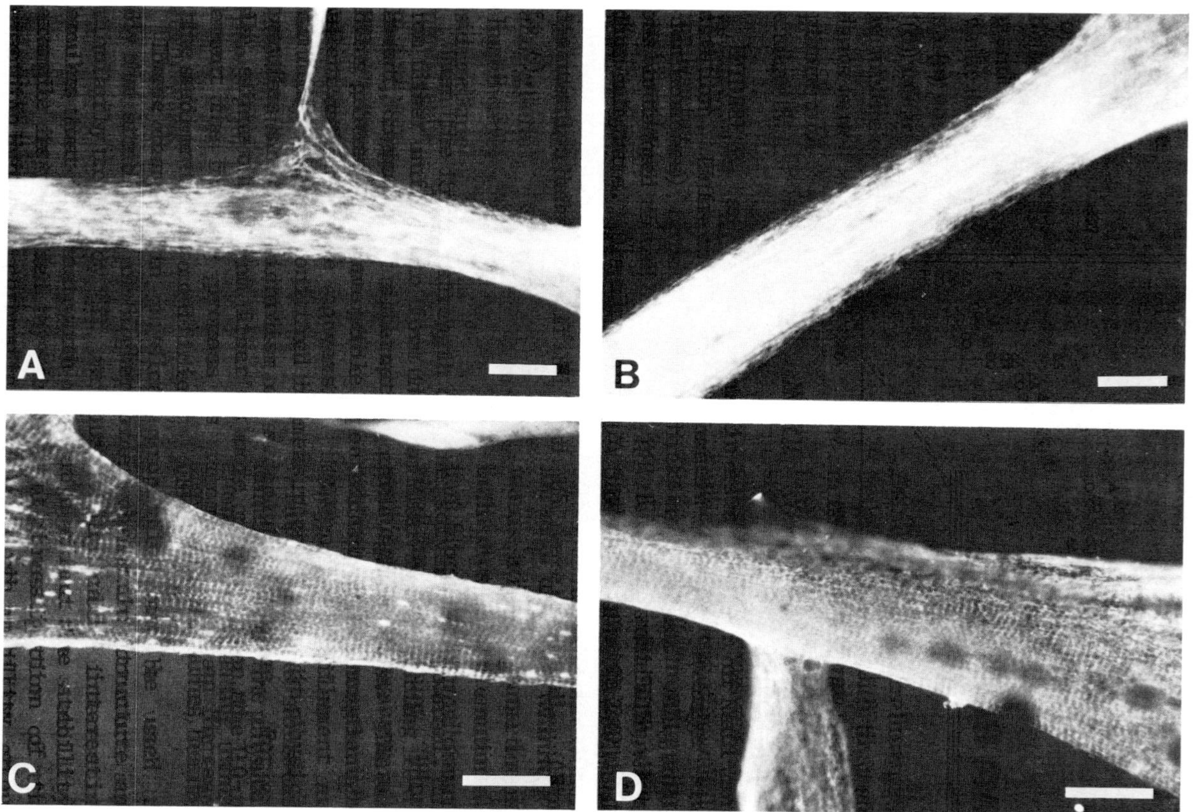

its accumulation during myogenesis (3). Several other phospho-proteins are present in altered amounts during myogenesis. For example, ^{32}P-labeled muscle tropomyosin is a very minor component in 6 h cultures, but increases substantially during myogenesis as the intracellular concentration of tropomyosin increases (3). Conversely, phosphorylation of another prominent cellular protein prevalent in early myoblasts is substantially reduced during the course of differentiation (Fig. 1, asterisks).

No changes in phosphorylation of desmin and vimentin are apparent in autoradiograms from isoproterenol- or 8-BrcAMP-treated myoblasts (6 h and 24 h) or early myotubes (Fig. 1E–G, isoproterenol result not shown). This is contrasted with the increased phosphorylation induced by either 8-BrcAMP or isoproterenol seen in 7–8-day-old cultured myotubes (see reference 7) and in 19–day-old cultured myotubes (Fig. 1H). These data suggest that the increase in IF protein phosphorylation stimulated by hormones or cAMP is restricted to mature myotubes (greater than 7 days in culture).

B. *Effect of BrcAMP and Isoproterenol on the Distribution of Desmin and Vimentin During Myogenesis*

Treatment of 4-day-old myotubes with 10^{-3} M 8-BrcAMP or 10^{-5} M isoproterenol had no effect on intermediate filament distribution as assayed by immunofluorescence with antibodies to desmin or vimentin (see Fig. 2A,B and Table 1). In these cells, antisera to desmin and vimentin reveal a dense filamentous network that fills the cytoplasm. Occasional longitudinal increases in filament density are seen, which may represent the initial association of intermediate filaments with assembling myofibrils (3). The lack of effect of 8-BrcAMP or isoproterenol on 4-day myotube IF distributions is consistent with the lack of cAMP stimulated increases in IF protein phosphorylation in cells of this age (7) (see also Fig. 1).

FIGURE 2. Distribution of desmin in control and 8-BrcAMP-treated myotubes. No differences in intermediate filament distribution were apparent between control (A) and 8-BrcAMP-treated four-day myotubes (B) examined by immunofluorescence microscopy with anti-desmin. The few myotubes from 8-BrcAMP-treated 8-day cultures that exhibit Z-line striations (D) with anti-desmin appear similar to untreated 8-day myotubes (C). Bars are 20 μm.

TABLE 1. *Effect of 8-BrcAMP and Isoproterenol on the Association of Desmin with Myofibril Z Lines During Myogenesis in Tissue Culture*

Days in culture	Analogue or hormone added	Concentration (M)	Number of experiments	Length of incubation (h)	Z lines with desmin (%)
4	Control		2		0
	BrcAMP	5×10^{-4}	1	2	0
			2	24	0
5	Control		2		6 ± 1
	BrcAMP	10^{-3}	2	24	0
6	Control		6		20 ± 4
	BrcAMP	10^{-3}	1	1	13
			1	3	0
			3	>16	2 ± 1
	BrcAMP	5×10^{-4}	1	1	34
			2	16	4 ± 2
	Isoproterenol	10^{-5}	1	1	20
			3	16	6 ± 3
7	Control		7		27 ± 5
	BrcAMP	10^{-3}	1	2	2
			1	12	2
			2	>16	5 ± 3
		5×10^{-4}	3	>16	6 ± 3
		10^{-4}	3	>16	4 ± 4
		1.3×10^{-5}	1	>16	0

	Isoproterenol	10^{-5}	3	>16	15 ± 7
		10^{-5}	2	48	25 ± 1
	Br Adenosine	10^{-4}	2	>16	31 ± 12
8	Control		9		43 ± 5
	BrcAMP	10^{-3}	2	2	26 ± 6
			2	5	44 ± 4
			2	8	32 ± 12
			4	>16	6 ± 5
		10^{-4}	7	>16	8 ± 6
		10^{-5}	2	>16	17 ± 11
	BrAMP	10^{-4}	3	>16	51 ± 4
	Br Adenosine	10^{-4}	2	>16	50 ± 1
	BrcGMP	10^{-4}	2	>16	32 ± 7
	BrcGMP	10^{-4}	2	>16	36 ± 10
	BrGMP	10^{-3}	2	>16	9 ± 4
	Isoproterenol	10^{-5}	1	8	0
			5	>16	10 ± 5
9	Control		6		46 ± 8
	BrcAMP	10^{-4}	2	>16	34 ± 1
		10^{-3}	1	18	49
			1	54	44
	BrAMP	10^{-4}	2	>16	55 ± 2
	Br Adenosine	10^{-4}	1	>16	50
	BrGMP	10^{-4}	2	>16	54 ± 3
	BrcGMP	10^{-4}	2	>16	53 ± 7
11	Control		4		52 ± 8
	BrcAMP	10^{-3}	4		47 ± 5
	Isoproterenol	10^{-5}	4		49 ± 12

By day 6, approximately 20% of the untreated multinucleate myotubes exhibit a striated pattern in immunofluorescence with desmin or vimentin antisera (Fig. 3 and Table 1). Pretreatment of cultures for 20-24 h with 8-BrcAMP or isoproterenol prior to fixation on day 6 results in a marked decrease in the number of cells with fluorescent striations (see Fig. 3 and Table 1). No other obvious changes in the distribution of intermediate filaments are observed in treated cells. In general, those myotubes in 8-BrcAMP-treated cultures that do exhibit desmin- or vimentin-containing striations appear similar to striated myotubes in control cultures (Fig. 2C,D).

As myotube differentiation proceeds, the percentage of myotubes in control cultures that exhibit striations with antibodies to intermediate filaments increases, reaching 40-50% by days 8 and 9. Eight-day cultures treated with 10^{-3} M 8-BrcAMP for 24 h prior to fixation show a marked reduction in cells with Z-line striations (see Table 1 and Fig. 3).

Antisera to α-actinin were used to ascertain the effect of 8-BrcAMP on Z-line assembly. No changes were apparent in the proportion of myotubes exhibiting α-actinin Z-line striations in 6- or 7-day-old myotube cultures treated with 10^{-3} M 8-BrcAMP (see Table 2). This result suggests that 8-BrcAMP acts at a stage subsequent to the assembly of α-actinin-containing Z lines, consistent with the proposed role of intermediate filaments in the the assembly of myofibril bundles (3).

The 8-BrcAMP-stimulated increase in phosphorylation of desmin and vimentin reaches a peak within 1 h after administration of 5×10^{-4} M 8-BrcAMP (7). However, this length of treatment has no noticeable effect on intermediate filament distribution in 7-8-day-old myotubes. We have found that a noticeable effect on filament association with the Z lines is clearly observed only when 7-8-day-old myotubes are exposed to 8-BrcAMP for periods greater than 16 h. In most of the experiments we have used 10^{-3} M BrcAMP. Even though we have not examined the BrcAMP dose response in detail we have observed that 5×10^{-4} M as well as 10^{-4} M 8-BrcAMP and 10^{-5} M isoproterenol will also inhibit the transition of desmin and vimentin to the Z disc in 7-8-day-old myotubes (see Table 1).

To investigate the specificity of 8-BrcAMP in inhibiting the association of desmin and vimentin with the Z line in 7-8-day-old myotubes, we examined the effect that other brominated nucleotide analogues have on this process. 8-BrAMP, 8-Br Adenosine and 8-BrGMP, when administered to 7-8-day-old myotubes at a concentration of 10^{-4} M and for periods greater than 16 h, have no observable effect on the redistribution of intermediate filaments. 8-BrcGMP at the same concentration (10^{-4} M) and the same length of time (greater than 16 h) has only a slight effect on this process but

TABLE 2. Effect of 8-BrcAMP on Z Line Assembly As Assayed by Double Immunofluorescence with Antibodies to α-Actinin and Desmin[a]

Days in culture	Analogue or hormone added	Concentration (M)	Number of experiments	Length of incubation (h)	Z lines with desmin (%)
5	Control		2		82 ± 3
	BrcAMP	10^{-3}	2	24	68 ± 4
6	Control		2		73 ± 3
	BrcAMP	10^{-3}	2	24	70 ± 4

[a] *In this set of experiments cells were stained in double immunofluorescence by the indirect-direct method with antibodies to α-actinin and desmin as described (3). Since control and 8-BrcAMP values for desmin are comparable to those presented in Table 1, only the α-actinin values are shown here.*

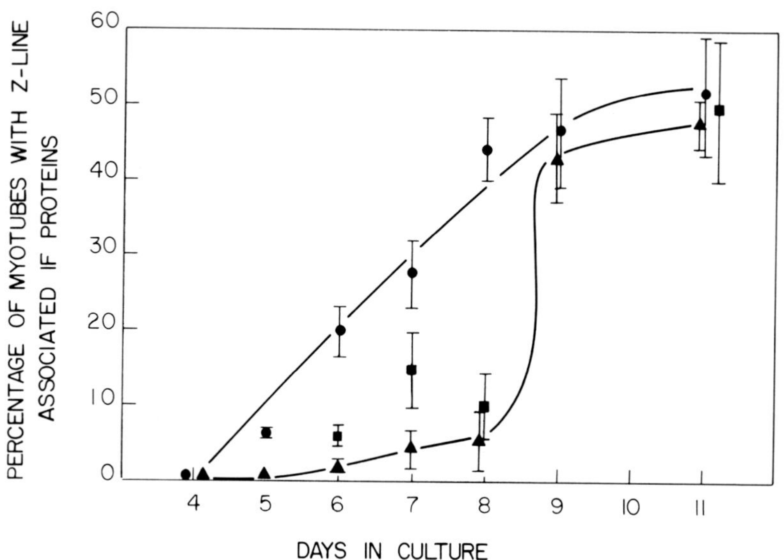

FIGURE 3. Inhibition of IF redistribution to the Z line during myogenesis by 8-BrcAMP and isoproterenol. Control myotubes (circles) were fixed for immunofluorescence microscopy on the day indicated. The proportion of cells exhibiting Z-line striations with antisera against desmin or vimentin was determined. At least 100 cells were counted on most coverslips (the number of coverslips examined is given in Table 1; mean and range are shown for each point). Identical cultures of myotubes were pretreated for 20–24 h with 10^{-3} M 8-BrcAMP (triangles) or 10^{-5} M isoproterenol (squares), fixed on the indicated day, and examined as above. Treatment of 5- to 8-day myotubes with these agents results in a marked decrease in the percentage of cells exhibiting Z-line striations by immuno-fluorescence microscopy. Little or no effect is seen with 8-BrcAMP on 9-day or older myotubes suggesting that once the IF proteins associate with the Z lines, this association cannot be reversed by 8-BrcAMP.

has a more marked effect in inhibiting the association of IF proteins with the Z line when used at higher concentrations (10^{-3} M) (see Table 1). In cells older than 9 days in culture, when approximately 40–50% of the cells exhibit association of desmin with Z lines, exposure to 8-

BrcAMP (10^{-4} M or 10^{-3} M, 16 h) or isoproterenol (10^{-5} M, 16 h) has no effect on the redistribution of intermediate filaments (see Table 1 and Fig. 3).

Inhibition of the redistribution of desmin to the Z line by 8-BrcAMP appears to be reversible. Six- to seven-day-old myotubes were treated with 10^{-3} M 8-BrcAMP to inhibit the redistribution of desmin to the Z line. When these cells were chased for 48 h in medium containing no 8-BrcAMP, the cells exhibited desmin in association with Z lines comparable to controls (see Table 3).

V. DISCUSSION

The results presented here suggest that 8-BrcAMP and iso-proterenol, when administered to myotubes differentiating in tissue culture at the time of the redistribution of intermediate filaments to the Z line, inhibit this redistribution. Prior to the onset of this redistribution (younger myotubes), this cyclic nucleotide analogue or this β-adrenergic agonist have no observable effect on the distribution of intermediate filaments. Similarly in fully differen-tiated myotubes where intermediate filaments have already transited to the Z line (9 days or older in our culture system), these two compounds also have no effect on the distribution of the filaments. These observations indicate that the effect of 8-BrcAMP or iso-proterenol on the redistribution of intermediate filaments is restricted to the time of their rearrangement and association with the Z line; once this process has been completed neither of these two components are capable of reversing it.

We have observed that maximal inhibition of the redistribution of intermediate filaments is brought about when either 8-BrcAMP or isoproterenol is present in the medium for periods longer than 16 h. Although we have not examined the time course of inhibition in detail the time required to observe this inhibition is much greater than the 8-BrcAMP stimulated increase in phosphorylation of desmin and vimentin observed previously at this time in differentiation (7). This 8-BrcAMP stimulated increase in desmin and vimentin phosphoryla-tion observed in 7-8-day-old myotubes reaches a peak within 1 h after administration of 5×10^{-4} M 8-BrcAMP (7). It is possible that the immediate effect of 8-BrcAMP on IF protein phosphorylation leads to a more time-consuming rearrangement of intermediate filaments. The data on hand, however, do not adequately address the question of whether 8-BrcAMP can reverse the association of IF proteins with the myofibril Z line, or simply inhibits any new associations. This latter hypothesis could also explain the length of time required for an observable effect (the time required to see 8-BrcAMP effects would

TABLE 3. *Reversal of Effect of 8-BrcAMP on the Association of Desmin with Myofibril Z Lines During Myogenesis in Tissue Culture*

Days in culture	Analogue or hormone added	Concentration (M)	Number of experiments	Length of incubation (h)	Z lines with desmin (%)
7	Control		2		28 ± 5
	8-BrcAMP	10^{-3}	1	24	8
	Chase				
9	Control		1		47
	8-BrcAMP-chased[b]	48	39		

[a]One plate from a 7-day-old 8-BrcAMP treated culture was chased for 48 h in normal grown medium without 8-BrcAMP and the number of Z lines exhibiting desmin fluorescence was compared to a control 9-day-old culture.

be dependent upon the rate at which the redistribution of filaments occurred), as well as explaining the decrease in effectiveness of 8-BrcAMP treatment observed in more mature (9-day-old) cells. It should be born in mind that the assay used in this study (immuno-fluorescence with desmin antibodies) can distinguish only these two states of intermediate filaments, namely as cytoplasmic filaments or in association with Z lines. Perhaps the use of a more sensitive assay or the use of monoclonal antibodies capable of detecting specific stages in the rearrangement of the filaments will help elucidate these hypotheses further. Nevertheless, from the results presented it is evident that this inhibition of the redistribution of intermediate filaments is specific to cyclic AMP analogues and β-adrenergic agonists and specific to a given time in myotube differentiation. These observations suggest that the levels of cyclic AMP play an important role in the association of intermediate filaments with the Z line. Similarly, the observation that this cyclic nucleotide analogue has no effect on the appearance of α-actinin containing Z-line striations or the general striated appearance of differentiating myofibrils further suggests that the effect of 8-BrcAMP is specific for this step in myofibril assembly.

Correlation of the inhibition of intermediate filament redistribution by 8-BrcAMP and isoproterenol with the previously reported increase in phosphorylation of desmin and vimentin stimulated by these agents (7) suggests that high levels of cAMP-dependent IF protein phosphorylation are associated with a less differentiated state, when intermediate filaments are dispersed throughout the cytoplasm. This hypothesis is consistent with the age-dependent sensitivity of myotubes to cAMP-stimulated increases in IF protein phosphorylation; maximal cAMP-dependent phosphorylation of IF proteins in immature myotubes would prevent further stimulation of phosphorylation by endogenous cAMP. The onset of sensitivity to exogenous cAMP corresponds in time to the redistribution of intermediate filaments to the Z line during myotube maturation, suggesting that a decrease in IF protein phosphorylation is associated with this event. Prolonged artificial increases in phosphorylation induced by endogenous cAMP then inhibit the normal redistribution of filaments during differentiation. At present, evidence supporting this model is incomplete. Further support could be obtained by measurements of intracellular cAMP levels, active cAMP-dependent kinase levels, and the total levels of IF protein phosphorylation during myogenesis. It is interesting that this model, which equals high cAMP-dependent phosphorylation with an immature state, is contrary to most reports that cAMP levels increase during differentiation in many cells (for example, reference 8). In this regard, increases in intracellular cAMP are noted prior to myoblast fusion, the earliest step in the process of myogenic differentiation (9) and such a rise in cyclic AMP has been implicated as being the critical intracellular change responsible for initiating events that culminate in myoblast differentiation 4 to 5 h later (10, 11). However, these experiments have also indicated that the hormone responsible for the positive regulation of myoblast differentiation in vitro is not acting through the β-adrenergic receptor (11). Thus the regulation of the onset of myoblast fusion by cAMP appears to be opposite to that of the regulation of the redistribution of the IF proteins to the Z line, since the latter event is most likely mediated by the β-adrenergic receptor and requires a lowering in intracellular cAMP levels.

The mechanisms by which cAMP-dependent phosphorylation of intermediate filament proteins could control filament distribution are unknown. To date several intermediate filament-associated proteins (as opposed to filament subunits) have been identified in muscle cells (12, 13), and undoubtedly many remain to be found. In particular one of them, synemin, has been shown to be phosphorylated by cAMP-dependent protein kinases in vivo (14) and has been implicated as functioning to crosslink intermediate filaments both in avian erythrocytes and in muscle cells where it is expressed (15, 16). Thus changes in the phosphorylation of synemin or other intermediate filament associated proteins may affect interactions of intermediate filaments with each other (e.g., increased synemin self-interaction and thereby increased filament crosslinking), or with other cytoplasmic structures (such as myofibril Z lines).

REFERENCES

1. Lazarides, E. (1980). Nature 283, 249.
2. Bennett, G. S., Fellini, S. A., Toyama, Y., and Holtzer, H. (1979). J. Cell Biol. 82, 577.
3. Gard, D. L., and Lazarides, E. (1980) Cell 19, 263.
4. Granger, B. L., and Lazarides, E. (1979). Cell 18, 1053.
5. Granger, B. L., and Lazarides, E. (1979). Cell 18, 1053.
6. O'Connor, C. M., Balzer, D. R., and Lazarides, E. (1979). Proc. Natl. Acad. Sci. USA 76, 819.
7. Gard, D. L., and Lazarides, E. (1982). Mol. Cell. Biol., in press.
8. Prasad, K. N., and Kumar, S. (1974). In "Cold Spring Harbor Conferences on Cell Proliferation" (B. Clarkson, and R. Basenga, eds.), Vol. I, p. 581. Cold Spring Harbor, New York.
9. Zalin, R. J., and Montague, W. (1974). Cell 2, 103.
10. Zalin, R. J. (1976) Develop. Biol. 53, 1.
11. Curtis, D. H., and Zalin, R. J. (1981). Science 24, 1355.
12. Granger, B. L., and Lazarides, E. (1980) Cell 22, 727.
13. Breckler, J., and Lazarides, E. (1982). J. Cell Biol. 92, 795.
14. Sandoval, I. V., Colaco, C. A. L. S., and Lazarides, E. (1982). J. Biol. Chem., in press.
15. Granger, B. L., Repasky, E. A., and Lazarides, E. (1982) J. Cell Biol. 92, 299.
16. Granger, B. L., and Lazarides, E. (1982). Cell, in press.

WORKSHOP SUMMARY: Hormone-Receptor-DNA Interaction

William T. Schrader
Department of Cell Biology
Baylor College of Medicine
Houston, Texas   77030
        and
John D. Baxter
Department of Medicine
University of California
School of Medicine
San Francisco, California   94143

The purpose of this Workshop was to discuss steroid hormone regulation of gene expression. Two lines of experimentation were addressed, which are summarized below. The workshop was convened by J. Baxter and W. Schrader, who presented for debate the opening general question of whether regulatory sequences in DNA necessary for hormonal control of genes have yet been either identified or localized. Since all studies to date have involved probing DNA regions relatively close to expressed genes (perhaps up to 1-5 kb away at the most), a corollary to this question was whether present knowledge yet allows one to concentrate efforts in these proximal regions. Eight speakers presented brief synopses of their efforts in this regard.

Three speakers described work using steroid hormone receptor complexes as probes for gene regulatory sequences. W. Schrader (Abstract #751) described further control experiments not covered by his plenary lecture on the interactions of purified chick oviduct progesterone receptors with DNA sequences flanking the chicken ovalbumin gene. Whereas a preferential receptor-binding DNA sequence located upstream of the cap site between -135 to -249 bp was identified, he showed that alternative DNA sites of high affinity exist both within the transcription unit and external to it. A DNA-cellulose nucleotide sequence competition assay using crude receptor reported by E. Mulvihill (Abstract #440) was described. The method has the advantage of not requiring purified receptor.

F. Payvar (no abstract) described DNA binding studies for mapping the sites on mouse mammary tumor virus DNA that bind purified rat liver glucocorticoid receptors. A set of strong binding sites was detected; apparently two lie between -150 and -400 bp upstream of the cap site in the left-hand long terminal repeat sequence of MMTV. At least two additional sites were found within the transcription unit, one about 30% of the distance downstream from the site of initiation of transcription, and the other about 80% of the distance. These internal sequences were tested for their biologic activity as regulators of glucocorticoid responsiveness. When cloned into recombinant vectors upstream from a marker gene not otherwise under glococorticoid control, these DNA sequences allowed glucocorticoid responsiveness of the marker in mouse L cells containing glucocorticoid receptor. Electron microscopic analysis of putative receptor-DNA complexes confirmed the direct protein-DNA binding data.

B. Littlefield (Abstract #959) showed work on purification of an oviduct non-histone chromosomal protein termed the "acceptor protein." This protein is necessary for the chick progesterone receptor B subunit to bind to chromatin. The acceptor was extracted from chromatin with 4.0 M guanidine hydrochloride, and further purified. Reconstitution of the protein back onto recipient DNA by gradient dialysis restored receptor-binding activity. The reaction preferred chick DNA compared to procaryotic DNA. No direct sequence preference of this acceptor for DNA is yet known; however, the protein is present in chromatin remnants after exhaustive DNase I digestion. Sequence complexity studies of this DNA show enrichment for reiterated sequences, leading to the possibility that the acceptor proteins may be located at or near these highly repeated sequences in situ.

These three discussants showed that an analysis of receptor-DNA interactions has the potential for defining putative regulatory regions. It is interesting to note that for both progesterone receptor and glucocorticoid receptor there exist high-affinity regions proximal to the start of transcription. The estimated affinity of these sites compared to total DNA is only on the order of 10 - 100-fold. Thus it is entirely problematical at this juncture whether the interactions described can account for the biologic responses in vivo. Little is known about the possible mechanism by which a receptor-DNA interaction might affect gene transcription; progesterone receptor DNA binding appears to interact at AT-rich sequences and to allow helix destabilization. These initial observations have not yet been extended to an in vitro DNA transcription system necessary to test such a hypothesis.

The second segment of the Workshop presented studies of regulatory sequence determinations made by insertion of recombinant DNA into recipient cells followed by screening for hormone responsiveness. Two workers studied estrogen-inducible chicken oviduct genes. E. Lai (Abstract #995) reported on expression of the chicken ovalbumin gene cloned in pBR322 when introduced into human breast carcinoma cells MCF-7. Transformants containing the gene copies showed a 3-5

fold induction of mRNA for ovalbumin or immunoassayed ovalbumin protein in response to estradiol administration.
Similarly, the 5'-flanking DNA of this gene was linked in
pBR322 to the bacterial XGPRT gene. Recipient MCF-7 cells
survived only when estrogen was present, indicating the
activity of the proximal 5'-end of the gene as a necessary
and sufficient signal for hormonal control.

R. Renkawitz (Abstract #982) used similar reconstructions
to examine hormonal control of the chicken lysozyme gene.
These workers microinjected recombinant genes into chick
oviduct cells in primary culture. When the gene was linked
to SV-40 T antigen sequences, its transcription could be
distinguished from host-cell lysozyme gene products. Estrogen induced T antigen transcripts several fold. Recombinants
using only the proximal upstream DNA sequences flanking the
gene were similarly effective at allowing estrogen induction
of T antigen product as detected by an immunofluorescence
assay.

In both of these presentations above, estrogen served as
the inducing stimulus. These chicken genes are stimulated
several hundred-fold in the intact chicken; the reports presented showed only about a 10-fold induction in the recombinant cells. Thus, the possible importance of the locus of
gene insertion remains unknown; the systems do not yet mimic
the in vivo sensitivity of these genes for regulation.

Three additional speakers dealt with regulation of genes
by other agents. P. Gruss (Abstract #993) described his
experiments using SV-40 derived "activator" DNA sequences to
enhance transcription. The cis-acting effect of these sequences was mimicked by other sequences obtained from the
Maloney sarcoma virus long terminal repeat DNA. The two
"activator" sequences have only a small region of homology.
This finding stressed the conclusion that functional assays
must be used to probe for such enhancer DNA activity; similarity of DNA sequence organization is not yet sufficiently
well understood.

R. Warren (Abstract #981) described his work studying
the mouse metallothionein gene and its flanking DNA function
in stimulation of this gene by cadmium. Flanking DNA from
the metallothionein gene was coupled to a Herpes simplex
virus (HSV) thymidine kinase gene plasmid and cadmium regulation of TK was monitored in mouse L cells. The proximal 250
bp of upstream DNA was sufficient to allow Cd regulation of
TK activity. Thus, this DNA region is sufficient for regulation of this function. Interestingly, regulation by glucocorticoids was lost in the fusion plasmid, suggesting importance
of more distal sequences for this hormonal control.

G. Pavlakis (Abstract #985) described similar experiments using the metallothionein gene flanking DNA fused upstream from human growth hormone (hGH) genes. Expression in cultured monkey cells of hGH protein was found to be under Cd^{++} control, but not glucocorticoids. Transient infection of the cells was also tested; the hGH expression was also affected by Cd^{++} during this acute expression experiment.

In summary, the Workshop showed two distinct lines of approach to mapping of regulatory sequences for hormones. It was the general consensus of the participants that DNA sequences proximal to genes are loci for regulation of their expression by hormones and other ligands and the data strongly suggest that steroid-receptor complexes can bind selectively to certain structures of DNA. Since heterologous fusion genes can also be regulated, it is doubtful that sequences within the transcription units themselves are obligatory for receptor-gene interaction. Future work combining gene-cutting approaches with receptor-DNA binding methods should no doubt extend the present level of understanding.

Index

Numbers refer to the chapter numbers.